From Whirlwind to MITRE

History of Computing
I. Bernard Cohen and William Aspray, editors

William Aspray, *John von Neumann and the Origins of Modern Computing*

Charles J. Bashe, Lyle R. Johnson, John H. Palmer, and Emerson W. Pugh, *IBM's Early Computers*

Paul E. Ceruzzi, *A History of Modern Computing*

I. Bernard Cohen, *Howard Aiken: Portrait of a Computer Pioneer*

I. Bernard Cohen and Gregory W. Welch, editors, *Makin' Numbers: Howard Aiken and the Computer*

John Hendry, *Innovating for Failure: Government Policy and the Early British Computer Industry*

Michael Lindgren, *Glory and Failure: The Difference Engines of Johann Müller, Charles Babbage, and Georg and Edvard Scheutz*

David E. Lundstrom, *A Few Good Men from Univac*

R. Moreau, *The Computer Comes of Age: The People, the Hardware, and the Software*

Emerson W. Pugh, *Building IBM: Shaping an Industry and Its Technology*

Emerson W. Pugh, *Memories That Shaped an Industry*

Emerson W. Pugh, Lyle R. Johnson, and John H. Palmer, *IBM's 360 and Early 370 Systems*

Kent C. Redmond and Thomas M. Smith, *From Whirlwind to MITRE: The R&D Story of the SAGE Air Defense Computer*

Raúl Rojas and Ulf Hashagen, editors, *The First Computers—History and Architectures*

Dorothy Stein, *Ada: A Life and a Legacy*

Maurice V. Wilkes, *Memoirs of a Computer Pioneer*

From Whirlwind to MITRE

The R&D Story of the SAGE Air Defense Computer

Kent C. Redmond and Thomas M. Smith

The MIT Press
Cambridge, Massachusetts
London, England

Set in New Baskerville by The MIT Press.
Printed and bound in the United States of America.

Library of Congress Cataloging-in-Publication Data

Redmond, Kent C., 1914–
From whirlwind to MITRE : the R&D story of the SAGE air defense computer / Kent C. Redmond and Thomas M. Smith.
p. cm. — (History of computing)
Includes bibliographical references and index.
ISBN 0-262-18201-7 (hc : alk. paper)
1. SAGE (Air defense system)—History. 2. Military research—United States—History.
I. Smith, Thomas M. II. Title. III. Series.

UG633 .R378 2000
355.4'5'0973—dc21 00-029228

Contents

Acknowledgments

We offer our thanks to all those who encouraged, supported, and assisted us in this historical recreation of the development of the SAGE air defense computer.

Those whom we interviewed gave generously of their experience and knowledge, permitting insights otherwise unobtainable. Jay W. Forrester and Robert R. Everett deserve particular mention in this regard. The contributions of many other participants, too numerous to be listed here, are identified in the notes.

We are grateful to MITRE archivists Edward Galvin, David Baldwin, and Frank Mastrovita, who met our requests for documents and illustrations with dedicated industry and cooperation. Patricia Pendleton Morgan read the manuscript with a critical eye, suggesting changes that led to superior organization and greater clarity.

Most of all, we are grateful to Robert R. Everett, without whom this book would never have been written. His support and encouragement never failed us, from the moment we first proposed writing the history of SAGE to his assistance in the final selection of informative textual material and illustrations.

From Whirlwind to MITRE

1

"It Worked!"

In 1951 a strategic breakthrough was reported by the electrical engineers who had been in charge of a mock military operation carried out in the skies above Bedford, Massachusetts. According to their routine technical report, "three successive trial interceptions were run with live aircraft under control of the WWI [Whirlwind I] computer."[1] Few of those who received the report were aware of its import, and most of them were preoccupied with its limited military significance. The computer's promise was not yet perceived outside a small circle of mathematicians, engineers, scientists, and administrators, most of whom had no idea they were involved in the beginnings of what would be, by the century's end, a multi-billion-dollar worldwide industry.

Technology sometimes has a way of masking its impact on human affairs. Its apparently dry and toneless technicalities can escape wider notice even while signaling historic changes. The report of the events of April 20, 1951, quite obscured the fact that a group of young MIT electrical engineers, working with a group of radar engineers from the Air Force Cambridge Research Laboratories, had just demonstrated the practicability of defending North America against a possible airborne attack.

As the international tensions of the Cold War increased, particularly with the Berlin Blockade of 1948 and the outbreak of the Korean War in 1950, U.S. policy makers were, with growing urgency, seeking an effective defense system.

What the Cambridge-based engineers achieved on April 20, 1951, seemed to show genuine promise. For the first time, radar data on an approaching "enemy" aircraft had been sent over telephone lines and entered into an electronic digital computer, which had almost instantaneously calculated and posted instructions directing the pilot of a "defending" aircraft to his target.

Not only did the report fail to convey the magnitude of the technical innovation; it gave no hint of the engineers' elation. "It worked! Almost the first time, it worked," electrical engineer C. Robert Wieser recalled proudly some 33 years after directing the test.[2]

"Three days later," Wieser recalled at a gathering of nostalgic computer veterans at Boston's Computer Museum 37 years after the event, "the decision was made to build the Cape Cod System."[3] This was to be an elaborate, multi-radar experimental system tied into Whirlwind I, an experimental computer being developed at the Massachusetts Institute of Technology. Its task would be to generate, under mock battle conditions, detailed operating information that would enable engineers to develop a continent-wide air defense system incorporating machinery so advanced that it had yet to be designed. The job would keep the engineers busy for most of the 1950s constructing the SAGE (Semi-Automatic Ground Environment) air defense system.

Bob Wieser and his fellow engineers were not the only ones who were elated by the April demonstration. The commanding general of the U.S. Air Force, Hoyt S. Vandenberg, sent a formal letter of congratulation to MIT Professor George E. Valley Jr., chairman of the ad hoc committee of the Air Force Scientific Advisory Board that had recommended the test and its underlying concept. Calling the demonstration a "remarkable achievement," Vandenberg hailed it as "a milestone in the development of a modern Air Defense System" and "a potentially great contribution to the national defense."[4]

In itself an unprecedentedly complex communications and data-processing system, SAGE was to be a subsystem of a larger continental warning system that would use radar and computers to report on and assist in directing military operations in the air. Its weapons systems were to include manned interceptors, antiaircraft artillery, and ground-to-air missiles.

Crucial to the SAGE system's projected detection and control functions were two pieces of equipment that, in 1951, remained to be designed, built, and tested. One would be "a high-speed electronic digital processing machine," to be located at each radar site, that would perform "real-time processing of radar signals" so that the data could be sent over telephone lines to the command-and-control center. The other would be the central computer, which would accept and process the data from the radar sites. Both machines would be at the very edge of the technical state of the art. Too novel in design to have any record of demonstrated performance, they would have to be tested as they were being built, and they would have to be built to such high standards of quality that their performance and reliability during shakedown operations would be capable of meeting the demands of normal operating conditions. And military operating and maintenance personnel would have to be trained.

With a software industry not yet in existence, programs to make the computer and the data-processing system operate would have to be designed, tested, debugged, retested, and brought up to the same standards of performance and reliability that were expected of the hardware.

Owing to the perceived urgency of the work, a typical American technical approach that had occasionally worked in wartime had been invoked even before the crucial Bedford tests: this was to be a "crash program." Emergency priorities were set, and virtually unlimited funding would be available from the federal government.

The almost inadvertent policy of ad hoc and sometimes frantic improvisation was attributable to international political, economic, and military tensions. There was nothing novel about the practice of establishing and carrying out national defense policies; however, the forms of defense had changed along with the technology and the goals of national leaders.

The Bedford tests were foreshadowed by two unrelated historical developments that had coalesced in 1950. One of these was Project Whirlwind,[5] an engineering research project for the U.S. Navy that a group of young graduate engineers, led by Jay W. Forrester at MIT, had undertaken during World War II. The other was the conduct of U.S. foreign and military policies during the preceding half-century, one small consequence of which was the appearance at the end of 1949 of an ad hoc technical committee made up of civilian scientists and engineers and chaired by Professor Valley. To understand how these separate enterprises joined together to produce the successful Bedford tests, which led to the building of SAGE, we shall look first at the separate, slower-moving current of historical circumstance that by 1950 had brought concerns about air defense to the forefront of national military policy. Then we shall turn to the precise circumstances that got the engineers of Project Whirlwind involved in these defense-related matters even before the nation's top policy makers could be sure that in SAGE they really had the appropriate technical solution.

At the core of the story lies the issue of the proper management of the scientific research and engineering development investigations that combined to form the intricate process familiarly known as *research and development.*

2

The Beginning of the Beginning

At the time of the Bedford tests, there was a pattern of sometimes friendly and sometimes hostile international relations. That pattern had roots in the nineteenth century and was dominated by tradition. On the other hand, a powerful technique was emerging that had been virtually absent from military thinking in the nineteenth century and which was still not fully understood: improving military weapons by means of research and development. "R&D" moved forward as rapidly as military funds and human ingenuity would allow, impeded only by traditional institutional inclinations to stand by tried and true practices. The conduct of foreign affairs and the pursuit of R&D proceeded separately as well as together, and long before the digital computer became important they gave rise to a debate about national defense.

A version of the debate, in the years immediately after World War II, centered on whether the United States should continue to rely primarily on a policy of massive retaliation by long-range bombers carrying atomic warheads if its vital interests should be threatened by the military actions of a foreign power. In particular, would the threat of massive retaliation be enough to counter the persistent Soviet threat while the technical capabilities of weapons systems were expanding? Massive retaliation had been discussed within U.S. military circles as early as February of 1946, but as time passed it was increasingly considered to be only a partial solution—and one that ignored the advantages of mounting a passive defense system.[1]

In August of 1949, the Soviet Union detonated its first experimental atomic device. As a stone breaks the calm surface of a pond and sends out ever-widening ripples, this technical event challenged Americans' confidence in their security, which derived from geographical isolation and regional political dominance. The danger posed by atomic bombs, intercontinental aircraft, and ballistic missiles suddenly became clear. Deeply disturbed, many believed that not just the security of the United States but the survival of humankind was at stake.

A major question in the second phase of the national debate concerned the question of what defensive measures, both political and technical, were appropriate. The technical problems raised by the Soviet Union's atomic capability were perhaps easier to solve than the political and military problems, for the latter were highly emotional and their consequences were by no means predictable. The two world wars apparently had persuaded Americans to accept as prudent a policy of maintaining a military presence in foreign lands and even waging war overseas if that should be necessary to contain Communist aggression.

But there was no precedent in the national experience for the possibility of an atomic strike on the homeland. Before the recent technological advances that had converted the skies into intercontinental routes of attack by weapons capable of destroying a city in a single strike, the American people had not suffered the fear of foreign attack that had long been the chronic lot of their European cousins, and American policy makers had not been in the habit of adjusting foreign policy and national defense policy in response to dramatic advances in weaponry. Indeed, the situation had been quite the opposite for so long that it was difficult for those in charge to recognize that there were changes in the wind that would require the nation to change its course. Since these changes were technological, they appeared irrelevant to traditional principles of foreign policy and military policy. For example, in 1898, when the Joint Army-Navy Board had recommended extending financial support to Samuel Pierpont Langley's experiments in heavier-than-air flight,[2] the officials who sat on that board had no way of knowing that by 1950 technological advances would be forcing American military leaders to look for new and appropriate ways to defend the country against intercontinental air attack.

Efforts to use aircraft in the "Great War" of 1914–1918 had had little lasting impact on air strategy. Theories of longer-range strategic bombing developed and tested in a preliminary way during that war were filed away. By 1922, military textbooks were offering the lesson that "tactical bombing is . . . a necessity, while strategic bombing is . . . a luxury."[3] The visionary pronouncements of such a perceptive theorist as General William Mitchell went unheeded; policy makers appeared not to have heard him point out how the concept of national security based on geographical isolation was being affected by the increasing range and speed of aircraft.

During the 1930s, even though the "Air Arm" of the U.S. military was interested in developing multi-engine long-range bombers, the government's military advisers remained unconvinced of the implications for defense. In July of 1934, a War Department Special Committee investigating and recommending solutions to current disputes over the status, the organization, and the mission of the Army Air Corps preferred to assess the current situation and to reaffirm the traditional faith in geographical isolation and regional dominance:

> The "air invasion of the United States" and the "air defense of the United States" are conceptions of those who fail adequately to consider the effect of ocean barriers and other limitations. Aircraft in sufficient numbers to threaten serious damage can be brought against us only in conjunction with sea forces or with land forces which must be met by forces identical in nature and equally capable of prolonged effort.[4]

Later in the 1930s, stimulated in part by the ominous military and political rumblings in the Far East and in Europe, U.S. defense forces began to give more active consideration to defense against air attack. Interception exercises were held in California in 1936, and 2 years later exercises were conducted to test the feasibility of employing civilian ground observers and commercial communication facilities to alert antiaircraft batteries and pursuit aircraft. The most significant technical gains, however, occurred in the arcane area of radar design. By the end of the 1930s, the Signal Corps was developing radars that had a range of 120 miles. But the secrecy surrounding that work was so great that operational forces were given no opportunity to set about integrating the new and advanced equipment into strategic or tactical practice. As the historian Robert Futrell pointed out more than 30 years later, even "the air warfare theorists at the Air Corps Tactical School were uninformed about [radar's] potential."[5]

The outbreak of war in Europe in the autumn of 1939 presumably increased Americans' concern about national security, but non-involvement in Europe rather than the state of the nation's defenses became the primary issue. In February 1940, responding to General Henry H. Arnold's suggestion that the air defense of the nation be studied, the War Department created the Air Defense Command, a small planning group with authority to study the situation but not to take corrective action. In the main, Americans continued to feel secure behind their oceanic moat.

This sense of security was given voice by the U.S. Chief of Staff when he testified before the Senate in May:

> What is necessary for the defense of London is not necessary for the defense of New York, Boston or Washington. Those cities can be raided . . . but . . . continuous attack would not be practicable unless we permitted the establishment of air bases in close proximity to the United States.[6]

The War Department did take precautionary action to counter a potential threat, directing "that commanders of armies and overseas departments prepare or revise plans for an aircraft warning service which would include provision for the use of detectors."[7] The Air Defense Command's planning group responded by drawing up plans for a warning system of radar units, ground observers, and processing centers, but at the same time it recommended strengthening overseas bases, on the ground that "so long as these were held, the likelihood of a serious attack on the continental United States was considered to be slight."[8]

In the autumn of 1940 and again in early 1941, exercises were held in California in which radars, civilian ground observers, pursuit aircraft, antiaircraft artillery, and information centers that processed the incoming data and issued instructions for counteraction were integrated into a single system.[9] These military efforts reflected the deteriorating international situation rather than any overhaul of U.S. foreign policy or any striking technical improvements in military readiness, and their ineffectiveness became evident on December 7, 1941.

Meanwhile, a two-way transatlantic flow of military observers had commenced in 1940. American observers (including Captain Gordon P. Saville, who in the postwar years was to play an influential role in air defense) crossed the Atlantic while the Battle of Britain between British fighter planes and German bombers was at its height. Their object was to observe Britain under attack and to study British countermeasures to the German air attacks. They learned "the well-kept secret that the British had developed a system of electronic early warning and fighter control."[10] (One group of observers arrived on August 15, 1940, the very day on which German Field Marshal Hermann Goering launched a massive air attack from across the North Sea and from across the English Channel. The German bombers suffered such grave losses against defending British fighter aircraft that the Germans described the day as "Black Thursday."[11]) Some of the American observers were given special missions; others were assigned to cover the situation in general. All reported favorably on the United Kingdom's ability to resist the Nazi onslaught.[12]

Later in 1940, a number of British scientists and engineers traveled to the United States, at first to exchange technical information with their American counterparts and eventually to work side by side with them on projects of common interest.[13] The first group to arrive was the British Technical Commission, led by Sir Henry Tizard. On September 8, 1940, this commission gathered in Washington with the intention of initiating an exchange of information with American scientists and engineers associated with the newly created National Defense Research Committee, directed by Vannevar Bush.[14]

The British brought with them their multi-cavity magnetron, described by Bush as the "heart of radar."[15] They demonstrated the magnetron and provided the information necessary for its duplication by American manufacturers. The discussions, which extended into November, were unrestricted in scope and content. The Secretary of the British Mission reported:

> A considerable amount of important and highly secret technical information was given to the U.S. authorities. . . . Many documents, drawings, and samples of secret equipment were handed over . . . in cases where the information required was not available here arrangements were made . . . for obtaining it from England.[16]

The Americans were also candid, reserving only the Norden bombsight.[17]

During the aforementioned discussions, members of the British Mission had suggested that the United States pursue the development of "microwave aircraft-interception equipment and a microwave position finder for antiaircraft fire control," and that research facilities be established for that purpose. After "detailed and careful study of military and civil microwave developments in the United States . . . , exhaustive examination of the corresponding British facilities and experience . . . [and] extensive conferences with the British Commission,"[18] the Microwave Section of the National Defense Research Committee approved such a project, and in the autumn of 1940 the Radiation Laboratory was established at MIT. The technical experience developed during the war years in the Radiation Laboratory (directed by Lee DuBridge) and in Gordon Brown's Servomechanisms Laboratory built up a pool of competence and expertise in military technology that was to serve MIT well in the postwar years.

In March 1941, the British-American scientific and engineering collaboration begun by the informal visit of the Tizard group gained official status with the enactment into law of Lend-Lease legislation, which committed the United States to provide all assistance short of war to the nations under assault by the fascist powers. Later that spring, earlier recommendations for the establishment of a permanent "Central Technical Office" to facilitate the exchange of information were implemented. Offices were opened in Washington and London.[19]

The pooling of technical information by Britain and the United States during World War II was not without precedent. During World War I, representatives of the scientific and engineering communities of both countries had crossed the Atlantic to consult and to lay the groundwork for exchanges of information related to military research and for development programs related to matters of shared interest and need. Scientific attachés had been assigned to U.S. diplomatic missions in the capitals of the Allied and Associated Powers to organize and facilitate the flow of information.[20] However, after World War II—in contrast with what happened after World War I—the collaboration carried over into the postwar decades; the two nations sought, both bilaterally and through the North Atlantic Treaty Organization, to contain the threat from the East.

Before the United States entered World War II, both Japan's aggression in the Western Pacific and the potential threat to the Atlantic approaches perceived in Germany's early victories in Europe had aroused interest in the state of U.S. defenses. During 1940 and 1941, the U.S. War Department, having considered experience gained in field exercises, information relayed by American observers in the U.K., and technical data obtained in the exchanges with British experts, took action to set up an aircraft warning system along the western and northeastern coasts.[21]

By the time of the Pearl Harbor attack, the Pacific coast from Seattle to San Diego was guarded by six early-warning radar stations; four more were added before the end of December. There were two stations in the northeast: one in Maine, one in New Jersey. These radar nets were reinforced by Ground Observer Corps stations, some 4000 along the East Coast and 2400 fronting the Pacific.[22]

The destruction wrought by the Japanese at Pearl Harbor (the keystone of the system defending approaches to the West Coast) exposed the United States to the danger of immediate offensive strikes. Two days after the attack, Secretary of War Henry L. Stimson reported to President Franklin D. Roosevelt that the Japanese action had "left the West Coast unprotected." The American people were warned that same day of possible hostile action against the continent. "Our ocean-girt hemisphere is not immune from severe attack," Roosevelt emphatically told the nation. "We cannot measure our safety in terms of miles on any map."[23] After the war, the historian Samuel Eliot Morison wrote that with one swift blow the Japanese had shattered the commonly held illusion that "it can't happen here" and replaced it with the fear that almost any hostile action was possible: "Even strikes on Puget Sound, San Francisco or the Panama Canal were not beyond the range of possibility."[24]

With an incomplete air defense net that was not adequate to resist a major assault, the War Department again turned to Great Britain for advice. The British Air Ministry sent Sir Robert Watson-Watt, a key figure in the development of British radar. Focusing on the West Coast, Watson-Watt found "dangerously unsatisfactory conditions . . . reflecting 'insufficient organization applied to technically inadequate equipment used in exceptionally difficult conditions.'"[25] In full agreement, Colonel Gordon Saville, then director of air defense for Army Air Forces Headquarters, described the report as "a damning indictment of our whole warning service."[26]

Efforts were made to improve the air warning service by incorporating new equipment and modifying older equipment, by expanding the radar warning net and extending it into Canada and Mexico, and by enrolling more volunteers in the Ground Observer Corps.[27] With the turning of the tide of war in the summer of 1942, however, continental defense was de-emphasized, and when Japan surrendered in 1945 the United States' wartime radar defense establishment was almost entirely dismantled.[28]

The war's end brought prompt demobilization, save for the forces needed to enforce the peace. In the euphoria of victory, the American people turned to the camaraderie of victory and the relaxations of peace. But this mood was not to last. By 1946 it was becoming inescapably clear that prewar ideological differences and animosities had been temporarily subordinated, not eliminated, and were again coming to the fore. The Cold War was well underway.

With the end of World War II, air defense of the North American continent had reverted to the status of a theoretical military problem. But in the latter half of the 1940s, as the Cold War began to intensify, the military began to pay more attention to the need for air-defense R&D, and a number of programs were undertaken. The Air Force, for example, actively supported the development of the BOMARC surface-to-air missile, and it encouraged several technical investigations at the University of Michigan's Willow Run Research Center.[29]

Higher echelons in the Air Force were reviewing their air defense policy options. At the Air Staff level, Vice Chief of Staff Hoyt Vandenberg sought the technical advice of a group of civilian engineers and scientists organized under the terms of the National Security Act of 1947. Known as the Research and Development Board (RDB), they were to assist the executive branch of the federal government by advising the Secretary of Defense. They would carry out technical investigations and make assessments for the military establishment, but only at the topmost policy levels. They were not at the beck and call of any of the armed services; rather, they operated at the highest level of policy within the Department of Defense: that of the Office of the Secretary of Defense. This arrangement was deliberate. It would insulate the RDB from direct involvement with the efforts of lower echelons of the armed services to implement decisions to introduce new weapons of advanced design. Perhaps inevitably, it also gave rise to the possibility that the RDB's concerns would become so theoretical that the RDB would become vulnerable to charges that some of its recommendations would have little practical value and no practical military results. Indeed, such a situation emerged during the late 1940s in the fluid, undecided realm of Air Force policy concerns over what to do about continental air defense.[30]

The course of events that involved the Research and Development Board in air defense problems was launched by a letter that the Vice Chief of the Air Staff sent to the chairman of the RDB, Vannevar Bush, in December of 1947. General Vandenberg expressed the Air Staff's apprehensiveness about the state of the nation's air defense and asked for an evaluation of the Air Force's R&D programs in that area. Deploring the lack of a national air warning system, Vandenberg explained that his greatest concern was with the "state of the electronics research and development program of the Services in connection with Air Defense and guided missiles problems." As matters stood, Vandenberg anticipated that there would be "no new developments of radically improved design available for production until after calendar year 1953."[31]

Vandenberg put three questions to Bush: Were new developments coming along at the best possible rate? Was the Army-Navy-Air Force program "properly balanced in this respect?" Were there "any serious deficiencies in this program which should be corrected without delay?"[32]

Bush referred Vandenberg's request to the Subpanel on Early Warning, which was part of the Radar Panel of the RDB's Committee on

Electronics. The subpanel found that in general the program was in proper balance, and that no serious deficiencies required immediate correction. These findings were passed up the RDB's internal chain of command.

The Subpanel on Early Warning seized the opportunity, however, to reiterate an earlier recommendation for "immediate action to establish in a limited coastal area a project for engineering and operational test and evaluation of the combined elements of a system." In January of 1948 the RDB was not receptive.[33] This recommendation was to be repeated in subsequent years.

How much pressure there was to get something done is evident from the subpanel's response to Vandenberg's question as to the rate of new developments: Though there were "ample plans and programs, representing considerable technical progress in radar early-warning equipment," the "limited funds and the difficulties of executing long-term research and development under two-year fiscal restrictions" did not permit the maximum rate of progress. If, however, personnel of "superior qualifications" were brought into the program, newly designed radar equipment could be production-ready by 1954. This would permit a "new improved radar early-warning system" to become "available in service within the period 1957–1959"—several years sooner than Vandenberg had anticipated. However, the early-warning system would have to be given immediate emphasis; otherwise, a new system would be "still years further off."[34]

In the meantime, another response to the question about air defense had been generated in mid December by Julius A. "Jay" Stratton, chairman of the RDB's Committee on Electronics, to be forwarded through Bush to the Secretary of Defense. This report warned that "accomplishment of a system completely effective against any and all offensive weapons should not be anticipated."[35]

Recognizing that more than the development of technically advanced radar equipment was required to counter future threats of attack by high-speed aircraft and missiles, Air Force leaders had already begun considering other means. One of these was automation. In April 1947, specifications had been drawn up for "automatic equipment to pick up and relay information to an air defense control system in an operation that would reduce the human element to a minimum." A year later, the Air Staff had developed serious doubts that "successful interceptions of high-speed attacking airborne objects" would ever be possible with "other than fully automatic means." This perception helps to account for the enthusiasm of Vandenberg's later response to the successful Bedford tests of April 20, 1951.[36]

In late 1948 and early 1949, the Air Force set about specifying military needs associated with air defense. In consultation with the Air Defense Command, Air Force Headquarters "reviewed the USAF operational requirements for the ground electronics portion of the Air Defense System . . . to establish a standard against which our current ground electronic research and development program can be evaluated." Out of this review emerged a ranked list of six "fields of electronic development for air defense to be emphasized." First on the list was airborne IFF ("identification, friend or foe") radar equipment; it was followed, in descending order of priority, by "modernization and improvement of post-war radars not in production," anti-jamming techniques, improved ground-control radar, an "Automatic Interceptor Director System," and an "Automatic Controller Computer System." The last two items are sufficiently important, in view of their novelty, their reflection of the technical state of the art at the time, and their acknowledgment of the technical difficulties that lay ahead, to warrant quoting the discussion of them in full:

> With respect to the automatic interceptor Director System, the air defense system requires that the delays and inaccuracies inherent in the human computation of interception courses and human command of the interception system must be eliminated. It is therefore required that an automatic course computer and guidance system for providing direction information to the defending interceptor aircraft or missile be developed. Additionally, for the complete intercept control system some means must be provided for the course computer to receive continuously information on the position and motion vector of the enemy target being intercepted. A command system for guidance of the interceptor aircraft or missile, which not only guides the defending weapon (aircraft or missile) but also verifies by another independent method that it is actually following the ordered course, is required. Any partial solution of providing a simplified course computer, whose output can assist the human controller, may materially improve the overall system's performance during the transition period from manual to semi-automatic to full automatic GCI.
>
> The Automatic Controller Computer is, with respect to time, the least urgent of these development projects, although it is an absolute must in guided missile operations when the delays introduced by human controllers can no longer be tolerated. This program should be undertaken on a study basis immediately, with development leading to a prototype completed in about 4 years.[37]

Automating the air defense system was not a new idea. In the United States and in Great Britain, the surge of computer development and application was beginning to crest. Air Force leaders were familiar with the use of analog computers during World War II, and presumably they were aware of R&D programs underway in both countries. The lines of communication between military leaders and the scientists and engineers involved on both sides of the Atlantic were open. Vannevar Bush was a key figure in the field of analog computers and was cognizant of the state of the art in digital technology. Karl Taylor Compton, when he resigned the presidency of MIT to assume the chairmanship of the Research and Development Board, carried with him to Washington in the autumn of 1948 a report for "distribution within influential official circles"; prepared for Compton by Jay Forrester and his associates in Project Whirlwind, it discussed the possible application of the digital computer to military needs.[38]

Air defense continued to be an area of persistent, modulated interest to the Research and Development Board and to the operational branches of the national military establishment. The RDB had established an Ad Hoc Panel on Air Defense, which, as was the custom, was expected to report on the theoretical investigations it was pursuing to the RDB's standing Committee on Scientific and Synthetic Analysis.

One member of the Ad Hoc Panel was Perry O. Crawford Jr., an early theorist in the application of the computer to command-and-control systems. During the war years, Crawford, having completed his graduate studies at MIT, had joined the Navy's Special Devices Center. While there, he had worked very closely with Jay Forrester and Robert Everett, the two young engineers who were directing, in MIT's Servomechanisms Laboratory, the project that was to develop the Whirlwind computer. Crawford had been instrumental in suggesting the digital computer to Forrester as the solution to a difficult design problem they had encountered. Shortly afterward, for Forrester and his engineers, the computer *became* the project.[39]

It was Crawford who also first brought to Forrester's attention the Ad Hoc Panel's interest in air defense. This occurred when the two men met in Washington in January of 1949. Both Forrester and Everett had already given thought to applying computers to air defense. Consequently, when the Ad Hoc Panel held its first meeting (on March 18, 1949, in Washington), Forrester sent one of his project engineers, Robert A. Nelson, to participate. Nelson had acquired experience with radar during World War II. He returned from the panel meeting with a conservative prediction: though the panel had not gone far enough into in its projected program of action to enable him to really judge the program, it appeared to him that progress would very likely be slow.

Nelson explained that, at the meeting, Crawford had presented a sample problem for preliminary theoretical analysis. It called for radar to be used to provide both "early warning and ground-controlled interception." A computer would be programmed to carry out a scenario of warning and interception. Nelson had difficulty with some of Crawford's presumptions about what might be expected from the state of the art, for philosophically the two men came from quite different directions. Nelson found the whole business "not very definitive," partly because there was so little agreement among the panel's members on how to define the problem and partly because he was left uneasy by the absence of engineering detail to support Crawford's general proposition that "in the next few years simulation devices used in conjunction with present knowledge we have about air defense and offense can lead almost automatically to a recommended systems development program." Perhaps Nelson's wartime experience with radar made him cautious; perhaps Crawford's vision of what the computer could do made him appear incautious. In either case, Crawford's bold anticipations of future hardware—not to mention software—could easily have been regarded as extravagantly theoretical and speculative, especially by an engineer. Nelson was unable or unwilling to follow what appeared to him to be Crawford's intellectual frolicking in the pitfall-strewn area of ends and means. "What Crawford would like," Nelson concluded in his report on the discussion, "is some little box labelled 'synthetic activation' tying in the aircraft's radar and armament to the ground radar. The discussion that followed was not very definitive."[40]

Nelson was happier to report that Harry F. Goode of the Navy Special Devices Center (who subsequently left that center to join the University of Michigan's air defense program) had "emphasized that the equipment should be flexible—something on a building-block idea—and that simulation equipment should serve in designing other equipment not before constructed." When the meeting broke up, the members of the panel agreed to study, for consideration at their next meeting, a problem that "has to do with the ground-controlled direction of a fighter intercepting and firing on a very high-flying B-29."[41]

Whatever the merits of this Ad Hoc Panel and its interest in simulated aircraft warning and control operations, the panel appeared to lack the internal consensus and the political or funding "clout" in policy-making circles to perform more than a subtle educational function for those exposed to its ideas. Although it met at MIT in late May of 1949, there was no perceptible effect on Forrester and his Whirlwind team—or, for that matter, on air defense R&D programs supported by the Department of Defense.[42]

The increased interest in air defense R&D was not limited to the policy-determining levels of the RDB and the Air Staff. It was also to be encountered at the technical policy implementation level, where Air Force scientific and engineering personnel routinely were engaged in projects related to air defense equipment, techniques, and procedures. MIT's Whirlwind group had become aware of this level of interest when John W. Marchetti (a scientist and an Air Force civilian administrator) and his colleague L. M. Hollingsworth had visited the project on May 3. As a part of their work at the Cambridge Field station (an electronics R&D facility operated by the Air Force), the two men were preoccupied with a theoretical problem of air interception. Marchetti and Hollingsworth envisioned a problem "involving 300 planes, buffer storage, and radar information with about the characteristics of that from the SCR 584 [radar], for controlling a number of anti-aircraft gun batteries through perhaps M-9 directors." Forrester understood the problem; he and Everett had addressed a similar problem in 1947 in a Servomechanisms Laboratory report (L-2, reissued in April of 1949 as R-158). "At first," Nelson reported in an informal note to Forrester, Marchetti and Hollingsworth "seemed quite skeptical that digital computation could accomplish quite efficiently some of their jobs in ground-controlled interception, as for example tracking while scanning for multiple targets," but their "interest in digital computation seemed to liven greatly" during the course of their visit.[43] Seven months later, Marchetti failed to bring the Whirlwind Project to the attention of an Air Force Scientific Advisory Board committee he had joined. The committee, chaired by George Valley, was looking for a computer with which to test an air defense concept it had in mind.

In December of 1949 (the month in which the "Valley Committee" was established), Norman Taylor of Project Whirlwind reported to Forrester on a meeting he and a fellow engineer, J. A. "Gus" O'Brien, had had with Air Force psychologists at Wright Field in Dayton, Ohio. They had discussed "display problems of the type we may need in Air Traffic Control" (a project the MIT engineers were working on under contract to the Air Force), and Taylor had learned there was also pressing concern over a different type of problem, not unrelated to air traffic control: "the Air

Figure 2.1
George Valley, circa 1995.

Intercept Problem." It appeared that only a small number of jet fighter pilots knew how to approach enemy aircraft from the rear. Jet fighters were so fast that they required the pursuer to begin his turn miles away in order to come in behind effectively. On training flights, Ground-Controlled Approach radio equipment was being used by radar observers, who would talk the pilot into position. Better techniques were sorely needed. "These people feel," reported Taylor, "if we can solve this one the Air Force has almost unlimited funds available. They are spending seven-figure money now and not improving techniques." If the fighter tried too sharp a turn, the "G-forces" imposed by the centrifugal force generated could cause the pilot to black out briefly—no way to wage victorious combat. The MIT engineers found the people at Wright Field "quite excited" at the operating speeds being designed into the Whirlwind computer. Forrester considered this information significant enough to log it into his administrative Computation Book, and a full report of the trip was published as a Servomechanisms Laboratory L-Note addressed to the "Whirlwind Planning Group."[44]

Even as Taylor was reporting to Forrester and the Planning Group, however, a revision of national policy—compelled by the USSR's demonstration of atomic capability in August of 1949—was imminent. Routine operational limits imposed by peacetime political and financial considerations were yielding to the formulation and adoption of a more aggressive and urgent air defense R&D policy by Air Force Headquarters. With the outbreak of the Korean War in the summer of 1950, the sense of urgency would become even greater.

The Soviet nuclear test of August 1949 and the intensifying Cold War had impelled Air Force Chief of Staff Hoyt Vandenberg to circulate among other members of the Joint Chiefs of Staff a memorandum noting the "desperate need" for a more effective continental air defense system. In November, Vandenberg's Vice Chief of Staff, General Muir S. Fairchild, referred Vandenberg's memo to a special meeting of the executive committee of the Air Force's Scientific Advisory Board, a high-level advisory group composed of civilian engineers and scientists. Two years had passed since Vandenberg had sought the help of Vannevar Bush and the Research and Development Board, and awareness of the lack of an adequate policy and an adequate system of air defense had increased.

The Scientific Advisory Board was the independent counterpart, at the Department of the Air Force level, of the Research and Development Board at the Office of the Secretary of Defense level. Chaired by a distinguished applied mathematician, Theodore von Karman of the California Institute of Technology, the Scientific Advisory Board advised the Chief of Staff of the Air Force on R&D opportunities for improving weapons.

Coincidentally, during November of 1949 a member of the Scientific Advisory Board, MIT physics professor George E. Valley Jr., was proposing in a letter to von Karman that the Scientific Advisory Board take definite steps toward offering technical solutions that would address the inadequacy of the air defense system.[45] Presumably, von Karman informally called General Fairchild's attention to Valley's proposal: at the special session of the Scientific Advisory Board's executive committee convened to discuss how they might respond to Vandenberg's memorandum, when Fairchild was requesting that "concentrated attention be given the subject," he cited Valley's letter as an example of the sort of action Vandenberg was seeking.[46]

Although it was not obvious at the time, Valley's letter and the Scientific Advisory Board's executive committee meeting brought to a close the beginning of the beginning—the array of historical causes, false starts, and real forward steps that had slowly brought U.S. air defense policy into sufficient focus to compel action. A massive, intricate, and leisurely conjunction of theoretical concept, practice, and accumulated experience had been occurring, brought to pass by profound shifts in the balance of power in international affairs since the nineteenth century, by reluctant changes in defense and military policies that had not been changed since the nineteenth century, and by an increasingly sophisticated twentieth-century scientific technology that had come to dominate the field of military aviation. This war-born scientific technology had already produced radar and a device still so new that it had scarcely been tried: the digital computer.

3

ADSEC

George Valley had worked in MIT's Radiation Laboratory during World War II, and the experience had given him a comprehensive grasp of the state of the art of radar, both its potential and its limitations. He was technically well qualified to be a member of the Air Force's Scientific Advisory Board, and by the end of 1949 his service with the SAB had made him quite aware of the United States' lack of an adequate air defense and of the heightened threat created by the Soviet Union's atomic capacity. Throughout the SAB's deliberations on the matter, Valley argued with vehemence and conviction that decisive action could be taken to strengthen the air defense system.[1]

In a letter to von Karman dated November 8, 1949, Valley discussed two things he was greatly concerned about as a result of his work as a consultant to the Air Force: a lack of "effective coordination between the operational and technical personnel" and a need for "technical guidance in systems research."[2]

In addition, Valley was disturbed by an article in *Air University Quarterly* (an official journal of the Air Force) that, he felt, presented "a discouraging view toward the possibility of providing an air defense."[3] The author of the article, expressing a conviction held by many in the Air Force, argued that an effective air defense system was an impossibility, and that the vast amounts of money that an improved defense system would require would be better spent on "(1) a U.S. strategic bombing force, (2) U.S. Navy and Western European naval operations, and (3) establishment of ground power in European and other nations." The writer concluded that the proposed defense system failed to "rate a high rung on the priority ladder."[4]

Valley contended that such "rumors and feelings of alarm should be allayed" either by proving them to be "unfounded" or by "getting some forceful and intelligent people to do something." Congress's growing interest in air defense made it imperative, in his view, that the Air Force act "to clear the matter up without the help of outsiders."

Valley suggested that the Scientific Advisory Board be allowed to set up an Air Defense Committee composed of members from several of its panels. The first phase of this committee's work would be a "Technical Investigation of Air Defense." Valley wrote:

> This would be a fundamental problem in the field of basic systems research as described in the Ridenour Report. The Committee would have at its disposal several ground radars and crews, together with a squadron of interceptor aircraft and a reasonably free hand to operate these facilities as it willed. The site for this should be near a large city so that the final product could be used to defend the city as well as to serve as a model for the other installations.
>
> I would envisage that the Committee proceed as though it were guiding a kind of thesis problem. It is important to proceed on this basis since the problem will take a long time to solve anyway, and we cannot get the required personnel to work on a full-time basis.
>
> For this reason I suggest that members of the Committee be chosen from the New York-Boston area and that the site be also in that area. I would propose that the problem be guided from week to week by weekly visits of the Committee and that the committee do its planning at informal sessions at convenience. It would also take the opinions of such other experts as it deemed necessary.[5]

The problem, Valley believed, was not unlike the problem the British had dealt with before the outbreak of World War II when they had set up radar stations facing the Continent. Immediate expenditures would be limited, for "the Committee would consult as private members, held together by their mutual respect." Finally, Valley would be glad to pursue the matter further if von Karman thought it desirable.[6]

Von Karman did think it desirable. Valley's recommendations were converted into an SAB proposal, which the SAB's chairman forwarded to the Air Force Chief of Staff 3 weeks later. Approval was prompt. An ad hoc committee was formed immediately, and the Chief of Staff charged it with determining "the operational development of equipment and techniques—on an air defense basis—which would produce maximum effective air defense for a minimum dollar investment."[7] In mid December, Vice Chief of Staff Muir Fairchild informed the Deputy Chiefs of Staff that a new committee to study the technical aspects of continental air defense was being organized by the SAB. Thus the Air Defense System Engineering Committee (ADSEC), chaired by George Valley, came into existence.[8]

The Air Staff was troubled not only by the inadequacy of continental air defense but also by the perceived lack within the Air Force of an appropriate R&D effort. Valley's letter had touched on the latter point, if only by implication. Air Force leaders were already taking steps to redress the imbalance they saw.

The Air Force had been made a separate service in September of 1947, and the Air Staff faced the problems associated with organizing and structuring a new and independent branch of the national military establishment. At the same time, the Air Staff had to secure agreement on the mission and the role of the Air Force amidst intra-service differences and inter-service rivalries. Research and development continued at the level set under the Army Air Forces. Not until early 1948 could the Air Staff give R&D the attention its potential for improvement warranted. The Air Force Scientific Advisory Board, created in early 1946, was elevated to the level of the Office of the Chief of Staff, and the importance of its recommendations was given greater emphasis. Later in 1948 the SAB's assistance was requested in the preparation of an "ultimate plan" intended to ensure that R&D would be "given proper leadership [and] staffed by competent military and civilian technical personnel."[9] In response, the SAB appointed a Special Committee on Research and Development Facilities, Budget and Personnel (commonly called the Ridenour Committee, after its chairman, Louis Ridenour of the University of Illinois, who had been the head of the Radiation Laboratory at MIT at the end of World War II). That committee first met in July of 1949. Deliberations continued until mid September, when it submitted its conclusions. Almost immediately, the Ridenour Report became the technical bible for organizing and conducting R&D within the Air Force.

The Ridenour Committee recommended the "establishment of a Research and Development Command, separate from the Materiel Command, and . . . some reorganization . . . in the [Air Staff] to set off the function of research and development from the logistics-procurement, mobilization, and supply functions." The time had come, the report emphasized, to focus, at least in part, "not on the Air Force in being . . . but on the Air Force of tomorrow." Pursuant to the dictates of managerial protocol, these recommendations were implemented on January 23, 1950, when the Office of Deputy Chief of Staff for Development was created and the Air Research and Development Command was activated. Major General Gordon P. Saville, who had been associated with air defense since the days of World War II, was appointed to command the new staff office, and Major General Donald L. Putt was transferred to it to continue his duties as Director of Research and Development. Both officers firmly supported an active Air Force R&D program, for they recognized the necessity of staying abreast of technological change. Several weeks later, General Putt told the Subcommittee of the House Committee on Appropriations that national security would "depend on . . . combining strategy and technology."[10]

The Air Defense System Engineering Committee came into being at an opportune time. It began its work while the technical services of the Air Force were being reorganized to give more emphasis to research and development. Before the end of January 1950, Vice Chief of Staff Fairchild was notifying the Commanding General of the Air Materiel Command that an Air Defense System Engineering Committee of the Scientific Advisory Board had been appointed, and that it was composed of seven civilian experts (including its chairman, George Valley). Citing "the urgency attached to the Air Defense question" and "the many beneficial contributions" that the new committee could make, Fairchild requested "full cooperation and expeditious action within your command in connection with all the work of this group."[11] Without waiting for Fairchild's announcement, Valley's committee had already begun to meet.

The new body rapidly became known by its acronym, ADSEC, although informally it was often called "the Valley Committee." Officially it was to report to the Chief of Staff through the Scientific Advisory Board, but most of the time it worked closely with Major General Saville.

From the start, ADSEC was an advisory body, with no responsibility for the implementation of its recommendations. That arrangement raised the possibility that the committee could be ignored if not for the political reality of its support at highest Command levels within the Air Force.

ADSEC met regularly on Friday mornings. Also present were a military secretary, Major Richard T. Cella (who also served as a liaison with the parent Board), a representative of the Continental Air Command, and invited specialists from fields relevant to the committee's investigations. It functioned informally avoiding written reports as much as possible; its requests for action were submitted through the Air Force officers who were present. Its purpose was to "do anything technically possible to assist the Air Force in carrying out its mission to defend the United States from attack by air."[12]

Five of ADSEC's seven members were drawn from the Scientific Advisory Board: Valley, a physicist; Allen F. Donovan, a specialist in aerodynamics; Henry C. Houghton, a meteorologist; H. Guyford Stever, an aeronautical engineer and an authority on guided missiles; and Charles Stark Draper, an aeronautical engineer. William R. Hawthorne, an expert on aircraft propulsion and a member of the SAB but not of ADSEC, worked closely with the group. The two non-SAB members, both radar specialists, were George Comstock of the Airborne Instrument Laboratories and John Marchetti, Director of Radio Physics Research at the Air Force Cambridge Research Laboratories. With the exception of Donovan, who was from the Cornell University Aeronautical Laboratory, all the members drawn from the SAB were on the faculty of the Massachusetts Institute of Technology.[13]

At its first meeting (December 27, 1949), ADSEC embarked on a task it was to shoulder until January of 1952, when it was dissolved by the Scientific Advisory Board. (By that time its creative work was done and the program it had initiated was being carried forward by Lincoln Laboratory.) All its meetings took place at the Air Force Cambridge Research Laboratories, at 220 Albany Street in Cambridge, Massachusetts (a location convenient for most of the members, all of whom were from the northeastern U.S.). The Cambridge Research Laboratories had been directed, in accord with Fairchild's request, to provide whatever facilities and services ADSEC might require. Thus began a relationship that was to carry on throughout the life of ADSEC and continue with its successor, Lincoln Laboratory.[14]

The Air Force facility in Cambridge was selected as a meeting place for ADSEC not only for its convenient location but also because its research laboratories possessed a technically experienced and competent staff, a consideration that did not escape Valley and von Karman when the committee was being formed. In 1945, as World War II approached its end, the Air Technical Service Command of the Army Air Forces had foreseen the need for continuing R&D and had set about retaining the services of scientists and engineers who had been engaged in wartime programs at Harvard and MIT. Unable to persuade them to leave the Boston-Cambridge area, military officers had set up a "field station" in Cambridge. Many of those who had accepted employment at the Cambridge Field Station (known later as the Air Force Cambridge Research Laboratories and still later as the Air Force Cambridge Research Center) continued to maintain a keen interest in air defense systems. Some had transferred in 1945 from MIT's Radiation Laboratory to the Field Station in order "to continue ground radar research." It made sound, practical sense to have John Marchetti on the committee: in addition to his administrative and technical ties to the Cambridge Research

Laboratories, his own directorate was engaged in investigations of potential significance to the committee's work.[15]

In accord with Valley's earlier suggestions, ADSEC's task had two parts: "(1) To make such recommendations as are necessary to bring the presently planned air defense system to its maximum inherent effectiveness" and "(2) To design a model of the best air defense system conceivable, under only the limitations of basic natural and economic laws."[16]

For the first several months, the members of the Valley Committee studied air defense in general, the current system, and the changes planned. The results of their investigations were "depressing but not discouraging." Visits to Ground Control Intercept (GCI) stations left the aeronautical engineers on the committee "astounded." The stations, they discovered, were equipped with "completely inadequate and antiquated means . . . for the control of the new interceptors." They quickly recognized that "the necessary improvements imposed conflicting requirements which, in the long run, would require some radical departure from the established concepts, both of usable techniques and of organization of the air defense system as a whole."[17]

In ground equipment and procedures, the manual approach that was in use was essentially the system that had been used during World War II. Consequently, it was unable to incorporate hardware of more advanced design or to accommodate equipment and techniques for communicating, filtering, displaying, or storing incoming information.[18] It made, the committee acknowledged, the "best use of the materials available," but it was technically inadequate and obsolescent. There was too much delay in the flow of information between the area commander and his field units, communication by voice led to confusion, and only a few of the aircraft detected could be tracked or intercepted. Also, the committee found a "very serious and fundamental technical limitation" in the fact that "large long-range ground radars" permitted "low-flying aircraft to remain undetected in some regions." The solution ADSEC proposed to the last flaw was to use great numbers of small radars that supposedly could operate unattended for 30 days at a stretch. This, however, compounded another problem: the system's "low traffic-handling capacity." Though the low traffic-handling capacity could be remedied by "improvements in communication and data display," the committee noted, the proposed solution to the low-altitude problem "would impose still greater burdens on the already inadequate data-analysis part of the system." Consequently, the committee concluded, "the traffic handling capacity of the system would have to be not merely improved, but improved by orders of magnitude."[19]

Such were the conditions and the prospects of the existing military air defense system, and it existed in a world altogether different from the realm of research and development in which MIT's Whirlwind machine was being designed and built. Consequently, when George Valley stumbled on Project Whirlwind, in January of 1950, he promptly recognized the disparity between current military air defense activities and those being carried on at the R&D level in the MIT project. He also saw that Whirlwind could provide no quick fix to the defense problem.

Eight months into its study of the air defense problem, ADSEC described the contemporary system as "lame, purblind, and idiot-like," adding that it made "little sense" to improve interceptor weapons without improving radar and without developing a centralized facility capable of reaching prompt and rational decisions.[20]

The answer, ADSEC declared, was the digital computer. It should be able to perform the "prime function of each Data Analyzer"—that is, to compute the "position of all the aircraft in range of its associated observation posts." There were additional benefits, for the digital computer was also "capable at no additional cost in complexity of performing any of the following functions as desired: (1) Automatic GCT (control signals from xmitters on posts) [,] (2) Automatic Target Evaluation (take-off signals by telephone lines by shortest route to best airport from pertinent Data Analyzer) [,] (3) Automatic Data Filtering[,] (4) Automatic Defense maneuvers *but only according to predetermined strategic principles.*"[21]

Increases in requirements for automatic functioning would adversely affect the traffic-handling capacity of the computer, but this matter could easily be handled by incorporating additional computers into the system. Initially, ADSEC estimated, each 2 million square miles would require about 100 Data Analyzers to be "commercially installed and maintained," at a cost of $500,000 each.[22]

To those abreast of the state of the art, ADSEC's recourse to the electronic digital computer was neither surprising nor novel. Although the prevailing expert view saw it primarily as a major aid to extended mathematical computation, theoretical progress in computer design had advanced far enough that the digital computer could speculatively be included in complex military command-and-control systems.[23]

In the planning realm of military affairs (not to be confused with actual defense operations), by June of 1947 the Army Air Forces had prepared a requirement for "a large scale digital type computer . . . capable of computing and printing in a matter of a few hours the major phases of a plan and of printing in detail the entire program in two to three days."[24] But this was all speculative and highly theoretical.

In July of 1948, at a meeting held on the University of California's Los Angeles campus, Perry Crawford discussed a more specific application of the computer to control aircraft and declared that "in the area of flight operations" digital computers should reach the peak of their application.[25] But this, too, was speculative and highly theoretical.

Early in 1949, a Major General in the Air Force was willing to list among the Air Materiel Command's priorities in air defense the development of "an Automatic Controller Computer System."[26] This also was speculative and highly theoretical. No such system existed or was under construction.

Programs for the design and development of the digital computer were well underway in Great Britain, and consideration was being given to its application to air defense. For "upwards of 10 years" the Telecommunications Research Establishment had been conducting a program to develop a "master computer" that would accept, store, deliver, and distribute information obtained from radar stations. Initially an analog computer may have been the focus of this effort, but in 1947 TRE had commenced development of TREAC, "the only *parallel* computer being designed in Britain at the time."[27] Even as Jay Forrester and his colleagues were preparing a special report on military applications of computers for MIT's outgoing president, Karl Taylor Compton, to take to Washington (where he was to serve on the Research and Development Board advising the Secretary of Defense), British Telecommunications Research Limited was preparing a report to the Ministry of Supply in which automation of air defense was discussed.[28]

Almost from the outset, Valley and his ADSEC colleagues had contemplated incorporating a computer into the aircraft control and warning system they envisioned, although initially, they admitted, only the "haziest notions were had of what the central computer would be like or what it would actually have to do." Automatic computation, the group reasoned, was the only way to cope with the technical and quantitative problems arising from the use of numerous low-wave radars. Once the "conception of a centralized computer with many small radars . . . began to take form," Valley recalled, the problem arose of where to find a computer capable of testing the concept.[29] Unfortunately, this problem did not suggest its solution. In the realm of military practice no such device existed, and in December of 1949 Valley recognized how wide was the gap between practice and promise.

On January 20, 1950 (3 weeks after ADSEC's first meeting), Perry Crawford dropped by MIT to visit with Jay Forrester in the Servomechanisms Laboratory, where Forrester and his engineering team were building the giant Whirlwind experimental computer under Navy sponsorship. Crawford wanted to discuss a Research and Development Board Air Defense Panel report he was writing. In the course of conversation, he passed on the news of ADSEC's recent formation. In his administrative diary, Forrester noted that Crawford had told how "a new committee had been set up by the Air Forces on Air Defense System Engineering with George Valley here at MIT as chairman." Crawford had mentioned that among the members of Valley's committee were "Guy Stever and Draper of MIT and Marchetti of the Cambridge Field Station and Comstock of Airborne Instruments Laboratory."[30] The information Crawford conveyed to Forrester that morning betokened a fundamental change in the fortunes of Forrester's computer project, but too subtly to suggest to either Forrester or Crawford how profoundly the future of the project would be altered.

Events were converging on Forrester and his project engineers—events in which they had taken no part. Before the end of January, forces had been set in motion that would dominate Project Whirlwind's future and would sweep aside whatever real-time application projects (such as air traffic control) the engineers had been pursuing for most of the preceding year. None of the participants was yet in a position to appreciate the fact that at last a mission had been found for the Whirlwind machine that would challenge its potential, absorb the attention of its engineers for a decade or longer, and demonstrate the power of digital computer R&D to strategically transform contemporary technology.

4

Valley's Discovery

The circumstances of Project Whirlwind's involvement with ADSEC appear to have been quite accidental. Whirlwind came to George Valley's attention as a consequence of an encounter on the MIT campus with Jerome ("Jerry") Wiesner, an electrical engineer and administrator at MIT's Research Laboratory for Electronics. Valley had described to Wiesner the concept of an air defense system that ADSEC had been developing since its formation a month earlier. As Wiesner recalled it in 1950, Valley then remarked on the difficulty the committee had encountered in its search for "an information-gathering and information-correlating center that could organize with extreme rapidity great numbers of diverse pieces of information." Wiesner suggested that Valley look into the digital computer project Jay Forrester was running in the Servomechanisms Laboratory.[1]

Wiesner brought Valley and Forrester together for lunch at MIT's Faculty Club. Valley needed a computer; Forrester needed a mission. Forrester wrote the following in his log book:

> On Friday, January 27, I had lunch with Jerry Wiesner, and we were joined by George Valley to discuss his committee work on Air Defense System Engineering. He outlined some of the shortcomings of the present air defense and early warning system and his plans for a research project involving the Cambridge Field Station, Draper's group, and possibly the Digital Computer Laboratory. Later he came to the laboratory to see Whirlwind I operating with test storage, and we discussed possible participation in his work. He said he would get together Draper, Marchetti, and I sometime early in the next week to discuss a possible project before he goes to Washington on February 1 or 2. He seemed quite interested in possible use of Whirlwind I for analyzing data from a chain of doppler radar stations which would give range rate only.[2]

After suggesting the second meeting, Valley left, carrying with him a select batch of Project Whirlwind reports, including Servomechanisms Laboratory Reports L-1 and L-2. Forrester and Bob Everett had written these reports—which described a hypothetical command-and-control center for antisubmarine warfare, built around a computer—in 1947.[3]

The following Monday, Valley returned to the Barta Building,[4] which housed the Whirlwind machine, with three members of ADSEC (Charles Draper, H. Guyford Stever, and John Marchetti) and Eugene Grant of the Cambridge Field Station. Forrester noted in his administrative notebook that they went over material he and Valley had discussed Friday. Walking among Whirlwind's rows of equipment racks, the men observed various components and subsystems being tested for performance. Power had been restored to the machine in mid January after a month's shutdown. The performance of the storage tubes was also being monitored with a television display.

After their tour of the laboratory, the men began discussing how Whirlwind might be integrated with some of the trial systems Valley's

Figure 4.1
The Barta Building.

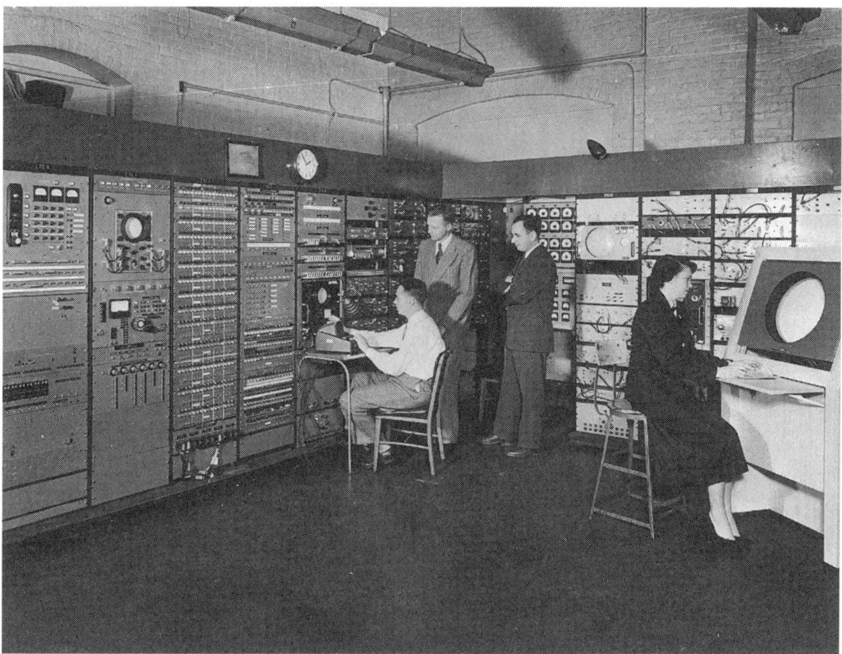

Figure 4.2
Left to right: Stephen Dodd at the test console of the Whirlwind computer; Jay Forrester; Robert Everett; Ramona Ferenz at a display scope.

committee had been considering. "All of the men seemed to be very enthusiastic about the prospect," Forrester thought. He was heartened to find they were realistically aware of the cost picture and "the fact that they would be called upon to share some of the basic $600,000 a year budget for the laboratory, plus additional charges for special work on their own project." It was not unreasonable to think of a level of effort by July that would be at a $200,000 per year rate. "The Committee is apparently meeting in Washington in the next few days to crystallize the matter further."[5]

By the middle of February, it was apparent that Valley was becoming increasingly committed to the idea of using Whirlwind to test out and demonstrate on a limited scale the feasibility of ADSEC's approach to the air defense problem. He phoned Forrester to invite him to the next meeting of the committee to "discuss the proposed system" (whose major components were the radar set and the Whirlwind computer), and he remarked informally that by March 1 the Air Force would approve their continuing to develop the system. The Washington meeting had evidently gone well, and rather than introduce obstacles it apparently expanded Valley's optimism. He suggested confidentially to Forrester that, if it seemed necessary, the Air Force might take over the entire cost of operating Project Whirlwind. Forrester recognized that this reflected Valley's enthusiasm and was not confirmed Air Force policy.[6] Valley then moved on to technical details. Could the computer handle predetermined flight plans of friendly aircraft in such a way that they would not be reported? The "situation board"—a wall map of military operations, continually erased and updated by trained personnel—should one day be replaced by a board electronically controlled by the computer and reporting selected radar data. Getting to more immediate matters, Valley told Forrester that ADSEC expected to have its own radar set operating by midsummer and turning out "information which we can start to make use of," and that "later they might add four or five Air Force radar sets within a 200 mile radius of here." Valley was impatient to get to work on the technical details, so that they might then move on to the tactical and strategic implications of the early-warning system he had in mind.[7]

Four days after Valley's phone call, Forrester brought two of his project engineers to an ADSEC meeting: his second in command, Bob Everett, and his laboratory administrator, Harris Fahnestock. It was a planning meeting devoted to technical details, as Valley had promised. The military radar station near Bedford had to be readied for tests that the committee had in mind. If these tests worked, then the Cambridge Research Laboratories' decoding system, designed to put the radar signals on a telephone line in pulse form, should enable signals from the Bedford MEW (microwave early warning) radar to reach the Whirlwind computer in coherent form. Also, Doppler radar stations had to be built, to provide more accurate range data. "We discussed the equations for Doppler station data reduction," Forrester noted in his logbook. How much of a computing load would they impose?

Forrester felt that Valley and his committee were placing higher expectations on Project Whirlwind without possessing the "hard" information on the Whirlwind machine's operation that would support such expectations. "We shall have to be careful," he warned, "not to be driven into a situation demanding more than we can deliver." He and Marchetti would go over the details in the coming 3 weeks and would "report on what seems feasible."[8]

Valley always presented his case with authority, confident that the Air Force fully supported his committee and its work. Whether discussing the matter with administrators of MIT's Digital Computer Laboratory, or later with Whirlwind's sponsor, the Office of Naval Research, or with upper-level MIT policy makers, he never showed any doubt that they would receive appropriate Air Force support. He was well aware that authoritative voices within the Air Force held his committee's purpose to be of paramount importance, warranting a free hand and full support. In particular, Major General Gordon Saville and Major General Donald Putt were firm supporters of R&D, and especially of air defense R&D.

Events were to justify Valley's confidence. On March 28 the committee's task was assigned a 1-A national priority, and shortly thereafter Air Force Headquarters directed that "no other 1-A priority projects at Cambridge Research Laboratories will have precedence over the ADSEC projects."[9] The Air Materiel Command mirrored this emphasis in Technical Instructions issued to lower operational levels on June 2, 1950: "The urgency of ADSEC's project cannot be over-emphasized since results of their work may alter the Air Force's basic concept of an Air Defense System."[10]

More tangible evidence of support was the financial undergirding that promptly implemented the recommendations affecting Project Whirlwind and the Air Force Cambridge Research Laboratories.[11] To meet immediate financial requirements in late March, a sum of $300,000 was allocated from fiscal year 1950 funds. That money was to be evenly divided among three areas of investigation, two of which—Very Low Altitude Cover Study and Investigation of Long Range Radar Technique—were already underway. Both types of radar, it was noted in the directive from Air Force Headquarters, were intended for use with a high-speed computer. The third area of investigation, Air Defense Systems Studies, was new; it would focus on "a study of the Ground Wave Radar, the exact extent to be determined by the Committee."[12] Inserted into the proposed Air Force budget for fiscal year 1951 was the sum of $1,650,000 for a single line item labeled Air Defense Systems Engineering Project. The monies would be disbursed through the Air Materiel Command "for expenditures as directed by the committee."[13]

The ADSEC funds were accorded special status. When reductions in the budget for fiscal year 1951 were being considered in the autumn of 1950, the Air Materiel Command's Engineering Division requested that the R&D budgets of the Command's various laboratories (including Geophysical and Radio Physics at the Cambridge installation) be pared. At Cambridge the bulk of the paring was to occur in the Geophysical section. In compliance, Robert Rader, Chief of the Air Defense Group in

Radio Physics Research, prepared, in order of priority, a list of the projects that were to suffer cutbacks. ADSEC's funding was spared.[14]

By the autumn of 1950, when ADSEC submitted its second and final report, the pieces were in place and the tests to determine the concept's feasibility could proceed. The system proposed was an electronically modified, advanced version of the manual system of observing, telling, and plotting that had been developed during World War II and carried over into the postwar period with some improvements in radar and weapons. The manual system, however, was unacceptably slow, and it lacked the "capacity . . . to handle mass raids quickly and effectively."[15]

Two remedies were needed. The first was the development of a way to automate the ground functions in order to permit rapid and accurate processing and display of essential data. Since these data would determine the command decisions that would have to be made, it was imperative that techniques be perfected to transmit incoming and outgoing data rapidly and accurately. Whirlwind pointed the way to accomplishing this. The second remedy was on its way to being achieved at the Cambridge Research Laboratories by a group, under the direction of John V. Harrington, that was developing techniques for processing digital signals at the radar sites for transmission over telephone lines. This project would provide another major technical accomplishment that would prove essential to the success of the overall air defense effort.[16]

The increased confidence engendered by the promise of these pioneering efforts was expressed in ADSEC's final report, issued in October of 1950, which clearly stated the conviction that operational decisions in air defense would be made at command centers and would be based on calculations performed by a digital computer:

> We envisage the data coming in from the analyzer to consist of three categories: interceptors, identified aircraft, unidentified aircraft. . . . In the initial operation of this system, and possibly always . . . judgments will be formed by men. . . . A possible situation board . . . would present . . . a three dimensional picture in which the three categories of aircraft are given different colors. This should be fairly easy to contrive, given digital computer signals to work with instead of radar signals. From this board all judgments affecting large groups of aircraft can certainly be made.[17]

ADSEC sought to create a superior air defense system either by improving existing defenses or by designing a replacement. As time passed, particularly after George Valley discovered the existence of Project

Whirlwind, ADSEC's efforts were concentrated more and more on the design of an experimental model system that might be operational by 1954 at the earliest. That date proved unrealistic, but the committee's concentration on the ideal of an experimental model eventually led to a number of major R&D actions, each of which became an almost independent project.

In view of the complexity of the projected model system, it made sense at the time to review and confirm ADSEC's concept by creating a new study group. Project Charles, organized in 1951 at the Air Force's request, is an indication of the confidence that was placed in ADSEC's work: instead of bringing in a group dominated by outsiders—an outside inspection team, if you will—the policy makers asked MIT to provide administrative oversight for it.

As a consequence of the Project Charles group's recommendation that ADSEC's approach be continued and implemented, the new Project Lincoln was staffed by many of the key members of the Project Charles team. Later redesignated Lincoln Laboratory, it was organized, sponsored, and administered by MIT at the request of the Air Force. Responsibility for upgrading the communication system was assumed by the Bell Telephone Laboratories, which the Air Force asked to undertake a program whose aims were to bring the Continental Air Defense System "to the best operating condition consistent with its design capabilities" and to "make recommendations for its continued effective operation."[18]

In 1967, Theodore von Karman remarked in his memoirs that "perhaps the most impressive single item in the accomplishments of the [Scientific Advisory] Board was the development of an air defense system for the North American continent."[19]

5

Managerial Misgivings vs. Technical Accomplishments

At the end of January 1950, when George Valley and his ADSEC associates discovered the Whirlwind computer, Jay Forrester and his Project Whirlwind engineers happened to be uniquely prepared to take on the task they had in mind. But in 1950 this was not something that outsiders could evaluate in a sufficiently informed way, nor could outsiders be sure that those on the inside were not overreaching, perhaps carried away by their enthusiasm. Indeed, there were administrators at MIT and in the Office of Naval Research who feared that the latter might well be the case.

In December of 1949, when ADSEC was created, it was not at all clear to the funding policy makers of MIT and ONR that the Whirlwind computer had a future. The new chairman of ADSEC did not yet know that it existed. Although it was being funded by the Navy, the Research and Development Board's technical advisers to the Secretary of Defense could find no military mission for it. Program managers within ONR considered its mathematical design parameters ill-conceived and its hardware design too expensive to develop.

It took until the spring of 1950 to turn the fortunes of Project Whirlwind around. The central policy problem was seen to be how best to manage R&D on limited funds—in this case, how to manage the novelty of digital computer R&D during the first decade of the electronic digital computer's existence. At the beginning of 1950, half a year before the Korean War broke out, Project Whirlwind appeared to some to be so beset with troubles that its funded level of effort would have to be curtailed and its activities brought to as inexpensive a close as upper-level administrators could devise. Both ONR and MIT policy makers had lost a great deal of faith in the project and were prepared to move it onto a side track, from which, at worst, it might later be inconspicuously derailed; at best, superseded and obsolescent, it might perform some modest functions away from the forefront of the larger computer R&D enterprise that would have passed it by.

Consequently, Valley's discovery of Whirlwind in January of 1950 was fortuitously opportune for ADSEC and for Project Whirlwind, because Valley brought with him not only a challenge that would test Whirlwind's untried promise but also an incidental solution to its funding problems.

As Project Whirlwind was progressing steadily toward technical accomplishment of its mission in the laboratory, it was becoming increasingly unacceptable to its funding sponsor, the Office of Naval Research. The dollar costs were simply too high; it was swallowing up most of the money available to ONR for computer R&D. Meanwhile, as 1950 approached, a number of external policy considerations at the upper levels of the executive branch of the federal government were incidentally darkening the outlook for military R&D funding in general and for Project Whirlwind in particular. Some of these difficulties concerned novel, broad, fundamental questions of how science and science-based technology should best be carried on now that World War II was over and the Cold War had begun. Others concerned narrower issues of how MIT should best go about encouraging computer R&D during the first decade of the electronic digital computer's existence. Both the broader and the narrower issues of managing R&D bore directly on the project's situation, on the funding and programming problems it was encountering at the time Valley stumbled on it, on the impact of Valley's involvement, and on the status and the future of the giant machine being constructed in the Barta Building.

First, the broader issues. Project Whirlwind had originated in the Servomechanisms Laboratory, which MIT Electrical Engineering Professor Gordon Brown had organized to meet the special needs of World War II, and the project was of course responsible to MIT's administration for the conduct of its activities. Since its major funding sponsor was ONR, it was accountable to ONR for how it spent the monies the Navy allocated annually for design and construction of the computer. The focus of responsibility and accountability was on Jay Forrester, the project's director.

The basic views held by ONR and MIT executives regarding the nature of R&D and the philosophy proper to its guidance were firmly rooted in their wartime experience in the early 1940s and in their professional assumptions about the aims of science and engineering. At bottom lay the wartime discovery by military policy makers that scientific technology could accomplish spectacular improvements in weapons, multiplying their versatility, power, and deadliness. An intensive application of science to technology had been innovatively directed during the war by the Office of Scientific Research and Development under the leadership of Vannevar Bush, a former member of MIT's faculty and administration.[1] Although the OSRD had demonstrated beyond any doubt the extraordinary effectiveness of coordinated federal support of military R&D, once the war was over the civilian engineers and scientists who had staffed that office returned to their peacetime activities, and the office was shut down.

It took no great feat of logic or imagination to share the common wisdom of civilian scientists, engineers, and policy makers in the pure and applied sciences that understanding natural processes opens the way to their control. Indeed, the belief was widely held among political and economic leaders after the end of World War II that, since European pure and applied science had been shattered and devastated by the political, economic, and military convulsions that had wracked the subcontinent between 1939 and 1945, the basic-research wellsprings of peacetime scientific technology and modern industrial wealth would soon run dry unless Europe's basic research tradition could be revived in university laboratories of science and engineering, its natural home. The linkage of basic science to scientific technology had a powerful influence on the thinking of the MIT and Navy executives who were committed to building this new, war-spawned device, the electronic digital computer.

Although the United States had not been ravaged by war, there was an obvious inference from the wartime experience that federal support of science would simultaneously advance scientific knowledge and lay the groundwork for advances in science-based technologies useful to American society. Accordingly, Congress created the National Science Foundation. In 1951 the NSF began to invest federal dollars in support of science and engineering research. Seven years later (when, coincidentally, the Air Force put its first air defense center into operation), the NSF was spending more than $30 million a year.[2] None of this NSF money was intended for military R&D, but it indicates how firmly U.S. policy was becoming committed to federal support and encouragement of scientific research, from which engineering development would naturally follow. It was a policy that was becoming "as American as apple pie," and it was a major factor in the emergence of computer technology during the later 1940s and through the 1950s.

As the growing belligerence of the USSR became evident in the first few years after the end of World War II, forward-looking U.S. military officers were unwilling to wait for Congress to organize federal support of science and technology. They began to initiate their own limited programs in military R&D. In 1947 the Navy established the Office of Naval Research, whose mission was to extend research grants in areas of interest to the Navy to college and university faculty members across the country.[3]

ONR happened to inherit Project Whirlwind when it took over the functions of the wartime Naval Office of Research and Inventions. Technical and administrative supervision of Project Whirlwind was assigned to the Mathematics Section of the Science Branch, which was organized according to scientific discipline and staffed primarily by civilian scientists. Since ONR's policy charge and bias deliberately emphasized scientific research over applied engineering, leaving the latter to Navy offices and programs more directly concerned with improving weapons, its policy makers in command of funding were more interested in the digital computer's scientific potential than in its military potential.[4]

The policy issues affecting military R&D funding in late 1949 and early 1950 inevitably affected the funding fortunes of MIT's Digital Computer Laboratory. And in 1948 the ONR's budget had been cut from $22 million to $11 million.[5] No one was expecting the Korean War, whose eruption in June of 1950 was to cause an abrupt and across-the-board reversal of military spending policies.

The MIT and ONR administrators involved in computer R&D understood that no sharp line existed in practice between basic and applied research or between science and engineering. Nor did ONR wish to antagonize MIT and other research institutions by arbitrarily and sharply curtailing certain programs while expanding its support of others. Nevertheless, it was not clear to the ONR program managers funding Project Whirlwind that MIT was undertaking the kind of computer R&D that merited their uncritical support, particularly when Whirlwind was consuming the greater part of their computer R&D budget.[6]

At issue here was not only a struggle of wills between Jay Forrester and Mina S. Rees, head of the Mathematics Branch of ONR (although that was a factor), but also a contest between rival conceptions of the proper way to carry on computer R&D. Nor were MIT's administrators lacking in sympathy for ONR's predicament. Jay Stratton, MIT's provost, happened to share a number of judgments with Mina Rees and her superiors at ONR regarding what constituted proper R&D in general and effective computer R&D in particular. Neither uninformed nor unreasonable, the leaders of MIT and ONR found Jay Forrester's philosophy of funding and managing computer R&D too expensive to be practical under the stringent federal budget circumstances that prevailed in early 1950. More telling, they feared that the Digital Computer Laboratory had fallen off the crest of the technical wave. As a result, Forrester, far from appearing admirably resolute, seemed to be becoming intractably stubborn in his refusal to live within a modest budget proportioned to the value of the modest contribution that ONR believed his project was likely to make to the future of the digital computer.

As January of 1950 gave way to February, and as MIT was preparing to confer with ONR about the Digital Computer Laboratory's R&D budget for the coming fiscal year, Stratton was turning over in his mind the question of how MIT might best manage computer R&D in ways that would enable short-run goals, such as the completion of the Whirlwind machine, to advance long-run policies.[7]

As has already been noted, Stratton's philosophical views were in substantial agreement with those of Mina Rees. Their academic backgrounds were similar enough; he had been a physicist and she a mathematician before each had become an administrator. Also similar were their views of the dependence of applied science on basic science. And both, as administrators, had wrestled with budgetary limitations.

ONR's misgivings about the Digital Computer Laboratory's operations were both well considered and well informed. In 1949, an inspection team composed of Mina Rees, Charles V. L. Smith, and Karl Spangenberg (all of ONR) and an outside consultant (Harry Nyquist of Bell Telephone Laboratories) had failed to allay ONR's concerns about the substantial funding it had been providing to the R&D enterprise that ADSEC later would discover. That team's report is significant for its detailed description of the state of the Whirlwind computer. Although it antedated by nearly half a year George Valley's discovery that Whirlwind existed, it described a situation that still prevailed in 1950. In August of 1949 the report was sent in the form of a letter to Jay Forrester and to three top-level MIT administrators: President James Killian, Provost Jay Stratton, and Nat Sage, the Director of the Division of Industrial Cooperation.[8] Though the report ended on a reassuring note, it mildly voiced several disturbing concerns in its discussion of some of the development problems the project was caught up in. Those technical problems gave ONR serious pause, partly because they required practical engineering solutions and partly because they reflected the Whirlwind computer's deficiency, in ONR's eyes, as a machine really useful for scientific research.

These issues are of particular interest to this story on several counts. They reveal how far along toward completion the Whirlwind computer had come. And, although neither the ONR administrators nor the MIT administrators could read the future, they were issues that would have to be resolved before the machine could be used for crucial tests of the feasibility of ADSEC's air defense scheme. They are of interest also because they reveal how Project Whirlwind was carrying on computer R&D. Finally, these problems were significant indicators of the state of the art of computer R&D in 1949–50 and in the next few years, when ADSEC and its successor, Project Lincoln, were testing the computer's usefulness as a command-and-control center for air defense.

The report of the ONR inspection team emphasized four areas of concern about the Whirlwind machine. First, internal storage was so severely limited (as a consequence of unforeseen chronic development problems with the electrostatic storage tubes) that additional "auxiliary storage" apparently would be needed to make the machine sufficiently useful at an early date. Communicating with the machine posed a second major concern, for the reliability of the reader-recorder developed by the Eastman Company was untested. The reader-recorder was intended to record "binary numbers on photographic film by . . . light from a cathode ray tube" and then read the recorded data by "scanning of the processed film with a cathode ray tube."[9] The inspection team therefore recommended that "detailed consideration be given to the question of input-output and auxiliary storage." The third concern was a fundamental weakness that ONR saw in the machine as a potential scientific instrument, and it was a concern that had been voiced before: "the limitations that will inevitably result from the present word length of Whirlwind I, and from the fact that it is a fixed rather than a floating point machine." These were limitations on Whirlwind's capacity to assist in scientific computation. Since MIT's project engineers and mathematicians were already studying the possibility of coding double-length-word operations, the inspection committee recommended that "routines for the coding of both floating point and double precision operations be established." The fourth concern had to do with the storage tubes used for the machine's internal "memory." Though the committee was encouraged by the quality of the project's tube development work, there was little likelihood at this juncture of determining "accurately how much it would cost in terms of time and effort to obtain the desired improvements" in the storage tubes' performance. Consequently, the committee recommended, "in the interest of obtaining at least a workable internal storage within the present Fiscal Year," that "the Project concentrate on obtaining a sufficient number of usable tubes."

The report ended on a courteous note, the inspection committee declaring that generally it was "favorably impressed by the thoroughness of the engineering effort displayed by the Whirlwind staff, and by the energy, enthusiasm and directness of approach." However, the committee was not satisfied that Whirlwind was as good a mathematical machine as could be obtained for the money, or that it would ever acquire a sufficiently large and reliable internal storage capacity to be worth the cost. And what use would it be without adequate input-output facilities? Although the inspection team was willing to put on a hopeful face and applaud the engineers' effort, the bulk of the report pointed out serious deficiencies. It did not take much imagination to wonder whether the $750,000 that ONR had allocated for the 1949–50 fiscal year was being well spent.[10]

The ONR inspection team was not alone in its criticism of Project Whirlwind. In December of 1949, just as ADSEC was getting organized under the top Scientific Advisory Board of the Air Force, a three-man Ad Hoc Panel (of which Harry Nyquist had become a member) issued to its parent Research and Development Board of the Office of the Secretary of Defense a report on the general state of military support of computer R&D. In this report the panel sharply criticized Project Whirlwind for the high military R&D costs it had incurred while developing a computer that had no explicit military mission.[11]

Thus, when George Valley and his committee appeared on the scene, with their particular technical needs and their different interpretive frame of reference, Project Whirlwind was a beleaguered and apparently wayward R&D enterprise in the process of having its funding cut back.

Against this background of events, January 1950 can be seen as a watershed month that separated the Navy's involvement during World War II and the immediate postwar years from the Air Force's involvement during the Korean War and after. But the Korean War was 6 months away, and it was not at all clear to the MIT and ONR participants that a fortuitous combination of events was impelling George Valley on an intersecting course with Project Whirlwind and its huge, unfinished computer.

To complete the picture of this critical juncture of events, it is necessary to look inside MIT's Digital Computer Laboratory and to examine Project Whirlwind's technical problems and its level of technical accomplishment. It is here, especially, that evidence is to be found of how uniquely and fortuitously prepared to meet ADSEC's need the Whirlwind engineers had become.[12] By 1950 the DCL had become a sturdy, vigorous R&D organization with a distinctive élan.[13] Most of its youthful, high-spirited engineers realized that they were working at the challenging frontier of engineering, mathematical, and physical knowledge. The first postwar computer conference we are aware of (a conference Jay Forrester had attended as a novice) had occurred less than 5 years before; it had been convened by R. C. Archibald, a Brown University mathematician then serving as chairman of the National Research Council's Committee on Mathematical Tables and Other Aids to Computation.

While exploring uncharted waters, Project Whirlwind was conducting itself unlike any contemporary computer development project. These circumstances affected both the data of experience and the philosophy that the engineers in the project acquired. At the same time, the state of the art made evaluation of these data and of the evidence provided by the project's efforts at once easy, disturbing, and difficult. The project was deceptively easy to evaluate by the standard of customary development projects—and thus disturbing. It was difficult to evaluate because its merit was obscured by the youthfulness of its managers and by the intricate unorthodoxy of their methods. Aside from the premium placed on technical knowledge in any competent technical evaluation, there was the matter of style. Forrester's style of R&D was to "go first class" or not at all, which meant that costs were subordinated to engineering needs rather than vice versa. Such a philosophy had been considered essential to winning World War II, whether or not the proposition could be proved. But it was not considered essential to peacetime R&D projects, and this was a major source of the philosophical differences that had developed between Forrester and the ONR administrators.

Another manifestation of Forrester's style was the elaborate information-exchange system he set up among the engineers of the DCL as they pursued their different personal tasks. Frequent written reports kept the investigators who had to write them, as well as their supervisors and colleagues, aware of what was going on, and these reports were generated in a critical, helpful atmosphere in which the engineers were encouraged to be frank about their technical problems and to avoid dissembling cover-ups. They had the luxury of being honest in pursuit of inquiry and frank in identifying problems, and they enjoyed it. This subtle but fundamental element of the high esprit de corps of the DCL affected the full-time engineers, the part-time graduate students, and the clerical assistants.

There were progress reports, trip reports, conferences and conference reports, proposals, master's theses stemming out of work done in the laboratory, technical memoranda, and discussions with visiting experts from industry, academe, and government. *Any* information relevant to the technical challenges being faced in the laboratory was considered worth setting down in a dittoed "M-Note," "C-Note," "E-Note," "Bi-Weekly Report," or other such report or note. The high level of communication and incidental documentation of the laboratory's technical affairs was one of the abiding unorthodox features of Project Whirlwind, and it continued after the project organization became Division 6 of Lincoln Laboratory, working exclusively on the air defense computer.

In December of 1949, as arguments were being readied for upcoming budget discussions with ONR, the Research and Development Board's Ad Hoc Panel was completing and circulating its adverse report, and (coincidentally) ADSEC was coming into being, physical construction of Project Whirlwind's room-size computer was well along. Of the four major parts of the computer, the arithmetic calculating element and the central control (which coordinated the functioning parts) were already in limited operation. The arithmetic element, assembled in January of 1949, and the central control, operating since June of 1949, were both undergoing extensive testing by being hooked up to the 32 "manual test storage" circuits, which were controlled by toggle switches, each of which could be flipped to signify either zero or one. When the switches were flipped to produce different binary sequences and patterns of one and zero, the information could be passed by the central control to the arithmetic element in sequences that tested the arithmetic element's ability to add and consequently to carry out the other major arithmetic operations (subtract, multiply, divide) that were essential to the machine's usefulness.

Although the 32 registers of manual test storage sufficed for testing purposes, they were not adequate for any useful work. To handle computation problems, the machine would require automatic internal storage and input-output equipment, which were still under development:. To finish these items would require careful planning of the levels of effort and budgeting of corresponding levels of funding.[14] Rows of racks, separated by service aisles, awaited this equipment.

The input-output devices were to provide data to the machine in the form of binary numbers, and they were to provide instructions regarding what the machine should do with the data—e.g., store them, or add two numbers and store their sum, or compare two numbers to see if they were the same (by subtracting one from the other and obtaining zero if they were equal or some numerical value if they were not). Furthermore, the input-output devices should present the results of the machine's computations, together with as detailed a record of the machine's operations as a user might desire in order to check on how the machine was obtaining its answers. Essential to these operations was the fourth major internal part: the internal storage, or "memory," which operated as a reservoir to hold the data and instructions fed into the machine, the intermediate results of calculations, and the final results. In other words, the internal storage served as a central "holding tank" for the numbers—the raw machine language that, in binary form, was transported back and forth between storage and the arithmetic element, where the numbers were operated on to produce new numbers according to the rules of arithmetic.

Since the storage element served as the central "warehouse" for both the information and the programs that processed the information, the storage also had to be connected to the input-output equipment. Without an operating internal storage of sufficient capacity, the computer would never be able to function the way it was supposed to. Forrester and the MIT and ONR administrators understood this, and shortly Valley would understand it too.

Forrester and Everett had become committed early to the idea of a rapid-access, rapid-retrieval form of internal storage, and Forrester had selected the electrostatic storage tube as the modular component to be developed. The "final design" Forrester had in mind late in 1949 would include enough tubes to handle 2048 (8×256) binary digits arranged in 128 16-digit words. The first bank of four tubes would probably be installed by the end of February 1950, he anticipated. By May they should be operating with the other parts of the computer, and by the first of July test results from the first bank should be sufficient to guide more rapid installation of the second bank of four more tubes. To build the second bank of tubes and install it in the computer should take about 4 months, so by November of 1950 the hardware capacity should reach 2048 registers—if the necessary funds were forthcoming.[15]

On December 15, 1949 (6 weeks before Valley first saw Whirlwind), the components that had been working were shut down to allow removal of temporary wiring and installation of the permanent high- and low-voltage lines of the computer's power distribution system. All the power feeders would then be in place, and only the input-output equipment and the machine's internal electrostatic storage system would remain to be hooked into the main power lines once the subsystem portions of the machine had passed extensive tests in isolation. The shutdown, which had been planned to last 2 weeks, actually lasted 4. The Central Control Room was enlarged, and test equipment for the various parts of the computer was transferred to nine new racks provided in a specially designated Central Test Control section. Air conditioning ducts had to be moved to accommodate rerouted wiring, a new Power Bay Fuse Indication Panel had to be installed, and several hundred cables for the electrostatic-storage section of the machine had to be laid.[16] This general "resetting of the stage" was intended to make the tests being carried on simultaneously in different parts of the giant machine more orderly. By improving both the physical and the electrical arrangements, the engineers expected to get more reliable operation, obtain more informative test results, and make more efficient use of their time.

A small indication of progress was the successful operation of the computer on the very first test problem when testing resumed on January 16, just 11 days before George Valley's unanticipated visit.

6

The Problem of "Internal Storage"

Because creative achievements in research and development usually arise from previous design activities closely tied to the state of the art, they are apt to have small beginnings. Such was the case with the design of the Whirlwind computer's memory.

For nearly 5 years before George Valley found them, Jay Forrester and Bob Everett had been deeply involved in the design of the Whirlwind computer. They were fully aware that the electronic digital computer was a new type of machine. Already conceived by others and under technical development in military secrecy during World War II, it was still far from being a practical device.

As early as the spring of 1947, Forrester was systematically running through in his head and setting down in his laboratory notebook possible hardware alternatives for storing electronically the data and the operating programs such a machine would need in order to operate effectively. This problem of internal storage—they hadn't begun to call it memory yet—was a tough one, especially when high reliability of recall was considered essential. Forrester considered internal storage crucial because he had in mind a variety of exotic (for those days) and exacting tasks that might be carried out by the general-purpose computer that he and his project engineers were designing and building for the Navy.

There already existed, Forrester knew, a number of alternative experimental devices for storing within a computer the information to be processed, the instructions for processing it, and the new information that would be produced. Some of these devices could store information in a linear arrangement; others stored it in a two-dimensional arrangement.

One workable linear arrangement was the mercury acoustic delay line, in which an electric pulse was received by a piezoelectric crystal at one end of a tube of mercury. The crystal vibrated upon receiving the pulse, and the vibration was transferred to the mercury, setting up a tiny acoustic shock wave, which took some time to travel down the tube of mercury until it struck a second piezoelectric crystal at the other end of the tube. When the vibration struck the second crystal, the crystal generated a corresponding electric signal. The signal could then be amplified and transmitted through a wire connected to the first crystal, setting it to vibrating a second time. This repetitive "holding pattern" procedure could be continued for as long as one wished to store the signal in the delay line. Storage was thus accomplished by a process that used the passage of time as a "receptacle." Forrester considered this storage device too slow to be widely useful, and he had no intention of installing it in the machine that he and his fellow engineers in Project Whirlwind were designing and constructing.

Two-dimensional storage made use of the electrostatically charged surface of, say, a disk in a vacuum tube. A number of insulated "spots" could be placed on such a surface, and each of these spots could be given a positive or a negative charge by focusing an electron beam on it. This "electrostatic storage tube," first tested successfully in England in June of 1948 by F. C. Williams on Manchester University's Mark I computer, used the two dimensions of a spot on the disk as its storage receptacle. But Forrester knew that the current experimental models were unreliable and required excessive amounts of time to gain access to their stored information. He was especially sensitive to these problems because his own computer project at MIT was then engaged in developing and testing such an electrostatic storage tube.[1]

By mid March of 1947, Forrester was reflecting on the pros and cons of storing a digital item of information in a separate circuit, as in a vacuum tube, and then multiplying that single circuit by hundreds and thousands to achieve a practical storage arrangement that would accommodate enough information to be useful. More efficient than this arrangement, he saw, would be one in which a single circuit could control many points of information arranged over the area of a surface. Yet such a two-dimensional array would make less efficient use of the volume occupied by the storage equipment than would a closely packed, many-layered, three-dimensional arrangement—if only a working model of such a design could be created. Forrester remarked to Electrical Engineering Professor Harold Hazen and to his friend and informal counselor Nat Sage that such an idea ought to be given serious study, and he put the idea informally to several of the engineers working on Project Whirlwind. The next new student in the laboratory, he remarked, might well be asked to undertake an exploratory study.

By April, Forrester had progressed in his own thinking to consideration of a three-dimensional arrangement consisting of parallel wires passing through perforated plates stacked in a specially designed gas-glow discharge tube. Theoretically, he reasoned, why couldn't one store binary information by taking advantage of the double-valued impedance associated with the low-pressure gas-glow discharges occurring at the perforations? He wrote up the idea in a Servomechanisms Lab technical memorandum titled Data Storage in Three Dimensions.[2] However attractive this idea looked in theory, such an arrangement did not then appear feasible, as a laboratory assistant, William N. Papian, would observe in 1953 when reviewing the state of the art.[3] Forrester himself recalled—13 years after the fact, when involved in patent litigation over magnetic-core storage—that he had decided not to pursue the idea further at the time because closer scrutiny had revealed "the unreliability and non-uniformity of the crystal diode and gas discharge cells."[4] An attractive idea in theory, it wouldn't work in practice.

The importance of suitable internal storage capacity cannot be minimized in a survey of the history of the digital computer. The full potential of the computer simply could not be realized without a reliable memory of great capacity and speed. Forrester and his associates were not alone in working on the problem. The Harvard Computation Laboratory had been working since at least the summer of 1948 on a design that made use of small, metallic, magnetic rings, or cores. The Harvard design arranged the cores to form a "static magnetic delay line."

The Harvard efforts bore early fruit in a paper that researcher An Wang presented orally in June of 1949. It was followed by a paper co-authored by Wang and Way Dong Woo, published in 1950.[5] After learning of the Wang-Woo paper, Munroe K. (Mike) Haynes, a graduate student at the University of Illinois, turned his attention to the subject. In the autumn of 1950, upon completing his doctoral thesis, Haynes moved on to IBM, where he continued his research.

In 1947, at the University of London, Andrew D. Booth, in a lecture not connected with Forrester's interest, discussed the possibility of using magnetic cores as an alternative to storage tubes. In the laboratories of the Radio Corporation of America, Jan A. Rajchman explored the possible use of the magnetic core as a storage device.[6] It was Jay Forrester, however, who realized the concept in practice—an achievement that the computer pioneer Herman Goldstine later described as "one of the basic technological discoveries in the entire computer field."[7]

Two years passed as Project Whirlwind continued to design and assemble its computer and to develop the electrostatic storage tube, encountering various problems of storage reliability and access speed along the way.

In the spring of 1949, Forrester's eyes fell upon an advertisement by the Allegheny Ludlum Company for a magnetic material called Deltamax. Taking a fresh look at the principle of three-dimensional storage, he examined technical data issued by the company and decided to test the magnetic properties of small toroids, or cores, formed of spools of metallic Deltamax ribbon and manufactured by the Arnold Engineering Company (an Allegheny Ludlum subsidiary).

In retrospect, the idea of reversing the direction of direct current passing near or around an iron alloy and thereby reversing the alloy's residual magnetic polarity appears to have been an idea whose time had come. But realizing it in practice posed severe and separate problems that would take months and years to solve. Since electrostatic storage was then further along toward a solution, magnetic storage appeared in comparison to be even less of a practical alternative. It offered only an attractive course for investigative research.

By mid June of 1949, Forrester was writing in his laboratory computation book about a two-dimensional design "using a rectangular hysteresis loop material with coincident-current selection" and about a "cancelling-sense winding method of storage readout." By the end of June, he had begun testing the electromagnetic properties of Deltamax cores. That summer, he was working out the theoretical details of a three-coordinate, three-dimensional system that could magnetize and repolarize any core at will.

By early August, Forrester's tests of the Deltamax material had demonstrated that the polarity of a core could be controlled by applying coincident currents to separate windings on the same core. But since it took 10,000 microseconds to switch the core's polarity, the Deltamax alloy was too slow for "real-time" applications of the Whirlwind computer such as Forrester had in mind (e.g., air traffic control or gunfire control). After conferring with the manufacturer in August and September, he obtained improved toroids that yielded a switching speed of nearly 30 microseconds—still too slow, but heartening.[8]

The preliminary results of his summer investigations satisfied Forrester that he had accumulated enough specific and unanswered questions to justify putting William Papian, an electrical engineering graduate student in the Servomechanisms Laboratory, to work on a paper for a fall seminar in electrical engineering. Forrester wanted Papian to search the literature and to prepare to continue the experimental and conceptual work that Forrester had begun. While Papian was becoming conversant with the problem that summer and fall, Forrester continued to pursue his inquiries into the properties of the Deltamax material, then broadened his study to include other metals and ferrites. Papian soon discovered that Forrester was contemplating a design that would employ a single turn of wire through a ring, or core, and would accommodate thousands of such cores "arranged simply and inexpensively."[9]

By the middle of January 1950, Papian had completed his research paper. It discussed certain theoretical and practical aspects of the electromagnetic behavior of ferromagnetic alloys; these might be useful, Papian noted, in the electronic digital computers then under design development in various electronics laboratories in the United States. Later in January, as a consequence of conversations he had been carrying on with Forrester, Papian formally submitted a proposal for a master's thesis research project in the same area, to be supervised by Forrester.[10] Papian would continue his search of the literature and would attempt some preliminary experimental investigations.

As an electrical engineering graduate student working part time in Project Whirlwind while pursuing an M.S. degree, Papian already had come to realize he was involved in a one-step-at-a-time field of engineering research in which the transformation of theory into successful practice was by no means automatic or certain. The challenging unknown lay ahead, and the point of being an engineer was both to invent and to unmask the unknown. Besides, Papian felt, he had a good introductory understanding of the immediate problems he faced. He was certainly under no illusions. His work on Whirlwind components in the laboratory had immersed him in the performance problems of flip-flop storage circuits and test storage ("memory") composed of rows of toggle switches set by hand. Though these made no use of ferromagnetic alloys, the familiarity he was acquiring with the operation of their circuitry was valuable; it exposed him to the theoretical and experimental realities he would encounter in his thesis research.

Forrester's criticisms in 1949 and 1950 of the one- and two-dimensional storage designs were based on several years' investigations and were shared by his close friend Bob Everett. In May of 1950, Everett revealed the depth of his knowledge and experience in a colloquium talk he gave for MIT's Electrical Engineering Department. After examining "the need for storage in computation, the effect of storage characteristics on digital computer design, desirable storage characteristics, and the characteristics of existing storage devices," he analyzed the projected performance and the design virtues and drawbacks of the linear-storage acoustic delay line, the two-dimensional electrostatic storage tube, and the two-dimensional storage surface of a rotating magnetic drum.[11]

Everett realized that the linear-storage limitations of the mercury delay line could be offset by installing a sufficient number of mercury delay lines in the storage element of a computer, thus placing all three types of storage on a two-dimensional par. Such a typical mercury-tank delay-line installation, Everett concluded, might store 43,000 binary digits and require 150 millionths of a second to provide access to any digit. It would cost $30,000 to design and build and another $3000 a year to maintain, he estimated. In comparison, a typical magnetic drum might store 120,000 binary digits and require 8000 microseconds of access time, on the average. It would cost an estimated $25,000 to design and build and another $3000 a year to maintain. An electrostatic storage tube might store 32,000 binary digits, require 10 millionths of a second access time, and cost $115,000 to design and build and another $5000 a year to maintain. Once developed, each additional tube would cost about $1400 to construct and could be expected to last about 4000 hours.

When Everett compared the estimated first-five-years costs of these three devices, he came up with $45,000 for a typical mercury-tank delay line, $40,000 for a typical magnetic drum, and $252,000 for a typical bank of electrostatic tubes. Obviously the cost of electrostatic tubes was way out of line. But this was not the end of the story. How expensive would the storage be to operate? What would be the processing costs of operating machine time when storage capacity was compared with access time to any digit? The latter comparison, Everett realized, would measure the efficiency of the storage and consequently determine the nature and performance of the entire machine. Comparing storage capacity to access time, Everett arrived at what he called a "performance unit" for each type of storage. He calculated that magnetic-drum storage would cost, when the machine was operating, $2650 per performance unit, versus $157 for delay-line storage and $79 for electrostatic storage.[12]

Since it is easy to overwhelm an audience with numbers, Everett focused on three basic requirements for storage in a digital computer: reliability, speed, and reasonable cost. Reliability was essential, he pointed out, since "a single error anywhere in the machine can destroy the entire calculation." A practical computer should tolerate no more than one error per 100 billion (i.e., 10^{11}) operations—equivalent to one wrong number every 2 years in a telephone network handling 130 million calls a day.

Another requirement was high speed, and the overall speed of any computer would be set by the time required for access to any item of its stored information. Everett traced the speed story briefly. During the early 1940s, Professor Howard Aiken had worked with IBM engineers at Harvard to build the Automatic Sequence Controlled Calculator (also known as Mark I), which used electrically driven counter wheels and punched paper tape. Then came the use of electromagnetic relays in the Harvard Mark II and in the Bell Telephone Laboratories machines developed under the direction of George Stibitz. Mark II and the Bell Labs machines, like Mark I, used paper tape for storage. But counter wheels and relays were expensive calculating and storage devices. J. Presper Eckert and John Mauchly's ENIAC, built at the University of Pennsylvania during the war, used vacuum tubes for storage and computation; although its Eccles-Jordan flip-flop (or ring counter) was faster than a relay, the cost per digit was higher. ENIAC's internal-storage capacity of 20 numbers was set up by cables and plug boards and subsequently by selector switches. It suggested possibilities of impressive speed. But ENIAC was not built to really exploit the storage principle it embodied.

The next desirable step was the development of fast, erasable storage devices operating at relatively low cost per digit, such as the three Everett was comparing (magnetic drum, acoustic delay line, electrostatic storage). But when used for storage, he pointed out, "these new elements lost, necessarily, their computing abilities. They became storage only."

Finally, Everett examined cost per digit. "The demand for high-speed internal storage is insatiable," he remarked. Since at least some 25,000 binary digits of storage would be required at the minimum, low cost per digit was absolutely essential. Alongside this cost there would be the overall cost of developing the entire computer. When Everett compared these with each other and with computer speeds, he arrived at what he called a "computer figure of merit." A magnetic drum machine should cost $100,000 to develop and achieve 100 operations per second at a resulting cost per operation per second of $1000. An acoustic tank delay line machine should cost about $300,000 to develop and achieve 2000 operations per second, at a cost of $150 per operation per second. An electrostatic storage machine should cost in the neighborhood of $1,000,000, judging by their experience so far with the MIT machine. It should achieve 30,000 operations per second, at a cost per operation per second of $33.[13] Although the conclusion was obvious, the fact remained that no such machine of proven reliability had yet been built.

Both Everett and Forrester saw two broad categories of applications for the computer. The first involved scientific problem solving, which would require large storage capacity. Here, said Everett, "the magnetic drum machine may hold its own, the operating speeds of the faster machines being reduced by" their need to keep referring to a slow auxiliary memory, such as magnetic or paper tapes. The second kind of application would require the computer to keep up with events in the physical world as they happened. For such "real-time work," Everett saw no solution except speed and more speed. Putting more than one computer to work simultaneously was not really a solution, Everett pointed out, drawing an analogy between computers and humans: "When the job gets too big for one man, an assistant can be provided. This is seldom as good as doubling the working capacity of the first man. When you have more than one person or one machine working together, you have the problem of communication. . . . The best solution to the communication problem is to avoid it."[14] Everett preferred speed, and this turned his thoughts toward three-dimensional storage. Such an arrangement, he felt, "would give the greatest promise for an economical high-speed storage and, in fact, such a storage is under development."[15]

Everett and Forrester agreed, on a strictly operational and technical level, with the engineers working under them on the MIT computer project. Forrester put it this way:

> The fundamental storage problem is not so much that of discovering an element which will retain information; it is the far more difficult problem of switching or selecting among the large numbers of individual elements which make up the storage system.[16]

This difficult problem had not yet been solved in practice in any computer R&D program Forrester knew of, so it was by no means clear in 1950, even to informed observers, that he and his colleagues on the obscure computer project in the MIT Servomechanisms Lab had the solution in hand. In fact they were getting it in hand, but matters of such technical detail were not that readily predictable, and at the time they were not the focus of attention of policy makers at either the Massachusetts Institute of Technology or the Office of Naval Research.

7

Project Whirlwind's Perspectives on Computer R&D and Air Traffic Control

During the late 1940s, Jay Forrester and his Project Whirlwind engineers faced other design issues and challenges besides that of achieving a reliable memory of large capacity. Crucial among these were the computer's reliability and the practical uses to which it might be put at reasonable cost. For Forrester, reliability and practical application went hand in hand, and cost varied as a function of expectations of availability and performance. As early as 1948 he was assessing these in ways that would strengthen his project's capacity to meet ADSEC's need. That September, he and four of his project engineers were responding to a request from President Compton of MIT to report on the state of the art and where Whirlwind stood in relation to its competitors. Their report, titled Forecast for Military Systems Using Electronic Digital Computers, reminded Nat Sage to point out to Compton in his cover letter that "the staff of Project Whirlwind has been concerned for many months about the issues raised in your letter." For some time they had been acquiring information through "visits and exchange of information with commercial and military laboratories" involved in computer R&D and computer applications.[1]

The reason the novel and relatively untried stored-program digital computer offered such rich promise even before any machine other than ENIAC was in full operation, Forrester and his colleagues argued in 1948, lay in the prospect that it would combine a suitably large storage capacity, versatile switching systems that could carry out operations of selection and choice, and the facility to perform basic arithmetic operations. The speeds at which it would operate should range between 1000 and 50,000 arithmetic and switching operations per second. If no human link were required in this computing chain, a human link could be dispensed with; thus, both automatic and semi-automatic operation would be available. Also, the speeds involved meant that far greater quantities of information could be absorbed, digested, and converted to usable form—such great quantities in so little time that it would be feasible for the computer to keep up with events as they happened in the physical world, determine what to do next, and exert control. The computer could not think, in the sense that humans do; however, if it were programmed carefully enough in advance, step by step, it could be made to calculate and choose among the alternatives provided in, say, combat operations or "the supporting activities of logistics, design, information centers, etc."[2]

Forrester and his associates saw no reason why a computer tied to an appropriate control system could not intercept supersonic missiles. They outlined a technical sequence of eight steps, then pointed out that no equipment then in use could accomplish them all—but that digital computing systems should be able to.[3]

As matters turned out, the dollar cost of developing computer technology over the next 15 years would be quite beyond the scope of any program that the Truman administration would allow ONR to fund in the spring of 1950.

In September of 1948, at a conference in President Compton's office with Vice-President James R. Killian Jr., Nat Sage, and Sage's assistant F. L. Foster, Forrester had spoken of outlays of $100 million a year for the next 10 years, "if the apparatus that people counted on getting was to be made available." In his view, estimates of the time and cost required were sometimes "hundreds of times too low."[4] Forrester and his colleagues suggested that contemporary expert opinion was dominated by "linear extrapolation of past laboratory programs." They countered this view by pointing out that "historically, similar developments have grown exponentially; and the nearest parallel is found in the easier problem of bringing radar to an operational status."[5]

To meet the design challenge that they saw ahead, Forrester and his associates proposed to spend $65 million over 4 years in order to arrive at an informed estimate of what sort of computer systems the military establishment might reasonably expect. Should it prove necessary to accelerate the development of computer systems, the least costly time to do so would be during the first 10 years. Experience had shown that, where scientific technology was concerned, R&D costs were lower than the production costs of items scheduled for inventory and use.

In addition to balancing costs against time, there would be the need to train engineers and operating personnel. In 15 years a suitable R&D staff of 7500 could be built up, and by the end of that period 4500 people a year would be emerging to variously develop or operate military computer systems.[6] They calculated $2 billion as the grand total for the first 15 years of computer R&D, provided no speed-up was called for. But if the first 10 years were reduced to 6 the amount would increase by $410 million, and should the speed-up be imposed during the last 10 years of the 15 the extra cost would rise to $1.36 billion, simply because of the exponential growth curve that characterized the shift from the earlier R&D phase of national programs to the later, more expensive production phase. When one added up, for the 15 years, nearly $1.156 billion for research, development, and training, $257.5 million for application studies and field tests (alias "experimental construction"), and $627 million for preparation for production, the grand total (rounded off) would be $2.04 billion.[7]

Such were the orders of magnitude that Forrester and his colleagues were presenting to Compton and Killian, and they did not include actual manufacturing and production costs that could be expected to follow at the end of the first 15 years. Their report divided these costs—totaling another $1.97 billion—among eleven categories, ranging from air traffic control and interception information networks to simulation, training, and logistics. A program of this detail and scope, they suggested, should be able to equip the armed forces with the computer systems they would need.[8]

The analysis of prospects submitted to Compton for his consideration as chairman of the Research and Development Board appears to have had no immediate impact on the military R&D community, but less than a year and a half later it would provide warrant and historical evidence of the strong conceptual base from which the Whirlwind engineers would greet George Valley's air defense challenge. And that strong conceptual base was becoming a practical reality, even as Jay Stratton of MIT and Mina Rees of the Office of Naval Research were setting their sights on more modest goals for Project Whirlwind for a variety of compelling administrative and fiscal reasons. To judge from the Air Force's level of interest in the problem of controlling air traffic around military airports, the goals held in view by the military establishment were also comparatively modest.

It was in this policy climate that, in March of 1949, Forrester's engineering and mathematical staff in the Servomechanisms Laboratory began a modest study of problems of computer-assisted air traffic control for the Air Materiel Command's Watson Laboratories, located in Red Bank, New Jersey. Here was another fount of technical knowledge and experience that would prove uniquely applicable to the problem of air defense that Valley would soon bring to Forrester's group.

From its wartime experience of handling hundreds of airplanes at a time in the skies over Britain, the Air Force was aware of the need to enable air traffic controllers to cope with great amounts of traffic. Although the railroads still dominated domestic commercial travel, it took no unusual imagination to anticipate what the years ahead would bring.

During 1948, Forrester began to tap the Air Force's cautious interest in financing a theoretical study of the technical problems involved in applying a computer to air traffic control. Such a study would be a sensible preliminary to specific practical application. Although not all experts yet agreed that the computer could be helpful in solving the air traffic problems they anticipated, Forrester was confident that it could. And his vision was not limited to air traffic control. Consequently, late in the autumn of that year, when rows of racks were being erected on the second floor of the Barta Building to receive the electrical cables and wiring, the circuit panels, and the functional modules of the Whirlwind computer, a new group went to work in the Servomechanisms Lab. Led by W. Gordon Welchman, a British mathematician who had participated in machine-assisted cryptographic work during World War II, this Applications Study Group began to study "the mathematical background and the coding procedures that will be put on Whirlwind when it is running." The general approach on which the group decided after much discussion, Welchman said in a memorandum written for newcomers, was "to form a group that would do original work on problems of computer application."[9]

In November of 1948, while the Applications Study Group was being formed, Forrester and Welchman were also engaged in the preparation of a proposal for a two-year study of air traffic control problems. On the last day of the 1948, MIT's Division of Industrial Cooperation formally submitted the Servomechanisms Lab's proposal, titled Research and Investigations of Electronic Computers in Air Traffic Control, to the Watson Laboratories.[10] As Forrester noted in a cover memo accompanying the proposal, this project would be the first phase of a longer-range enterprise to provide an "interim," then a "target" air traffic control system. Testing of the interim system might occur as early as 1952, but the immediate contract proposal would be limited to such necessary preliminaries as a general system study of the entire problem, an examination of tasks that computers might carry out, and a "related study of auxiliary equipment and computer design applications."[11]

In Forrester's view, while some staff members of the Servomechanisms Laboratory already were devoting time to a contract with the Navy for

Project Whirlwind, others would be able to initiate "a study of the air traffic control problem in relation to knowledge, already available at this laboratory, of the nature and potentialities of digital computers." They would familiarize themselves with present air traffic operations and then proceed to automatic equipment applications that might be practical. Navigation, radar, and communications equipment present and pending would be examined for possible use with computers. Information to be fed to a central computer would have to be generated in suitable form, and transmission and use of information from the computer would have to be accomplished. So far, ways of carrying out these operations were either unknown or undemonstrated.

Computer speed and storage characteristics would be determined by sample program-coding procedures. As mathematical formulations of parts of the overall problem were worked up, they would have to be discussed with "interested persons" and then programmed for practical trials. Forrester took it for granted that, because air traffic control and computer design were both in flux, early investigations would be oriented to "the type of computer which is most familiar to the laboratory staff"—that is, Whirlwind. Conversion of coding for the later interim systems, Forrester estimated, would be "a simple matter." Eventually, laboratory work would be done.[12]

Forrester emphasized the magnitude of the ultimate program. Air traffic control presented a prospect so vast that one could expect only to "help map the field of actual work" in the near future. Specifically excluded from the MIT proposal were construction and installation of operational equipment, which Forrester considered to be beyond the scope of the intended research. Nevertheless, laboratory trials using Whirlwind would explore coding problems and their solutions.[13]

The cost of the investigation, Forrester estimated, would be just over \$9000 a month, or slightly more than \$216,000 for 24 months. Two groups of investigators would be involved. Those in the first group would make a substantial portion of their time available; those in the second would function in an advisory capacity. Forrester described the qualifications and backgrounds of the five men in the first group: W. Gordon Welchman (who would be the project's supervisor), Harris Fahnestock, C. Robert Wieser, William Linvill, and Theodore Hildebrandt. The three members of the "advisory" group were Forrester, Everett, and Philip Franklin of MIT's Mathematics Department. Franklin was already directing "mathematical aspects of digital computer research" associated with Whirlwind.[14]

On January 12 and 13, 1949, while awaiting a response to the air traffic control contract proposal, Forrester visited Washington to discuss possible computer applications with other government agencies. He spent an afternoon with Perry Crawford, who had first convinced Forrester of the extraordinary potential of electronic digital machines. As Forrester noted later in his record of the trip, Crawford "expected active computer financial support to become available from one or more of four places." These included the Office of Naval Research, the Air Navigation Development Board, a possible program to spend half a million dollars on combat-information-center evaluation, and "the air intercept program."

Forrester was sharply aware of Crawford's informed and thoughtful understanding of the state of the art of computer technology and his keen nose for trends. Although neither of the young men was clairvoyant, it was Crawford who correctly saw the shape of things to come. Crawford was right, too, about where R&D support was likeliest to come from, but another year would pass before Forrester would be able to stop looking. Meanwhile, Whirlwind's near future appeared less than promising.[15]

Upon his return to MIT, Forrester inquired informally of I. J. Gabelman, a program manager at the Watson Laboratories, whether any decision had yet been reached on the air traffic control proposal. Though nothing had been decided, it appeared that MIT's cost estimate was high; if a contract were made, costs would probably have to be negotiated. Subsequently, the Watson Laboratories, after reviewing proposals from several organizations (including the Franklin Institute, the Raytheon Corporation, and the Teleregister Company), awarded Contract AF28(099)-45 to MIT's Servomechanisms Laboratory.

Within the Servomechanisms Lab, the efforts of the Applications Study Group were absorbed into the new project, which acquired the project number 6673 from the Division of Industrial Cooperation. The project engineers realized that, just as they were new to practical air traffic problems, air controllers were unfamiliar with the digital computer. Their undertaking would begin to redress this imbalance. They realized, too, that the general character of high-speed computer programs was not widely known—a situation that was to be expected, since the computer was so new and untried. The way such programs solved problems could best be explored and understood by engineers, physicists, mathematicians, and air controllers alike if they were to begin with simple, practical programs, then test them, and then bring them into reliable working condition.

Nine months later, on the eve of Valley's visit, the air traffic control project's mathematicians and engineers had published detailed descriptions of how to program, step by step, an ideal theoretical computer of the Whirlwind design, telling it how to control the positions of up to 100 aircraft more than 10 miles away by means of private-line communication and a rotating antenna.[16] These descriptions were directed at readers who had an engineering mathematics background but were otherwise not acquainted with encoding and programming a computer. A basic introduction to the available "machine language" techniques was provided: "The operations fall into five groups as follows: Transfer operations, denoted by cy, ca, cs, cm, . . . Arithmetic operations, denoted by ad, su, . . . Shift operations. . . . Scale factor operation. . . . Change of program. . . ."[17]

Eight days before George Valley's first visit, on January 18, 1950, I. J. Gabelman went to the Barta Building and reviewed Project Whirlwind's present and future objectives with the mathematicians and engineers there, "together with allied air problems of a military nature that seem to be suitable for computer application." "As a result of this discussion," W. Gordon Welchman noted in the routine biweekly report to project members, "we were asked to begin to look at the problem of tracking while scanning."[18] And Forrester noted in his private administrative log book that they would likely soon be adding to their workload "some problems of greater military value, such as the use of a digital computer for scanning while tracking on a radar set." "Gabelman," he wrote, "suggested that we do this in a preliminary way immediately."[19]

Most members of the Air Traffic Control study group in the Servomechanisms Lab did not realize it at the time, but the script was about to be slightly rewritten. On April 27, 1950, the work was redirected to the needs of the Valley Committee. Putting airplanes *on* collision courses had pre-empted air traffic control.[20]

8

Do It Right, Cut It Off, or Temporize?

George Valley made his formal appearance on the Whirlwind administration-and-funding scene at the MIT-ONR budget conference of March 6 and 7, 1950. He joined the conference and submitted ADSEC's request to participate in funding Whirlwind on the afternoon of the sixth. Underscoring the conference's original reason for being, that morning's session was attended by three representatives from each side. MIT Provost Jay Stratton was accompanied by two of his deans, Thomas K. Sherwood of Engineering and George R. Harrison of Science, ONR Chief Scientist Alan Waterman by Mathematics Branch Head Mina Rees and Boston Office representative Carl F. Muckenhaupt.

When Stratton asked the ONR delegates what they thought of MIT's plans, there were no surprises. The ONR delegates responded courteously upon hearing that MIT had surveyed its computing activities and was thinking of pulling them together under a single director without disturbing present departmental affiliations. "They feel that the strength of their applied mathematics together with related interest in science and engineering," Waterman noted later in his administrative diary, "make it important to continue an active program in computing." By making necessary adjustments, they should "not get in the way of any government program and in fact would offer a good opportunity for training of students as well as research. They wish to inquire how such a move would be viewed by the Office of Naval Research."[1]

As Stratton well knew, Waterman and Rees were not hearing anything they did not already know, nor was ONR about to tell MIT how to organize its internal affairs. The parties were in quiet agreement that it was time to put Project Whirlwind on a side track. What mattered, of course, was who the director of the new enterprise would be. "The ONR representatives stated that they would welcome such a move provided a suitable head of the program could be chosen."[2] Both sides agreed that either Professor Henry Wallman or Professor Philip Morse would be an excellent choice. Each man had the academic credentials and the applied mathematics orientation that was desired.

Then the discussion turned to money. ONR indicated its inelastic budget limit ($300,000 in new funds) for Whirlwind for the coming fiscal year, and the MIT administrators agreed that the Institute "would see that the Project would live within that budget if it were not possible to secure additional funds."[3] Although Valley had not yet appeared, both sides knew it was by no means unlikely that additional funds could be obtained from another source. But that was not at issue here. What really was at issue was whipping Project Whirlwind into line by applying agreed-upon MIT and ONR policies regarding the funding of any computer R&D associated with that project.

At the afternoon session, George Valley was invited to meet in Jay Stratton's office with Stratton, Thomas Sherwood, Nat Sage, Jay Forrester, and (from ONR) Charles Smith, Alan Waterman, Mina Rees, and Carl Muckenhaupt.[4] Acting in his capacity as chairman of the Air Defense Systems Engineering Committee, which was attached to the Air Force's prestigious Scientific Advisory Board, Valley explained why his committee was seriously interested in assisting the completion of Project Whirlwind and making use of the machine. Smith later wrote that ADSEC, having examined the problem of defending North America against possible attack by fleets of enemy bombers, had concluded that "some steps should be taken immediately toward integration and presentation at a central point of target information received from a network of early warning radars." Smith also noted that, since a digital computer could operate on a time-sharing basis while processing target data, "Dr. Valley's subcommittee believes that Whirlwind offers many advantages for an experiment in such system integration." And Smith noted three reasons why Whirlwind should be so employed: it was physically convenient to the committee (most of whose members had been deliberately chosen from the greater Boston area), "it is very fast, and thus could handle information from a considerable number of early warning radars," and it was not far from finished. "The proposed experimental system," Smith continued, "would start by merely using Whirlwind to process the search information received from MEW set at Bedford Airport." Subsequently, two or three CW (continuous wave) or MEW (microwave early warning) radars could be added. Forrester interjected that Whirlwind ought to be ready "by this summer" to prepare for such tests, but that its internal storage would have to be expanded to handle the load as radars were added to the system.[5]

The group then turned to the administrative and funding implications of George Valley's news. Waterman recalled it this way: "ONR was asked what its reaction would be to bringing the Air Force in on Whirlwind, to accept this program. ONR suggested that this would be satisfactory, provided Navy interests . . . were protected and provided the Air Force would be willing to transfer funds to the contract."[6] Smith's account, less diplomatically worded than Waterman's, was more informative:

> Dr. Valley was quite sure that upwards to half a million dollars could be obtained from the Air Force without much trouble. Dr. Rees and I were very much of the opinion that the administration of the Project should be with ONR, and that the Air Force should transfer the funds to ONR. It had previously been agreed that one-half of the computing time of Whirlwind should be used for scientific problems. The remainder could be devoted in considerable part to the Air Force project, to the proposed anti-aircraft fire control study, and to any other military problems which may be forthcoming.
>
> I am heartedly [*sic*] in favor of Dr. Valley's proposal; it at once helps support the Whirlwind Project, makes use of Whirlwind in a control system, which after all is the reason for designing great reliability at great trouble and expense, and answers the objections in the Fink report by providing Whirlwind with a very definite use. I am only concerned that control should remain in ONR, which has provided the money all through the weary and discouraging period of research and development, and that the cost should not be excessive.[7]

Valley's request and his news did not alter Stratton's and Rees's agenda for the budget conference, and the following day was budget-cutting time. Those who had attended a morning session in the Barta Building (Rees, Smith, and Robert Bergemann of the Boston office of ONR; Forrester, Everett, Hugh Boyd, and Fahnestock of Whirlwind) were joined in the afternoon by Stratton, Sherwood, and Sage, and specific cuts were made to pare Forrester's proposed budget of $505,000 down to ONR's targeted $300,000 commitment and thus bring Whirlwind into line with the Navy's funding realities.

In a move exemplifying the degree to which a funding sponsor who has gained the upper hand may influence managerial policy, Rees first disclaimed any intention of telling MIT how to conduct its internal affairs and then suggested that the Institute "carefully consider how large an organization they wished Whirlwind to be." "In any case," Smith's "summary of conference" continued, "nothing further could be done until the Air Force decided how much, if any, money could be made available." Stratton responded that he hoped they would know in 3 weeks.[8]

Waterman of ONR noted that Valley would "try to persuade the Air Force to transfer money for the purpose" of funding Whirlwind on a half-time basis. Administrative support from within the Air Force would have to come from General Gordon Saville. "In order to bring it about," Waterman concluded, "it will be necessary for Valley to persuade the Air Force of a definite approach to the Air Defense problem involving a central headquarters for handling field information. This is where Whirlwind enters. All agreed that this would make an excellent solution to the entire Whirlwind problem. It would provide additional financing, help the Air Force in their high priority project, provide opportunity for scientific use of Whirlwind and make it available for Navy problems as needed."[9]

Presumably, if Valley's oral promise of half a million dollars had substance, Valley's committee would be getting the Air Force assistance it sought. Forrester, held closely accountable by Stratton, would get a reasonable budget and would be able to carry on some R&D too. And perhaps ONR, with MIT's help and now ADSEC's too, would see the Whirlwind machine brought to fairly rapid completion at acceptable dollar costs, in keeping with the limits imposed on ONR's funding resources by the Truman administration.

ONR's position was not at all unreasonable. Within the funding constraints it suffered, it was able to espouse a conventional R&D policy that subordinated expensive development to less costly research. Stratton shared this policy view. Forrester, it appeared, had been corralled.

Valley's information and the budget-pruning session with Rees and Smith that Stratton later joined had generated a need for a redrawn budget that would incorporate the program changes on which they had agreed. Forrester and three of his engineers proceeded to put together a revised budget for Stratton's provisional, informal approval.

A week after the ONR-MIT budget conference, Forrester, Everett, Boyd, and Fahnestock met with Professors Harold Hazen and Gordon Brown, Dean Sherwood, DIC Director Sage, and Provost Stratton to discuss the revised budget for the 1950–51 fiscal year. This time the provost would see to it that his young project director was receiving explicit advice and guidance. At the end of their session, "everyone seemed reasonably well satisfied with the program outlined and with the cost figures," Forrester thought. And on the issue of operating the laboratory he headed, he observed that "earlier doubts about the supporting organization items seemed to be largely dissipated after a breakdown was available to show their composition."[10] All the men were quite aware that what Forrester might be allowed to do for ONR need not limit what he might be allowed to do to satisfy the Air Force's aerial defense problem, so long as appropriate contractual funding limits were adhered to.

At the sixth page of the revised budget analysis, immediately after the "budget summary" sheet, Forrester had inserted a short essay he had written in response to a question that Stratton had put to him while they were reviewing the program with Rees and Smith: Why was it costing so much to bring the Whirlwind computer into operation? It all hinged on reliability, Forrester explained in the essay (titled Reliability in Digital Computers). On the next 7½ pages he set forth an "article of faith" that would prove so fundamental and enduring over the course of the Whirlwind-SAGE project that it explains the "go" of many subsequent events inside and outside the laboratory.[11]

Forrester was committed to "going first class," with expensive, painstaking R&D based on not an egocentric, aristocratic idiosyncrasy but on a carefully considered analysis and a gut feeling, born of several years of experience, of what the enterprise would need in order to succeed. Whether he was right is of less historical significance than the fact that he and his laboratory lived by the commitment, and his ideas prevailed for a while. In the spring of 1950, however, Forrester's was an engineering creed that neither Stratton nor ONR nor the RDB Ad Hoc Panel on computers could accept; they found it too expensive. Thus, the outcome was in doubt. Forrester responded by redoubling his efforts to persuade the decision makers.

Admitting that high reliability of the MIT approach and its associated cost in man-hours and dollars had become controversial, Forrester argued in Reliability in Digital Computers that "preliminary successful results" at MIT, in combination with the failures dogging other R&D efforts in which such precautions had not been taken, were already producing "a noticeable trend." In view of this, Forrester felt, it made practical sense to "set down some of the considerations which lead to the policies on which I have insisted."[12] He offered three reasons for "placing dependability of equipment above all other aspects of our work." First, a fault or a flaw that most pieces of electronic equipment could tolerate would prove catastrophic in a computer, reducing its instructional sequence to what computer "hackers" would later call "garbage." Second, computers in this first decade of their existence were acquiring a reputation for unreliability that was discouraging potential users. Third, regardless of the problems raised by these first two difficulties, Forrester and his engineers were proposing, he wrote, "digital machines for the field of real time problems where human life and large financial investments are dependent on completely error-free operation."[13]

When computer R&D was just getting started, no one had any cause to believe that such extreme reliability as Forrester was committed to achieving could be accomplished. For Forrester, however, the issue was drastically simple: "The only procedure known then or now is to identify every source of error and take precautions to eliminate all. This has accounted for a large part of our time and cost." It was not the principles that were the problem; it was their realization. Though the computer's value might be demonstrated philosophically and on paper, its actual and ultimate usefulness could be demonstrated only by actual successful operation, and that could be accomplished if Forrester's three main points were to be accepted.

Expanding on his three main points, Forrester noted that to appreciate just how unique a piece of electronic equipment a computer is one should remember that a radio could accommodate some static, a television some flicker, and a teletype machine some error without destroying a message, but a digital computer could tolerate no such error without destroying the message. Because every single pulse carried a unique meaning, a single error could produce chaos. Paradoxically, the basis of the computer's distinctive logical power was the source of its greatest weakness.

Whirlwind was being built to compute under its own instructions and to alter its instructions simultaneously, and thus it would be vulnerable to machine-generated mistakes in its instructions, quite apart from human or data errors. Not only could answers go wrong; instructions and even the entire program could be devastated.

Forrester noted that, relative to Whirlwind, earlier computers were awkward, inflexible, and difficult to program:

> None of the following machines can compute control orders during a machine sequence: ENIAC, BTL Mod V, IBM, SSEC, Harvard Mark I and Harvard Mark II. So far as I know, the Harvard Mark III and BTL Mod VI are equally limited. The Eckert-Mauchley [*sic*] Binac and the University of Pennsylvania Edvac were equipped to compute control orders but are not being operated due principally to failure to follow design and administrative policies which we support. A new class of computing machine is coming up. Whirlwind should be the first of these to operate successfully. All of them

> (Whirlwind, Raytheon, Univac, IAS, Bureau of Standards, California at Berkeley, California at Los Angeles, etc.) will be vulnerable to electronic disturbances like nothing else before them. We hope that the very best we can do may be good enough for a start.[14]

Forrester felt that using two machines simultaneously to obtain independent confirmation of results was a poor solution. As he surveyed the "track records" of current machines, he found further reasons to be apprehensive and to insist on greater reliability instead:

> Even the twin Binac machines, doing simultaneous computation, have not solved the reliability problem. Based on latest information, the Eniac which has been operated for some six years is still making errors at intervals. . . . Various reports come from the IBM, SSEC machine in New York; the only first hand information is from Dr. von Neumann and the Carter Oil Company. These people are completely discouraged with the SSEC after being able to get only 11 hours of successful performance out of a three-week period. The Harvard Mark III has not yet provided any data and probably will not soon if it is to be moved to Dahlgren.[15]

Regarding the demands imposed on a computer by real-time control activity, Forrester drew the conclusion to which he had been long committed when he considered using the computer for, say, air traffic control, fire control, or aircraft interception: such applications would have to insist on utterly dependable equipment. An air intercept experiment marred by computer malfunctions "could only do more harm than good."

Reflecting on the state of the art, Forrester concluded that reliability was "by far the most important unknown quantity."[16] Thus, it was absolutely necessary that Project Whirlwind avoid making the mistakes that had booby-trapped other efforts. To this end, the project had routinely been taking a number of precautions. The first was "to place no confidence in the benevolence of nature and hope that trouble will not occur in a computer part just because its operation is not known or understood." Forrester was amazed, he said, at the "prevalent and devastating tendency" to assume that "untested and unverified" components would operate satisfactorily. Rather than follow this course, Whirlwind engineers checked the performance of all purchased items, often testing them with equipment superior to that used by the manufacturer. Sometimes the standards they imposed were more stringent than the prevailing military specifications. A small but significant indicator of the value of this approach had been their avoidance of any problems with the several thousand coaxial cable plugs and connectors on the machine, "much to the surprise of those with extensive experience in electronic equipment."[17]

The performance of vacuum tubes, and any defects that they might develop from prolonged use, had come to the Whirlwind engineers' attention, and they began imposing tests more stringent than anyone had ever tried. As a result, Forrester explained, they had "anticipated by many months" a rash of tube failures that afflicted pulsed-circuit devices of diverse kinds. The Whirlwind engineers had been able to demonstrate that failures that had long been laid to "cathode poisoning, loss of emission, presence of contact potentials, etc." were actually caused by deterioration of cathode interfaces, and these findings led them to approach "nearly every large American tube manufacturer for advice and consultation."[18]

Finally, there was the technique of marginal checking, which Forrester had invented almost on the spur of the moment in 1947. Calling it "by far our most important, principal and basic contribution to computer reliability," Forrester described this technique as follows:

> The system of marginal checking finds deteriorating components and permits their replacement before they reach the threshold where they can cause random errors. Experimental results show that marginal checking can easily decrease the random error rate by a factor of 40 to 1. Marginal checking, along with properly designed check problems (which have taken several staff-years to develop) can cross-section the machine so that defective parts can be readily isolated. Work is not yet sufficiently complete but results to date amply prove the wisdom and success of these methods to our satisfaction.[19]

Forrester admitted that he did not expect immediate and widespread endorsement of his strenuous and expensive reliability policy. He fully expected more computers to fail. But over time, he expected, observers and builders would become alert to the "dangers of blind faith in large electronic computing systems." The fundamental issue of dependability, he concluded, involved "much more than mere development work; it is research of a very basic nature in a field largely governed by superstition, prejudice, rumor and trial and error."[20]

As he waited for the Air Force to respond to George Valley's request for funds to test the computerized air defense scheme that ADSEC was proposing, Forrester gave little indication that he was inclined to temporize.

Hindsight suggests that the Whirlwind engineers were both luckier and better prepared than they realized at the time. Forrester and Everett had been thinking since 1947 (perhaps longer) of a machine such as Valley was seeking, and they were intent on designing and building it as a reliable general-purpose computer that would be capable of serving, for example, as a command-and-control center for anti-submarine warfare. This was a role similar to what Valley and his committee had in mind for air defense.

Also worthy of mention here is the habitual way the Whirlwind engineers had been carrying on their technical affairs and acquiring experience and expertise, day by day, week in and week out, for several years. The trend in the laboratory between 1947 and 1950 demonstrated a consistency of commitment and a continuity of purpose that contributed to their technical preparedness. But whether their accumulation of experience and engineering insights would suffice remained to be seen.

9

The Bedford Tests

Changing the script formally took 2 months, but before the end of March 1950 the Watson Lab of the Air Materiel Command (funding sponsor of the Air Traffic Control program) received official notice that a higher-priority Air Force program was pre-empting the ATC project, taking over the technical effort to which the MIT engineers under Jay Forrester had been devoting themselves, and redirecting it to a specific defense goal. In a letter dated March 21, John Marchetti of ADSEC courteously indicated what would have to be done. There was no question who had the upper hand. "The Valley Committee would like very much to have this contract changed" to air defense, Marchetti wrote to Ralph Cole of the Watson Lab. "No change in the contract is required other than to have the contracting officer write MIT and give permission to reorient the purpose. It is the plan of the Valley Committee to put 500k of '51 money in on this job and take over a considerable portion of Whirlwind time."[1]

The change in goal from controlling friendly aircraft to detecting and destroying hostile ones had little immediate effect on the day-to-day operations of Project Whirlwind. Either goal would require the same mathematical formulations and derivative digital programs to achieve the required track-while-scan ability, and either would require the Whirlwind computer to be connected to radar equipment. The radar, together with apparatus that would convert the radar pulses to telephone-line signals and send them to the computer, would be provided by the Air Force Cambridge Research Laboratories (AFCRL).[2]

The engineers and applied mathematicians in the Air Traffic Control group did not wait for the legal formalities to be executed before taking action. Computer programs for the first Bedford experiments were prepared in the spring and early summer of 1950. The first experiments called for an Air Force "microwave early warning" (MEW) radar installation at Bedford Airport, just northwest of Boston and Cambridge, to send its data over telephone wires to Whirlwind in the Barta Building. The data would consist of binary numbers in a frequency-multiplexed form, with the 1 in any of nine digit columns identified by an accompanying audio tone. Narrow-band filters inserted at Whirlwind's end of the line would demultiplex the signals to render them suitable for Whirlwind's circuits. Since the circuits of the radar and those of Whirlwind would be operating independently and asynchronously, buffer storage of some kind would have to be provided to receive the signals and hold them until they could be introduced into Whirlwind's circuits. "Further," the Summary Report stated, "buffer storage must be such that the computer has access to it in a very short time, preferably a few microseconds."[3]

The Bedford radar had some severe limitations. Ground clutter made it difficult to distinguish significant radar returns for aircraft less than 10 or 20 nautical miles away. The range of effective operation was restricted to "between 70 and 80 miles for moderate-sized aircraft flying at altitudes less than 10,000 feet."[4] The site was poor. The radar set was located in a shallow valley with a range of hills about 40 miles to the west that confined tests to a triangle with one apex at Concord, New Hampshire, one at Sanford, Maine, and one at Newburyport, Massachusetts (all north and east of Bedford Airport).[5]

Since only one radar installation was involved, the engineers felt that they should be able to use test equipment that was already standard for Whirlwind. Accordingly, one of the five flip-flop storage registers already built into Whirlwind (and antedating electrostatic storage) would be employed to receive the radar data; other test equipment would convert the data to signals compatible with Whirlwind's 0.1-microsecond pulses, making sure that read-in of a signal did not encounter either register-readout or register-restorer pulses. The engineers would need only a few weeks, they believed, to adapt the computer to carry out these functions. They realized that they were moving into unexplored experimental territory.

The frequency-multiplexed telephone-line signals would be recorded on a Magnecorder tape recorder fitted with an auxiliary spooling mechanism that would provide an uninterrupted hour of preserved data. By employing such a reusable source of incoming data, the engineers opened up for themselves a variety of experimental test options, perhaps the most useful of which was to test some of the equipment whether or not the radar happened to be operating "on line" with the computer. They would be able to adjust the demultiplexer and the read-in equipment at will; the computer and the radar could be operated and tested independently, with neither holding up the operation and testing of the other. As the engineers pointed out in a progress report, "recorded data will allow controlled experiments with the computer program with identical input data and still not require an excessive number of aircraft missions."[6]

They would want to keep track of what was going on by selecting among the radar blips that would appear on the 12-inch-diameter long-persistence cathode-ray tubes that were to be used as display screens. Until they could obtain a couple of these, they would use the 5-inch-diameter oscilloscopes already available. Reception of stationary targets would not be suppressed. The men realized they should expect "the same target will frequently be reported with two consecutive values of azimuth or range." Although storms might overwhelm the data link between the radar and the computer, this problem could probably be avoided in the tests by restricting the radar's range to 40 miles and using a 15-second sweep cycle.

In prospect, a reasonable approach would be to track a selected target with "some method analogous to the 'light-gun'"—for example, to have the computer generate a movable spot on the screen and provide it with locating coordinates by "a short sub-program in the computer and controlled by means of a joy stick." The joystick would be used to shift the spot until it coincided with the radar blip that had been selected as a target. Then the computer, having "identified" the target, would pursue it automatically. The joystick was expected to be ready for use in a few weeks.[7]

It made sense to keep the first tracking programs for the tests simple and to use the computer at the start to "prepare synthetic (r, theta) data representing a number of aircraft behaving in the desired manner." Punched-tape terminal equipment then being installed in Whirlwind would be used to feed in synthetic data that Whirlwind would be unable to distinguish from data arriving from Bedford or from the magnetic tape recorder. While the computer would be converting all incoming signals to (x,y) coordinates and displaying these on one scope, certain targets would be selected manually with the computer's help and would be presented on the second scope.

The ATC (now "Air Defense") investigators would build more complex tracking procedures on these simple beginnings. The details are too technical to examine here at length, but a partial sketch of the intricacy and sophistication of the procedures will serve as a reminder that the computer programmers were moving onto untrodden ground. According to Summary Report 6: "Considerable thought has been given to operational problems of this type and a procedure known as 'manual intervention' has been formulated in which the behavior of the computer at certain specified points in a program is controlled by instructions that can be manually inserted by switches without stopping the computer."

They also were working on automatic acquisition of all targets within a selected range, on avoiding loss of one target when two came close together, and on superior methods of "smoothing" data that would go beyond "theoretical discussions of data smoothing" to examine particular situations and conditions. Echoing the discussion of a technical problem that Jay Forrester and Bob Everett had presented in Report L-2 in 1947 (reissued as R-158 in 1949), they recognized that there would be radar errors which they would have to estimate "from an analysis of the raw data corresponding to a tracked aircraft." "Such an estimate," they noted, "cannot be obtained by a direct method because the true position of the aircraft cannot be ascertained with sufficient accuracy."[8]

Late in the summer of 1950, the engineers hooked up to a module of 32 registers of test storage the computer in-out equipment that was to accept radar data and present scope displays. Then the whole was brought up to a level of operation that would render the machine ready to receive radar data. The first signals and test patterns from the Bedford radar were recorded on magnetic tape and then fed into Whirlwind, whose test storage had been given "a simple display program . . . to display the data on an oscilloscope as a B-scan."

Though test patterns proved easy to read, actual radar signals were accompanied by so much ground and cloud clutter that it overloaded the electronic data relay link between the radar and the computer. The AFCRL engineers realized they would have to go to work on the relay link in order to extract true data and weed out spurious signals. But even with these problems it was "possible to observe on the display scope the tracks of moving targets."[9]

In mid August, members of the group had visited Bedford to observe the radar in operation and to discuss the problems that had been encountered by the AFCRL staff responsible for incorporating the radar and its ancillary equipment into the test system. The MEW radar was not suitable for the task to which it was being put, and the problems were severe enough to persuade the Bedford group to "switch their attention to a reconsideration of the entire design of a radar system for this special purpose." Nevertheless, it appeared to the Cambridge group that the set, as it stood, would provide ample opportunity for them to experiment and to determine the effectiveness of the computer programs if ranges were greater than 40 miles under favorable weather conditions.[10] The MIT engineers were not, however, optimistic about the character of the data they expected to receive from Bedford.[11]

Meanwhile, the programmers in the analysis group (as distinct from the engineering group) were working on the ATC-become-Bedford test problems while looking ahead to the day when they should have programs ready to feed radar data into the interim electrostatic storage being readied for Whirlwind. They recognized that, since the object of tracking was to locate aircraft targets for interception, they would have to begin with preliminary experiments in tracking and in directing interceptors. These would have to make use of an interim storage no larger than 300 registers. Such minimal storage meant that, for purposes of scope display, they should make a special effort to find simple approximations to necessary trigonometric functions locating a point (x,y) in any quadrant.

Furthermore, the programmers realized that, rather than spend the time it would take to optimize the preliminary experimental procedures, they would do better to attempt an experimental interception under favorable conditions. Simple tracking and guidance procedures should be sufficient for that purpose. They could then concentrate on selecting the best methods of "smoothing" data that would sharpen their techniques. Also, they could begin introducing for study and solution special tracking problems, such as those that would be encountered when two tracks crossed or when an aircraft made a sudden turn. And the requirement of accomplishing automatic acquisition of targets always lay ahead.[12]

When they learned late in the summer that AFCRL had modified the operation of Bedford Airport's radar so that it could provide data "without clutter suppression at 4 rpm or with clutter rejection (MTI) at 2 rpm,"[13] the programmers realized they would have to get to work on programs to handle both modes. Though the slower mode provided clutter suppression, which was greatly to be desired, they needed to know whether genuine aircraft would escape notice in consequence and whether (as it appeared) 2 rpm might not be too slow for proper tracking.

After constructing a joystick and discovering how slow and awkward it was to zero in on a selected target, the engineers built a "bread-board 'light gun'" as "an alternative device for manual target acquisition." Their Summary Report described it as "a photoelectric device which is placed over the desired spot on a display scope. The next intensification of the selected spot produces a pulse in the light gun, and this pulse is fed into the computer to select the proper subprogram. Satisfactory signals have been obtained from a multiplier-type phototube."[14]

Reminiscing with a group of former SAGE engineers several decades later, Norm Taylor attributed the invention of the light gun to Bob Everett, "simply because he told me he invented it." Everett, a member of the group, said: "If I'd known it was that simple, I'd have invented lots of things!" He pointed out that he had "invented the light gun in the sense that I didn't know [whether] anybody else had done so. That's really how I invented the use of light guns in digital computers. There are a lot of cases like that; people had invented photoelectric cell pickups for other purposes."[15]

As 1950 drew to a close, the men were beginning to track commercial aircraft, getting such "good radar data on miscellaneous air traffic . . . at various times" as to perceive that "there are periods when the radar and relay link work well." But they had no luck with four special flight tests they set up, which involved bringing in a target at altitudes up to 10,000 feet from the north, west, and south sectors of the compass while operating the radar at an antenna speed of 2 rpm and using the Moving Target Indicator. Although they could not ascertain the precise cause of the trouble, "the target signal was not encoded by the digital relay link consistently enough for positive identification by a human operator viewing the Bedford digital PPI [plot position indicator] which displays the data sent over the radar link."[16] The conclusion they reached in Cambridge, especially in light of Whirlwind's "improved operation," was that the trouble lay in Bedford. Consequently, Everett, Taylor, and Wieser visited the Bedford site to discuss operation of the radar and to witness for themselves the difficulties which accompanied its operation. Their visit made clear to them that the problem truly lay in the radar, not in the relay link or the computer. The major source seemed to be the unreliable performance of the Moving Target Indicator.

At a subsequent meeting with members of AFCRL held to discuss the problem of radar data, the engineers decided to run tests without the Moving Target Indicator. Unless solid improvement was evidenced, the MEW would be taken out of service and overhauled during the waning winter months, so tests could begin again when weather permitted.[17] Two flight tests were made on February 10. Although the results were the best to date, they confirmed earlier suspicions. Thus, the radar was taken out of operation for maintenance and repairs in the hope of obtaining more consistent performance when the flying weather was expected to be better.[18]

To set up enough flight tests in the near future, the engineers sought to supplement AFCRL's aircraft pool by utilizing the MIT Radar Weather Project's aircraft and by arranging to be informed when the MIT Instrumentation Laboratory was flying aircraft it had at its disposal. They would be interested in setting up trial interceptions with crossing and parallel courses involving at least two aircraft at the same time.

For the planned interception tests to be significant, the engineers would have to obtain an interceptor that was considerably faster than the aircraft being interdicted. Furthermore, the interceptor would have to be able to tune in to AFCRL's transmitter frequency, since continuous and direct ground-to-air communication would be indispensable. So far they had located only one such "interceptor": a twin-engine B-26 (originally a Douglas A-26 attack bomber, developed during World War II) operated by the Instrumentation Lab and equipped by AFCRL with a crystal tuned to the AFCRL frequency. The best plane AFCRL had to offer was a slower C-45 that would be eligible for target duty.

Looking further into the future, toward the day when an automatic ground-to-air communication system would be installed, the Digital Computer Laboratory engineers were collaborating with AFCRL to connect Whirlwind to the system AFCRL was designing. They had agreed that the new system should be compatible with a new input-output register intended for installation in Whirlwind in the summer of 1951. The radio transmitter for the system would be located in the Barta Building with Whirlwind. If permission was granted, the 150.1-megacycle channel that Wright Field had assigned to the Instrumentation Lab would be used.[19]

The engineers were still using the 5-inch oscilloscopes, but they had discarded the joystick. By the end of 1950 they were using an improved model of the light gun to select targets and interceptors and to present them on a second oscilloscope. AFCRL had provided the DCL engineers with a 12-inch digital PPI screen equipped with decoders. It displayed the relay-link data directly, and they found it "very useful for study and editing of recorded data without tying up the computer." Its larger screen allowed quicker identification of target trails. Meanwhile, work was progressing (though not as rapidly as they would have wished) on "16-inch magnetically deflected scopes for displaying WWI output information."[20]

During the latter half of 1950, the Air Defense analysis and programming group working on the Bedford tests had been laying the essential computer-programming groundwork for intercept tests by writing the programs, testing and debugging them on the computer, and thus warranting their reliable internal operation in the computer. A successful intercept test would entail three major actions: tracking the target and the interceptor, computing the guidance instructions to give the interceptor, and communicating these instructions to the interceptor. By January of 1951 the men were reasonably confident that the machine programs and the procedure they had laid out would work. Only actual flight tests were lacking.

Following their customary engineering approach, the men had set up as simple an initial procedure as they could devise. The target and the interceptor would fly straight courses at the same altitude. The speed of the interceptor but not that of the target would be preset and known, and all incoming radar data would be converted from their "r, theta" radar values to (x,y) coordinates for display on a scope.

Tracking would be initiated manually by applying the light gun to a selected target on the main scope. So far, use of the test patterns indicated that the light-gun technique should work. Holding the light gun on the location of the target long enough to detect the computer's next scanned display would transfer the target's display to the second scope. Doing the same to the interceptor's spot on the first scope would select it too for display on the second scope. From that time on, the two selected blips would be tracked in isolation on the second scope, without further need for the light gun. Their courses would be predicted on the basis of the history of preceding sightings. A collision course would be computed, proper heading instructions for the interceptor would be displayed, and the scope operator would pass on the information by voice to the pilot of the interceptor.[21]

To produce the heading instructions, the computer would translate its internal calculations into a binary-coded decimal number, which would be displayed by the indicator lights of one of the flip-flop storage registers that Whirlwind used for test storage. The scope operator would read the lights and would convey the numerical compass heading by voice over the phone to Bedford, where he would already be patched into the radio transmitter there that would carry his words to the pilot aloft in the B-26. At least, that was how it was supposed to work.[22] But there were problems with radar transmission, and Whirlwind's new electrostatic storage performed erratically.

Whirlwind's interim electrostatic storage became available in mid November of 1950, and by the end of January the engineers had used nearly 100 hours of computer time. "A large part of the time was spent in wrestling with equipment failures," said the Summary Report. At the start, the digital radar relay link and the punched-tape equipment were the major identifiable sources of trouble, but as these were fixed the new electrostatic storage (ES) became the principal culprit. Since the engineers had been expecting "teething troubles" with ES, they had made preparations to "go manual" in their control of the computer in order to monitor the operation of any portion of ES while a program was working. They could even change the contents of any particular storage register manually. "This preparatory work," they reported in January of 1951, "proved to be extremely valuable, as it provided means of locating errors in storage and of making corrections to get a program running again." It also set up a further "tumbling dominoes" sequence in which a problem generated a solution and the solution promptly generated the next problem:

> The experience gained in error chasing led to several improvements in the techniques employed in programming and in operating the computer. For example, when the errors in electrostatic storage became too bad, the only reasonable course of action was to run the program in again from punched paper tape for a fresh start. As this was a frequent occurrence, it became an intolerable nuisance that the time required to feed the program in again and to start the computer on the program was about seven minutes. A concentrated attack on this problem led to a number of improvements as a result of which the total time for feeding the program in again and restarting is now about one minute. Part of this saving in time was achieved by reducing the amount of manual switching, and this in itself is an improvement in that it reduces the danger of errors. Incidentally, the read-in time will be further reduced later on by the introduction of photoelectric tape reading equipment.[23]

While the "dominoes" in this row were falling, a derivative branch was set tumbling. Since the electrostatic storage was so unreliable, the Whirlwind engineers had to find out why. They came to the conclusion, as the evidence mounted, that "a drift in the deflection pattern on the storage surface" inside any ES tube could be causing the errors, and that this situation was generated by the accumulation of an electric charge on the interior surface of the tube. Since the tubes were being constructed in the Barta Building, they would coat the interior surface with a compound called Aquadag before sealing a new tube (in order to prevent or at least retard the buildup of any electrostatic charge), and they would change the electrode voltages.

Despite all these difficulties, it became apparent to the engineers within 2 weeks after they began to use the interim electrostatic storage in November that all the program elements of their first intercept experiments could indeed be fitted into the 256 registers that constituted the storage. Fitting them required more work, but out of these efforts emerged a small, useful library of routines for computing square roots and trigonometric functions.[24]

While coping with their day-to-day problems, the engineers were making plans for the coming weeks and months. The most important tests would be the experimental intercepts, for these would demonstrate to the military policy makers whether centrally controlled, computer-assisted interception was practicable enough to be worth pursuing. Almost on a par with the flight tests would be finishing and testing the programs designed to handle multiple simultaneous targets and those designed for automatic target acquisition. Then there was the unfinished business of computing all the details of sophisticated collision courses for interceptors and their targets. To calculate such a course, rate tracking would have to be measured, for the computer "must not only be able to follow the target but must also compute the components of the target's velocity." Range and azimuth values were quantified at Bedford before being sent to Whirlwind, where the accumulation of such data rendered it susceptible to "smoothing" in order to produce more accurate course predictions.

But no matter what computational formulas and program procedures made practical sense for two aircraft, these would be inadequate for the simultaneous tracking of a larger number of aircraft. As the number of aircraft grew, particularly when they were coming from the same direction, there would be too little time for the computer to carry out the necessary calculations for all the aircraft, unless computation time were stretched out and provision were made for the insertion of fresh data from time to time. A longer interval must be allowed for the calculations than the scheduled 1/50-second interval between the arrival times of the information traveling in linear sequence over the telephone line between Bedford and Cambridge.

Furthermore, the computations would have to be programmed so as to allow for interruptions when fresh data arrived from the radar. Fortunately, the time required for the arrival of new information would occupy but a small fraction of the 1/50-second interval, so the computer would be able to make effective use of the segments of calculation time thus provided. Nevertheless, the programs to handle these interceptions would have to be new, complex, and radically different from those handling just one or two aircraft at a time.[25]

As the engineers looked ahead, they perceived the wisdom of pursuing several theoretical studies if time and the work load would permit.

Further careful study was needed of "methods of smoothing radar data for tracking, for prediction, and for the derivation of instructions for the guidance of aircraft."[26] In another direction, a general study should be made of what was involved in hooking up a computer to a network of radars, each of which would be sending in altitude data along with "r, theta" coordinates. Also, a survey should be made of the general problem of connecting a computer to a network of continuous-wave radars supplying "either azimuth or elevation, but no range or range rate" data.[27]

The failure of the January intercept tests put the preceding 10 months' work in question and sharply disappointed the MIT and AFCRL engineers. Out of their discussions and further transmission tests, the regretful conclusion emerged that they had no choice but to shut down the Bedford MEW radar for a trouble-shooting overhaul. This was done from February 12 to March 12, 1951, which turned out to be a spell of extremely poor flying weather. Drastic changes were made to the digital radar relay (DRR) equipment at Bedford and to its companion DRR multiplexer in the Barta Building. As a part of these repairs, a digital storage tube accounting for nearly one-third of the electronic equipment in the DRR was removed, thus reducing and simplifying the DRR's capacity to store target coordinates before sending them over the phone to Whirlwind. Other modifications to the radar produced a significant improvement in its output as judged from a video plot position indicator at Bedford, but the DRR remained disappointing, yielding erratic variations of range data. These problems were not permanently corrected until mid April.[28]

Bob Wieser and Gordon Welchman took advantage of the pause in intercept flight tests to restructure the growing aggregation of part-time graduate students, engineers, and mathematicians that had come to be called "the DIC 6673 group." This group had been formed to investigate the avoidance of midair collisions, but it was now seeking near-collisions. At the beginning, problem analysis had been a preliminary that had been closely tied to the writing of computer programs, and the programming had been recognized as a closely related and equally essential preliminary to putting the computer into operation. In consequence, there had been little separation of these activities within the group. Successful operation and debugging of programs had been dominated by trouble-shooting efforts that required both an intimate knowledge of the structure of the programs and an understanding of the mathematical analysis that underlay the programs. As their understanding of programming and their mastery of the computer had been improving, they had been able to develop the early programs they needed for the Bedford tests. Now the urgency to ready the programs was over, and they were waiting for the radar to be overhauled. Administratively, matters were complicated by the fact that seven new program writers, all lacking in hands-on experience with Whirlwind, had joined the group.

After discussing plans with Everett and Forrester, Wieser and Welchman took advantage of the lull to divide and specialize. "In the future," read the order they issued in the middle of February, "there will be analysts and programmers and operators. The analysts will leave the detailed preparation of programs to the programmers, and the running of the programs on the computer will be done largely, if not entirely, by the operators."[29]

The phased changes that went into effect were indeed called for. Month by month the computer was operating with greater reliability and was in running order more of the time, so the engineers would have more time to run their programs. But arrangements would have to be flexible as they moved from the old ways of doing things to the new ones. The rate at which the computer would continue to improve, avoid unexpected "down time," and increase its storage capacity could not be known in advance, nor would the vicissitudes of flight testing be under the engineers' control.

Once the radar was back in operation, top priority would, of course, be given to the intercept tests. In the meantime, David Israel, Jack Arnow, and Robert Walquist would undertake some necessary theoretical studies. Israel, Walquist, and J. Salzer would become the Problem Analysis Group, and they could expect much of their time to be taken up by efforts that would require them to use the computer. At the same time, there would have to be concerted efforts to develop more efficient operating procedures governing the use of the machine, to determine what kind of knowledge of computer operation would be most useful for a programmer, and to "use the computer with the object of finding out how to use it better."[30]

Any machine time allocated to air defense work that was not taken up by the Problem Analysis Group would go to a new Computer Operations Group, which would be responsible for spelling out operating procedures, breaking in new staff members, and combining more experience in using the machine with greater understanding of how the machine and its programs worked. The members of this group would have to coordinate their efforts with those of such veteran members of the Problem Analysis Group as Charley Adams, Steve Dodd, and Dave Israel in order to avoid wasteful duplication and overlap and to make certain that an intelligent, coherent, efficient program of using the computer and training new personnel would be developed.

The object of all these efforts was not simply a list of codified routines. The engineers needed to reexamine their activities in order to appraise both the ideal and the practical relationships among operators, programmers, and analysts. When they had established what these should be, they should be able, Wieser and Welchman wrote, to create an efficient, "fully established computer laboratory." "What happens after the next few months," Wieser and Welchman continued, "will depend on individual aptitudes and preferences, and on a growing appreciation of what is wanted."[31]

There was still plenty of unfinished work to be done on the computer in Cambridge in preparation for further flight testing. While the radar was shut down, the MIT programmers took a closer look at their programs for computing the headings of aircraft. A lack of precision in the calculation of heading angles in certain circumstances impelled them to switch to expected-time-of-interception calculations. With indicator lights displaying time to intercept in binary form, they felt, the operator would be able to read time to intercept over the phone to Bedford; it would then be relayed to the pilot of the interceptor.[32]

Partly as a result of the installation of better storage tubes, more useful computer time became available, and the computer time devoted to applications rose from 40 percent to nearly 90 percent. When the radar was operating again and flight tests were resumed, the equipment at both ends of the telephone line was found to perform much better. It appeared to be time to attempt tracking and guidance maneuvers that would produce an interception. The engineers set the radar to scanning at 4 rpm and the computer to both tracking a single aircraft in straight line flight and computing heading instructions to guide an interceptor to a selected point correlated with geographic landmarks.

But the first tests revealed more troubles with the digital radar relay. While tackling these difficulties, the engineers found that the last 10 miles or so of the interceptor's travel to the selected guidance point were afflicted with fluctuating guidance instructions. When they decided to ignore observed positions and use calculated "smooth" positions, better results were obtained.[33]

Satisfied that the computer's guidance program was better and hoping that the radar information would also be better, the engineers prepared to resume flight tests. They were less than a month away from April 20, the day they would never forget.

Viewed from the operating level, these events demonstrate how the MIT and AFCRL engineers achieved success one step at a time. For example, when the Bedford relay link resumed operation during the last week of March, several days of poor weather prevented the Whirlwind engineers from determining how effective their repairs had been.

Meanwhile, plans were made to resume flight tests the first week of April. First, guidance tests would be run in which the computer would direct the pilot. A preselected and arbitrary fixed point would be established, the coordinates of which would be inserted in the computer. In addition, the pilot would have a visual flight plan of the course, so that he could report the effectiveness of the guidance. Once these preliminaries were in place, the engineers would "run an interception program on a stationary radar target, such as an isolated ground-clutter echo."[34] This program, which would employ only one airplane, would be useful for introducing the guidance procedures to new personnel.

On April 2 there was heavy fog, and on the next day there was a low overcast with rain. On April 4, one guidance test was run in the morning and one in the afternoon. Although the plan had been to have the B-26 fly at 6000 feet, clouds forced a decrease in altitude to 4000 feet. Also because of the clouds, the course that had been fed into the computer was used: from Hanscom Field to Concord, New Hampshire, then to Sanford, Maine, then to Cape Ann and Plymouth, Massachusetts. "Results of the test were obscured by frequent range coding errors in the DRRL." The miscodings were "too small by 1 to about 20 miles." Nevertheless, the B-26 was approximately on course until it drew close to the preselected simulated target. "Near close-in, the heading instructions over-corrected the aircraft and caused it to turn away before reaching the target point." Because the data were erratic, the engineers could not immediately put their finger on the cause of the overcorrection. Also, the communication between the aircraft and personnel on the ground was faulty. Success seemed far from imminent.

On April 6, the engineers had two Instrumentation Lab aircraft at their disposal. The B-26 was dispatched along the same course, and a familiar problem reappeared: range-coding errors. But in spite of these errors, they were able this time to direct the B-26 to "within 3 to 5 miles of the target point." The second aircraft, a T-6, was flown from Bedford to Concord to Sanford to Portland and back to Bedford. Since the radar was able to track it to nearly halfway between Sanford and Portland, the engineers realized that they now could now use it and the B-26 in their tests. Later that day they attempted a guided interception between the two aircraft. No luck. It appeared to be the same old problem. The range errors were so bad that the light gun could not acquire the target. The first interception attempt had failed.[35]

Tests were suspended until further efforts could be made to get to the bottom of the range-coding problem. The engineers in charge—Taylor, with his intimate knowledge of Whirlwind; Wieser, with his understanding of the target-acquisition programs; and Jack Harrington at Bedford, with his knowledge of the radar installation—agreed to a schedule that would give them regular access to the ever-busier Whirlwind computer. Mondays would be devoted to equipment maintenance; the other four afternoons of the work week would be committed to radar data from Bedford.

To prevent clutter or nearby targets from blotting out a more remote target, a "manually controlled inner range gate" was installed at Bedford. Displayed on the video plot position indicator, it would require the man monitoring it to adjust the gate so as to eliminate nearby interference.

The problems with range coding were finally solved at the end of the second week in April, and on Friday afternoon the engineers were able to guide another aircraft. On Wednesday, clouds and rain generated so much clutter that they could not be sure they had eliminated the range-jump problem. When they monitored local air traffic on Thursday afternoon, the range-jump problem was still there, as Harrington saw for himself on a visit to Cambridge. Further repairs made to the radar link on Friday morning left them ready to try again, using a revised computer-guidance program. To their delight, the overcorrection upon close approach that they had noticed on April 6 was gone, and "heading corrections were very smooth."[36]

The following Monday was a maintenance day. On Tuesday afternoon, when they took their scheduled turn at the computer, the engineers found that the radar and its relay link were communicating properly with Whirlwind I, although no flight tests were involved. On Wednesday afternoon they conducted guidance tests with the B-26, directing it to Sanford and to Cape Ann. They again encountered operation problems with the relay link. At one point it stopped sending for "about an hour," and at another it stopped for "about ten minutes," Wieser noted. "While this did not prevent useful work," he added, "it is inconvenient."[37]

Thursday was a holiday. Friday was the day, nearly 15 months after George Valley discovered Whirlwind, that opened up the future. That afternoon, April 20, the engineers had three aircraft at their disposal: the Instrumentation Lab's B-26, a T-6 from the Servomechanisms Lab, and a

C-45 from the National Guard. The B-26 was the interceptor; the other two planes were targets. There was one 10-minute interruption, caused by the DRRL. An unusually large amount of interference was detected, and a "blind sector," apparently caused by "close-in clutter," briefly blocked the radar view of the target planes in one direction while they were flying at the prescribed altitude of 6000 feet.[38]

Since the radar did not have the ability to report altitude information, aircraft altitudes were pre-assigned and computer control was exercised in two dimensions. Three interceptions were run in succession. All three were stunningly successful. Wieser, with careful understatement, wrote in his formal report three days later: "The results were encouraging, and experiments of this nature provide valuable experimental experience in preparation for operating WWI with the Cape Cod multiple-radar system."[39] Wieser then set forth the technical details of the historic event. The pilot of the B-26, he said,

> estimated that on each interception he was guided to within 500 to 1000 yards of the target. He had had experience testing the Instrumentation Lab's airborne intercept radar and autopilot system and he expressed the opinion that our midcourse guidance would have permitted completion of the interception by the airborne equipment.
>
> The interception technique is, briefly, as follows: The target ships are given predetermined courses (we have no direct communication with them) in the northeast area 30 to 60 nautical miles from Bedford. Targets and interceptors are detected by the Bedford MEW, their coordinates are sent to WWI over a telephone line. They are identified on PPI scopes at WWI, and tracking of the interceptor and one target is initiated by means of a "light gun" on the PPI. After initiation, tracking-while-scanning of both ships is automatically done by WWI, and the proper magnetic headings are computed to guide the interceptor on a collision course. Headings are computed once per antenna revolution (15 seconds) and displayed as binary-coded decimal numbers on a set of indicator lights. A man reads the headings off the lights and relays them by telephone to Bedford, where another man relays them by voice radio to the pilot of the interceptor.
>
> After initiation of tracking, the interceptor was guided for about 10 minutes (about 35 miles) to the interception point. The interceptions took place at a range of about 40 nautical miles. The whole system worked well, and more experiments of this nature will follow.[40]

A year had passed since the Air Force had redirected the Air Traffic Control project to "application of high speed computers to tracking while scanning in accordance with the needs of the Valley Committee."[41]

With the three interceptions on April 20, the Bedford phase of testing the experimental systems began in earnest. It would continue until May 15, 1953, when the MEW radar was dismantled to permit expansion of the Bedford runway. The dismantling of the radar, Dave Israel noted in his biweekly report, terminated "the associated MEW flight-test activity which [had] continued at increasing tempo since the first successful guidance and interception tests early in 1951." Flight intercept tests continued after May 1953, of course, as the project moved toward bringing the next experimental system into being. During the interim period, needed data would be provided by a radar unit located in North Truro.[42]

The Bedford test program was the first of several experimental test programs that were put together and operated after Jay Forrester and George Valley joined forces early in 1950. Out of it would evolve a more complex and sophisticated system: the Cape Cod System, a multi-radar network on a larger scale, which would feed data to the Barta Building for processing by Whirlwind. And out of the experience and the data accumulated from setting up and operating the Bedford and Cape Cod systems would emerge a still larger system: the Experimental SAGE Subsector (ESS), which would be linked to Whirlwind's successor, the XD-1 prototype of the operational SAGE computer.

ESS flight-intercept testing would be the final program of the developmental phase of the proposed continental air defense system. It would make possible the Air Force's successful activation of the first direction center, at McGuire Air Force Base in New Jersey. From McGuire, the system—no longer experimental—would spread across the United States, expand into Canada, and ultimately lead to the formation of the North American Air Defense Command, which in later years, as long-range bombers were supplemented with intercontinental missiles, would fuse SAGE with kindred systems to produce a continent-wide command-and-control system centered under Cheyenne Mountain in Colorado.

10

Project Charles

Although George Valley's Air Defense System Engineering Committee was created by the Air Force Scientific Advisory Board to find a scientific and engineering solution to the problem of air defense, it was never empowered to organize the program that would be needed to put the solution into effect. It was, after all, a part-time committee of civilian scientists and engineers assembled to investigate, to advise, and to undertake initial experimental inquiry, but not to commit the national military establishment to a course of action.

For the purpose of taking action and selecting the precise air defense policy to be pursued, two steps were taken during the autumn and winter of 1950. Crucially involved in these were General Hoyt Vandenberg, in his capacity as Chief of Staff of the Air Force, and MIT physics professor George Valley, in his capacity as chairman of ADSEC. The outbreak of the Korean War in June of 1950 was not the cause of their actions; it simply added impetus to the quest for an answer to two emerging policy questions: Who should assume responsibility for continuing the air-intercept feasibility tests that ADSEC and Project Whirlwind were undertaking? Who should simultaneously supervise the design and development enterprise that would produce the elaborate and unprecedented air defense system already being envisaged by Valley and his associates?

In Cambridge, some veterans of MIT's Radiation Lab who were working with Air Force Headquarters officers witnessed a demonstration of what had been achieved with the Whirlwind computer and promptly suggested that a permanent laboratory be formed to "improve the air defense system, using the digital techniques outlined by ADSEC."[1] Informal suggestion became formal recommendation when the Air Force's Chief Scientist, Louis N. Ridenour, a Radiation Laboratory "alumnus," pointed out to General Gordon Saville, Deputy Chief of Staff for Development, that verifying ADSEC's results would require an appreciable expansion of laboratory and field efforts. There was general agreement, said Ridenour, that the work should be performed by a contractor in the Cambridge area who would work closely with ADSEC and AFCRL. The Massachusetts Institute of Technology, he added, had indicated a willingness to consider the work.

Under these circumstances, perhaps inevitably, General Vandenberg turned to MIT. In December of 1950, in a personal letter (drafted by Valley at Ridenour's insistence) to MIT President James R. Killian Jr., Vandenberg explained that the time had come to "implement the work of the part-time ADSEC group by setting up a laboratory which will devote itself intensively to air defense problems."[2] This was a task, he said, for which MIT was especially and uniquely qualified. The formal proposal would have to come from the Air Materiel Command, and the details of it should be worked out by the AMC and MIT. That Vandenberg wanted MIT to undertake this supervisory task was clear in his final plea:

> The air defense problem which faces the Air Force is of great importance to the people of this country. The problem is technically complicated and difficult. The Air Force must urgently increase its research and development effort in this area and in this we ask your help. I sincerely hope that you will be able to give this matter serious consideration.[3]

No other choice appeared so attractive to Vandenberg and the Air Staff, even though the ADSEC-MIT group had not been the only one on the scene. Indeed, some studies had preceded that group's work, others had paralleled it, and others would follow. Two parallel studies completed at about the same time as ADSEC's confirmed and reinforced ADSEC's findings and added urgency to the Air Staff's conviction that prompt remedial action was needed. One of these studies, done by the Weapon Systems Evaluation Group, reflected the apprehensions of the Department of Defense; it found the nation's air defenses woefully inadequate. The other study was directed by Undersecretary of the Air Force John A. McCone; dubbed by him "Project Nutty," it emphasized "the need for a fresh approach to the technical problems of air defense," and it expressly called for the mobilization of scientific talent for this purpose.[4] All those concerned agreed that the problem posed formidable technical and political complexities.

A number of circumstances already supported Vandenberg's formal request to Killian. The Whirlwind computer was already engaged in feasibility tests. The knowledge and experience acquired by the Valley Committee was also a factor, as Vandenberg's letter pointed out. Just off the MIT campus were the Air Force Cambridge Research Laboratories, with their administrative and technical resources and with the experience their staff members had gained from assisting ADSEC. No other U.S. institution possessed a comparable pool of experience and competence. A nationally respected leader in education and research, MIT had developed substantial experience in military electronic technology through the work of its Radiation Lab and its Servomechanisms Lab during World War II.[5]

Vandenberg's December letter did not catch President Killian by surprise. In October, Provost Jay Stratton had told him of conversations in which Louis Ridenour had described ADSEC's work as "our brightest hope in the field of air defense" and had expressed the conviction that it would be "necessary to establish a laboratory in connection with the project."[6] Stratton told Killian that it appeared "inevitable" that the Valley Committee's work would be expanded and that MIT would be asked to become directly involved.

The scenario Ridenour saw in his remarks to Saville and to Stratton was not the one that Killian and his MIT policy makers saw. The MIT administration was ambivalent. MIT was a peacetime center of higher learning that had only recently divested itself of the Radiation Laboratory. Its leaders were not excited by the thought of starting up another such enterprise. Even Jay Stratton and Gordon Brown questioned undertaking the project.[7] But no one could ignore the threat that the war in Korea, now scarcely 6 months old, might expand, and they were seeking to establish a policy that would permit MIT to be ready to undertake emergency military research and, on the other hand, would maintain the proper emphasis on MIT's primary mission as an educational institution engaged in fundamental research.

Shortly after the eruption of the Korean War, Killian had stated in a memorandum to the faculty that MIT should be able to accommodate limited numbers of "short-term" projects or studies, but that to make a practice of organizing and managing another major research center like the Radiation Laboratory ran counter to MIT's enduring educational mission and should be undertaken only if "required to solve some particular problem and if the Institute possesses special advantages for such an establishment."[8]

Clearly, air defense was such a "particular problem," and MIT did indeed possess special advantages conducive to pursuing a solution. The pressure on Killian and his associates to accept the task was great. Even though several other educational institutions were engaged in air defense research under government contracts, the Air Staff believed MIT to be the only institution capable of implementing the concept proposed by the Valley Committee. Both the Chief of Staff and the Secretary of the Air Force were insistent that MIT assume the burden; indeed, Air Force Secretary Thomas K. Finletter was said to be "annoyed" by MIT's reluctance. Ridenour, too, urged Killian to undertake the responsibility. While the two men were discussing the problem of air defense in a Washington park, the Air Force's Chief Scientist argued that the action would push MIT to the forefront nationally in electronics. Even more compelling was his argument that MIT would be performing a task which was vital to the national interest and which the Department of Defense was not qualified to shoulder.

MIT's leaders capitulated only after they had negotiated specific terms, identified first in intensive discussions within MIT and then in external discussions with General Gordon Saville, with Louis Ridenour, with Ivan Getting (the Air Staff's Assistant for Development Planning), and with Air Force Undersecretary John McCone. The MIT administration stipulated that the project should encompass the "entire field of Air Defense," and that it should be undertaken under the joint authority of the three military services.

The air defense project would be carried out in three phases. First, an appointed study group, Project Charles, would conduct an intensive and comprehensive analysis of the overall problem of air defense. Second, ADSEC's recommendation of an experimental radar net would be implemented under the "interim management" of MIT's Research Laboratory of Electronics. Third, an R&D laboratory would be brought into operation to correlate and continue the work begun in the first two phases.

At first, the Army and Navy members of the Scientific Advisory Committee to MIT's Research Laboratory of Electronics were concerned that the air defense project might erode their influence elsewhere within MIT. But their fears were allayed and the Advisory Committee's approval secured when Provost Stratton and F. Wheeler Loomis, the man selected to head both Project Charles and the new laboratory program, explained the conditions MIT had set.[9]

Owing to the urgency felt by all parties, the formal legal action needed to certify the agreement lagged behind the technical and administrative progress. A letter contract was issued on January 30, 1951; it was confirmed after further negotiations by Basic Agreement AF18(600)-11, reached on August 6, 1951—five days *after* Project Charles submitted its formal report of its findings.

After contractual negotiations were completed, the formal contract went into effect in February of 1952; designated AF19(122)-458, it was backdated to January 30, 1951. This contract would remain in effect until January 1, 1959, when the obligation was transferred and responsibility for completion of the task shifted to The MITRE Corporation, a non-profit organization newly formed out of Lincoln Laboratory's Division 6 and other members of the Lincoln Lab staff who had been associated with the SAGE program.[10]

During the 12 months between the letter contract of January 1951 and the formal contract of February 1952, several important policy actions were taken. The Project Charles study reaffirmed ADSEC's concept of air defense. The concept's feasibility was demonstrated by the ADSEC-Whirlwind team in the Bedford tests. Project Lincoln was organized to begin converting the concept to an operating system, and plans were prepared to build an Air Force R&D installation to be operated by MIT. The new laboratory would be incorporated into the master plan for research facilities to be constructed in the Bedford-Lexington area for the Air Force Cambridge Research Center.[11] Until the new facilities were erected, both groups would remain in Cambridge.

The first two phases of the three-phase program that MIT and the Air Force had agreed to undertake ran concurrently. In the middle of February 1951, while ADSEC (under George Valley) and the MIT Digital Computer Laboratory (under Jay Forrester) were engaged in their joint effort to test the feasibility of hooking a radar up to the Whirlwind machine, the Project Charles study group commenced its investigation in earnest. The third phase, building and organizing the new laboratory, began the following July, shortly after Project Charles completed its study and as it was preparing its report.

It was half true that, as a member of Project Charles suggested several decades later, the study group "blessed" ADSEC's earlier proposal, "sprinkling holy water on it, so to speak,"[12] but both MIT and the Air Force expected more from Project Charles than that, and they got it. The composition of the study group and its course of inquiry reflected the fact that MIT and the Department of Defense were girding themselves for a more ambitious enterprise than ADSEC had ever been expected to carry out. Not surprisingly, the earlier recommendations of ADSEC and the later recommendations of Project Charles were cut from the same cloth, to nearly identical patterns, by the same designers.

The course of action already embarked on was the course that continued to be followed.

When Killian had discussed with faculty members and key administrators how MIT should respond to Vandenberg's request for help, Jerry Wiesner and Jerrold Zacharias (a professor of physics) were especially vocal about the need for a broad-based study group. Killian agreed; he also insisted that MIT's examination of the air defense problem and its commitment to organize a laboratory be explicitly authorized and supported by all three military services. Killian and his advisers were well aware that, although the Air Force had been assigned primary responsibility for defense against air attack, the Army and the Navy were determined to play important roles. Given this political reality, Killian and his associates realized that MIT would be in the strongest position and would secure the widest latitude for action if it served not just one branch but the entire military establishment.[13]

Killian and his advisors, deliberately seeking diversity, drew the members of the Project Charles group from other organizations and other geographical areas. They would, of course, continue to make full use of the large pool of experience and talent available in the Boston area, but they deliberately selected Professor F. Wheeler Loomis of the University of Illinois to be chairman. Loomis was familiar to MIT because of his earlier work as Assistant Director of the Radiation Laboratory. Highly respected within his discipline, he would meet no difficulty in securing the cooperation of all the major American physics laboratories. His selection signaled that the study would not be dominated by parochial institutional or regional biases.

Adopting no obviously predetermined policy position, Loomis was able to stand above the fray that surrounded the air defense issue. He was willing to return "on loan" for a while to an institution, a locality, and associations that the MIT policy makers were confident he had enjoyed greatly during his tenure at the Radiation Laboratory.[14]

Even though Project Charles was a diverse group, it was deliberately weighted with individuals affiliated with MIT and with ADSEC. Of its 28 members, there were 11 from MIT and one each from Cornell, Harvard, Illinois, Rochester, and Wisconsin. Two members of MIT's faculty (George Valley and H. Guyford Stever) were members of ADSEC, and two others (Jay Forrester and William Hawthorne) had worked closely with that body. John Marchetti of the Air Force Cambridge Research Laboratories and George Comstock of the Airborne Instruments Laboratory were also members of ADSEC. Other government agencies were represented too. Two members were from the Weapon Systems Evaluation Group in the Pentagon, one was from the Research and Development Board, and one was from the Atomic Energy Commission. In addition, there were seven members from private industry: two from Polaroid and one each from the Airborne Instruments Laboratory (Comstock), the Bell Telephone Laboratories, Eastman Kodak, Hughes Aircraft, and Trans Sonics.

Project Charles also sought the advice of 16 consultants drawn from seven educational and research institutions, one government laboratory, and one industrial concern. These consultants were to take into account various economic and sociological implications of the fundamentally new national air defense policy. Three of the consultants were economists; they were called upon to examine and report on the "economic consequences of passive defense." The scientific and engineering consultants included Louis Ridenour of the University of Illinois, Lloyd Berkner of the Associated Universities consortium, and John von Neumann of the Institute of Advanced Study in Princeton.

The international dimension of air defense planning was not ignored. Project Charles solicited the input of a North Atlantic Military Liaison group composed of members from the American, British, and Canadian armed services, the British Ministry of Supply, and the Canadian National Research Council. And, following protocol, during a briefing session in February on "development trends and program of the services in air defense" the Project Charles group heard from British and Canadian members of the Military Liaison group, as well as from American representatives. "Comments by the British Delegation" concluded the briefing sessions. The precedent established by wartime collaboration during the early 1940s was being continued.[15]

On February 19, 1951, at the Pentagon, the members of Project Charles held the first of 16 briefing sessions. The protocol reflected the gravity of the occasion. At the first meeting, "Senior Service Representatives" delivered addresses of official welcome. At the February 23 meeting, British and Canadian delegates spoke. After the Pentagon briefings, the study group proceeded to the MIT campus, and from there they went on to tour selected research laboratories and military installations "to witness demonstrations of systems and equipments, countermeasures, missiles, etc." Then they returned to Cambridge for the final briefing sessions, held March 12 at MIT. The entire series of briefing sessions took 4 weeks. Varied and comprehensive, the briefings encompassed "lectures and discussions on policy, general systems, individual components, and specific areas of interest to the Army, Navy, and Air Force" of the United States and also to the armed forces of Canada and the United Kingdom.[16]

Thus intensively briefed, the members of Project Charles launched an investigation into the details of the "general problem of defense against air attack." The sense of urgency they felt was shared in other quarters. In the Senate, Henry Cabot Lodge Jr. called the air defenses of the United States "so feeble as to almost invite attack," and there were those who feared that the Soviet Union's atomic capability presented the United States with "a really serious threat of a devastating attack by a foreign power."[17] To these forebodings were added anonymous intelligence reports to Project Charles speculating on the USSR's technical ability to mount an international attack if it chose, even though these reports did not provide any detailed substantive knowledge of Soviet intentions or plans. The members of the study group felt they had little choice but to accept the possibility of a "sudden sneak attack" to which the nation's air defenses would be "highly vulnerable."[18]

In retrospect it appears that the assumptions about Soviet weapons and delivery systems were unfounded. But at the time, the civilian technical experts of the study group were unquestioning in their acceptance of the grimly plausible conclusions offered, and they included in their final report a separate section in which they described a surprise attack as an "enduring threat" to the United States and its allies, as well as to the Strategic Air Command's "retaliatory forces." As they saw it, it was imperative that this threat be responded to quickly. Procedures for identifying aircraft should be improved. The rules of engagement should be revised. Earlier warning of any approaching hostile aircraft should be facilitated by the use of radar-equipped picket vessels and airplanes, land-based radars in Canada and Alaska, and ground observers in Northwest Canada.[19]

Early in their study, the members began to realize that the scope initially set for the investigation was "too broad for a group their size within so short a time span"; thus, they narrowed the focus of their study. Although "fleet problems and air defense of forward bases" were not ignored, emphasis was given to defending the continent against air attack. Concluding that detection and control of the air theater were crucial, they devoted their attention to the issues they regarded as the most difficult and pressing: those involving "the complex, difficult problems of air surveillance and control."[20]

To reduce their work load to manageable proportions, the members of the study group distributed their efforts over three major areas, to be examined by four "working groups," each assigned primary responsibility for a specific investigation within one of the three principal areas. Two groups would be responsible for "aircraft control and warning systems," a third for "passive defense measures," and the fourth for "interceptors and other weapons for action against attacking planes." In their final report they argued that this division of labor defined and delimited the work loads of individual members and permitted each to concentrate his effort in an area of special interest and competence. Also, it formed a direct link between the experimental work that ADSEC had initiated and the planning that Project Charles was charged to accomplish; thus, it was no happenstance that George Valley chaired Committee A, which was responsible for "Aircraft Control and Warning: Long-Term Program." To handle the short term, Committee B was formed under Professor Jabez C. Street of Harvard. Nicknamed the Quick Fix Committee, it was responsible for "Aircraft Control and Warning: Early Improvements." Committee C (under Jerrold Zacharias) undertook "Passive Defense," and Committee D (under Gordon Brown of MIT's Engineering Department and Servomechanisms Laboratory) undertook "Air Defense Weapons."[21]

To ensure that the final report would "represent a fair consensus of the entire group," a committee of the whole reviewed and revised the findings and recommendations of each of the committees. On August 1, 1951, they submitted their final report in three volumes under the title Problems of Air Defense.[22]

The large problem of aircraft control and warning (AC&W), current and future, received the most attention, followed by air defense weapons and passive defense in that order. "Electronic Warfare," "Manpower in Defense," and "Meeting a Surprise Attack" were the subjects of short sections. Passive defense received the briefest and perhaps the most theoretical, long-range treatment; it consisted chiefly of recommendations for the dispersal of new residential and industrial construction, provision of shelters in urban areas, and plans (to be implemented by training exercises) for rapid recovery from an attack and for prompt conversion of nonessential manufacturing facilities to war production.[23]

Defensive weapons received extensive scrutiny. Current programs to improve interceptors were "reasonable in view of . . . technical limitations," but dismay was expressed that the interceptors scheduled for service by the spring of 1953 would be "inadequate to counter the threat expected by 1956." More emphasis should be given to developing technologically advanced aircraft, especially single-seat, all-weather interceptors with superior power systems, ordnance, and fire-control equipment.

Design of surface-to-air missiles, especially those intended to provide local defense against low-altitude attack, should be encouraged. Further R&D on long-range guided missiles ought to be pursued—particularly work on the turbojet-powered BOMARC. Such an R&D program must be "coordinated with long-term plans for the AC&W system" that Valley, Forrester, and their engineers were working on. Navigation procedures and equipment should be pushed to the extent that the state of the art would permit, then adopted and installed promptly. Facilities should be created to develop and retain U.S. military predominance in countermeasures and counter-countermeasure equipment and procedures. Finally, air defense information networks should expect to make maximum use of noncombatants and civilian personnel in order to free combat forces for action.[24]

The members of the study group were certain that the sine qua non of an effective air defense system was a "competent aircraft control and warning system." For the Air Force, this meant a system that embraced the "functions of early warning and ground control of interception (GCI)," along with the "operations that take place in Combat Information Centers (CIC), Traffic Control Centers, Air Defense Control Centers (ADCC), GCI stations, and Antiaircraft Control Centers," including the "necessary electrical apparatus." Defensive weapons would be substantially useless without "complete, accurate, and up-to-date knowledge of the air situation and adequate control capacity"; so important was this requirement that it had to be met even if the cost amounted to "a significant fraction of the entire air defense system."[25]

The deplorable state of the nation's defenses prompted Project Charles to propose two simultaneous actions. One would be the "quick fix" program to raise the existing manual defense system to as high a level as could be achieved by "modest improvements and extensions of the existing equipment and systems."[26] The other would be the long-term program that was the centerpiece of the Project Charles report. "Not necessarily bounded by present equipment and concepts of operation," it should, in the coming 5 years, produce a technologically innovative and superior AC&W system that could cope with and control the more sophisticated offensive and defensive weapons looming on the immediate horizon.[27]

This program reiterated and confirmed the concept that ADSEC had developed. Its first phase was validated in a preliminary way 3 months before the Project Charles report was issued when the Bedford tests began to achieve successful aerial intercepts.

Project Charles opposed any pluralistic R&D approach to the national defense problem. "Considerable duplication of effort" by the three military services was cited. The services were pursuing programs similar in "functional conception as well as in many structural details," and they were following similar timetables. These programs, the committee's members emphasized, should be consolidated "under a single management to work toward a single integrated system" that would be "capable of application in any other theater." The shortage of qualified electrical engineers and the fact that only one nationwide system could be installed made a consolidated effort imperative.[28]

In addition to the dangers of the present and the near future, there were longer-term dangers. Project Charles urged that attention be focused on the sort of optimum system that R&D organizations would have more time to devise without being distracted by emergencies. Such a system would provide ongoing improvements and sophistication of the centralized AC&W system that the proposed new air defense laboratory would develop. The ideal prospect that Project Charles offered was one of seamless, efficient, considered improvement, beginning with a centralized AC&W system having radar systems and digital computers as its primary components.

The centralized system would have to be able, day or night and under adverse weather conditions, to collect, integrate, and display "up-to-date knowledge of the position, velocity, and identity" of aircraft in any selected airspace. This surveillance would provide the defensive forces with the early warning they needed. It would allow time for positive identification, and the defenders would have time to select appropriate countermeasures. It would also furnish navigation data and guide interceptors back to their bases. Civil aviation would benefit, receiving navigation assistance, air traffic control, and help in rescue operations. There would, however, be a larger volume of information received and dispatched than ever before in aviation experience, so it would be imperative to develop communication techniques more sophisticated than voice radio. Finally, Project Charles warned, the system would have to be "highly flexible" and "capable of growth and on-the-spot alteration."[29]

The study group called attention to a glaring technical weakness that ADSEC had noted earlier: vulnerability to low-altitude attack. Not only was low-altitude radar coverage inadequate, but it was difficult to detect and to track aircraft when ground clutter, false signals, and radar echoes were present. The Ground Observer Corps, used effectively during World War II, was the only traditional instrument available to detect low-flying aircraft, but it was not up to the job of providing the kind of advance warning needed to intercept planes flying at high speeds. Nevertheless, the study group advised that until something better was available a Ground Observer Corps should be brought to maximum effectiveness through organization and training. Meanwhile, it was important that "automatic aids for spotting, reporting, plotting, and filtering be investigated."[30]

Attackers flying low over land could be detected and tracked, the Project Charles report argued, by "relatively small radars closely spaced and mounted close to the earth's surface." Since transmission and processing of data should be automatic, a primary need would be the rejection of "unwanted signals." Since radar output should be suitable for "automatic transmission," it was necessary that "sufficient data processing be accomplished at the radar." Since a large number of "small, short-range, continuous wave (CW) radars" would be needed, they would have to be "simple and cheap" and require minimal maintenance.[31] Radar coverage over water, which would have to be "satisfactorily uniform and reliable," could be "obtained from a network consisting of pairs of blimps and picket ships anchored about 100 miles offshore."[32]

Adding many small radars to the system introduced a new problem: air defense operations could not be "organized about a great number of small individual GCI stations." The solution lay, the study group was confident, in the concept of the centralized system then being developed and tested by Valley, Forrester, and their ADSEC, AFCRL, and Project Whirlwind colleagues. Their approach differed in its organizational structure from existing and proposed systems, for it combined "all the GCI station functions in a single place which is also the ADCC." Radar sites would be sources of information, nothing more. Designing automatic operation into the centralized system would increase the system's ability to handle large volumes of traffic, respond rapidly to emergencies, and use a variety of weapons, including interceptors, antiaircraft artillery, and guided missiles.

It was the electronic digital computer, the report emphasized, that would give the Air Defense Control Centers the capacity to receive, correlate, and process large quantities of data flowing in from many sources. It was the computer that would then display the data needed to make the tactical decisions that would secure command over and retain control of the battle scene. It was the computer that promised the flexibility of operation, the speed, and the storage capacity that would allow "more sophisticated handling of information in an area of a given size."

The experiments using Whirlwind had already demonstrated the computer's ability to handle and process data, "including interception calculation of data within a 15-second scan of time." Nor would reliability be a serious problem, for this could be achieved if three electronic processing systems were operated in parallel, "to permit comparative checking of results and to allow for the required maintenance time of the equipment." In addition, there was marginal checking of the working components of the computer—"a technique for detecting the deterioration in the performance of electronic circuitry before the failure threshold" of the circuit components was reached. Marginal checking had already been developed and incorporated in the Whirlwind machine, and its effectiveness had been demonstrated: over a summer, Whirlwind had operated for 90 percent of its scheduled time without shutdown, in spite of variables introduced by newly installed components and additions.[33]

By the end of 1956, the Project Charles report overoptimistically predicted, if the project were given firm support in both program priority and funding, "a full-scale centralized system with an automatic-processing center and some 100–200 radars could be installed" in an air sector covering some 150,000 square miles. Furthermore, the margin of cost between the existing system and the proposed one would be relatively slight. The capital cost of the inadequate existing system was approximately $34.5 million, and the annual operating cost was $21.55 million; the estimated capital cost of the proposed AC&W system was $38.625 million, and the annual operating cost was set at $28.4 million.[34]

The members of Project Charles advised that two actions be taken to implement their proposed long-term program. First, the feasibility testing already in progress at Bedford should be expanded to link several radars in the Cape Cod area to the Whirlwind computer. Second, a prototype of the proposed continental air defense computer should be designed and built without further delay. It should possess special features "uniquely suited to the air defense problem," and it should take full advantage of "the design and operating experience with the Whirlwind computer, the experience with other computers, the new digital computer components that have been developed in the last few years, and the operating experience and systems study going with the [proposed] Cape Cod air defense system."[35]

The problems that Project Charles explored were neither strange nor unknown. There had been other studies, and there would be more before October of 1953, when the policy debate over how best to provide for the air defense of the continent would be solved by presidential fiat.

And the solution that Project Charles recommended was basically not one that the study group had created. Even so, Project Charles was seen as a necessary, not a redundant, exercise, for it gave the process of updating national policy the endorsement of a group of expert and respected engineers and scientists, and it laid out a specific course of action.[36] Perhaps even more important, it prepared the way for the next phase: organizing an air defense laboratory to finish the complicated R&D job that ADSEC had experimentally begun. It was no coincidence that Loomis, Hubbard, Hill, Valley, and Forrester, who had participated intensively in the deliberations of Project Charles, would contribute prominently to running the new laboratory.

11

Lincoln Laboratory

Lincoln Laboratory was the entity that converted the ADSEC–Project Charles proposal from a concept to a practical reality. At its inception, in 1951, it was "Project Lincoln." In April of 1952, as it was moving into its new permanent quarters in Lexington, the name was changed formally to "Lincoln Laboratory, Massachusetts Institute of Technology," to indicate its enduring and special status.[1]

Project Lincoln was not merely an ad hoc managerial reflex to the recommendations made by Project Charles. In 1950, MIT had begun to negotiate with the Air Force the terms under which it would assist in the national air defense effort. By the autumn of 1950, Chief of Staff Hoyt Vandenberg had raised within the Air Force the question of how that service should organize an effective research and development program that would draw on expert civilian talent. Vandenberg's Vice Chief of Staff, General Nathan Twining, had presented the question to the Air Force's Scientific Advisory Board. In response, in January of 1951, the SAB appointed a special working group, chaired by Louis Ridenour of the University of Illinois, to "explore ways whereby the Air Force might attract people in universities and civilian laboratories to help with the current rearmament effort" (an effort given impetus by U.S. involvement in the Korean War). Ridenour's group was beginning its examination of the Air Force's R&D policy at the same time that Project Charles was addressing the overall technical problems and the associated managerial problems of air defense R&D.

In February of 1951, without waiting for the Ridenour group's recommendations on long-range policy, Air Force Headquarters issued internal interim guidelines on the Air Force's relationships with civilian laboratories. Recalling the military's opposition to independent, civilian-controlled R&D organizations during World War II and the difficulties that had ensued, the statement pointed out that Air Force sponsorship of laboratories under university operating control should avoid a repetition of that experience.[2] The Air Force was seeking to avoid giving such labs the freedom of action that Vannevar Bush's Office of Scientific Research and Development had possessed during World War II while exploiting the nonmilitary technical resources of the nation as fully as could be done. The Air Force should, said the February statement of policy, allow the following:

> Maximum latitude and flexibility in the interpretation of procurement regulations in writing contracts to cover work done in these laboratories. . . .
>
> Fullest availability to each laboratory of the problems, developments, and plans of the military departments in the laboratory's field of interest. . . .
>
> Reasonable freedom for each laboratory to initiate and manage specific projects in its program area, [with] unnecessary duplication . . . prevented primarily by insuring that each laboratory is well informed about work already being done in its field.[3]

While these internal guidelines were being formulated, the Air Force and MIT were negotiating over how MIT should continue to assist the Air Force with its mission of air defense. By the end of March, Air Force Headquarters and MIT had agreed that MIT would have maximum freedom in operating the proposed air defense laboratory. No effort would be made to "prescribe or attempt to influence in any way [the] internal management" that MIT would develop to operate the laboratory. To evaluate the laboratory's work for purposes of making payment and approving future plans, a technically competent board would be set up by the Air Research and Development Command of the Air Force and would act with "the advice and counsel of the Scientific Advisory Board."[4] In this way, ARDC would become the Air Force agency responsible for interpreting and applying to the air defense project the general guidelines that had been issued in February.

In May of 1951, Ridenour's group recommended in an interim report that the Air Force sponsor the creation of "more weapons system laboratories in order to properly integrate the rich variety of new developments achieved in components since the end of World War II."[5] The object was to provide "a method of conducting military research and development distinctly different from that normally followed in peacetime."[6]

Whatever were the Air Force's intentions in making ARDC its administrative agent and managerial point of contact with the air defense laboratory that MIT was to set up, MIT's top administrators held different views. Determined to avoid the development of any kind of positive operating control by any one of the armed services, MIT had insisted that the new laboratory be sponsored by all three services, even though its primary mission was air defense, to be formally implemented by a prime contract with the Air Force.[7]

But ARDC was itself a newly activated command that had been given managerial responsibility without either control of the funds (which still belonged to the Air Materiel Command) or control of the air defense technical program (which belonged to MIT). Under these circumstances, as MIT's new laboratory and its air defense program began to take shape, the nature and the extent of ARDC's authority with respect to Project Lincoln's authority generated chronic difficulties and tensions, and these were exacerbated by the fact that in 1951, when Air Force Headquarters delegated full responsibility for the administration of the air defense project to its newly formed technical command, it failed to define with sufficient precision the duties ARDC should perform or the authority it should exercise.[8]

The issues between Project Lincoln and ARDC were symptomatic of a deeper level of organic hostility between two traditional and distinctly opposed systems. Peacetime practices and regulations, which still prevailed in the uneasy days of the Korean War, tended to give increasing authority to the holder of the purse strings. Consequently, in defense contracts the military services often exercised intrusive operational influence and even control. For organizations that operated contractually within the military command structure (as Lincoln Lab did) yet were not integral parts of the military chain of command, this practice posed special difficulties, and MIT, Air Force Headquarters, and ARDC all had to contend with such problems in the course of the effort to mobilize engineering and scientific resources. Air Force Headquarters was seeking to work out policy solutions to such problems, and it was seeking to do so in a cooperative way by calling upon its civilian Scientific Advisory Board and the Ridenour working group. Unfortunately, the arrangement involving ARDC and Lincoln did not work as smoothly as high-level civilian and military policy makers hoped.

In the summer of 1951, MIT's commitment to air defense shifted from policy planning to administrative action. While ADSEC was working with the Project Whirlwind computer group in MIT's Digital Computer Laboratory, seeking to convert the interception achievement of April 1951 to a normal operating procedure, Project Charles was spelling out the course of future actions. Three of the actions spelled out were taken: drawing up a binding legal contract between MIT and the federal government, drawing up the charter under which Project Lincoln would operate, and spelling out the dimensions of the first year's operating budget for the project.

While policy guidelines were being forged to ensure that MIT would have operating control of Project Lincoln, legal representatives from Air Force Headquarters and MIT turned their attention to preparing binding, formal, contractual work statements on which both parties could agree. Six such requirements emerged from their negotiations and were incorporated into the Schedule to Contract No. AF19(122)-458:

> 1. Investigate, do necessary research, develop, and design working systems in the field of air defense.
>
> 2. Develop or have developed components necessary to implement (A) above.
>
> 3. Seek improvement of all associated components and equipment.
>
> 4. Produce or have produced initial test quantities of equipment for field trial by special test detail or appropriate agencies.
>
> 5. Furnish necessary procurement information and consulting services to appropriate military services to enable procurement of end items.
>
> 6. Furnish project reports to the Contracting Officer at intervals of approximately three months.[9]

Events would determine whether these would remain pious statements of intention or become muscled definitions of practice.

The formal contract superseded all previous agreements and established the legal basis for the national air defense project. It defined the respective responsibilities of the two parties, described the work to be performed by MIT, set the schedule for initial accomplishment at March 1, 1952, and estimated the initial cost (about $7.1 million). However, it did not include the special arrangements to which both parties agreed when MIT committed itself to carry out R&D for the armed services. Those arrangements were codified in a more basic legal document: the Charter for the Operation of Project Lincoln, dated July 26, 1951.[10]

The charter declared the intention of the three armed services to "establish, under the management of the Massachusetts Institute of Technology, a program of research and development to be known as Project Lincoln." Although the charter went through several drafts, the approved version was strikingly similar to one that Loomis had prepared and submitted on June 4. The relatively minor changes made were in response to criticisms by the Army.[11]

Both the charter and the contract demonstrated how reluctant the Department of Defense and the Air Force were to impose stringent controls and risk being accused of hamstringing MIT's conduct of Project Lincoln's activities. At the start, Air Force Headquarters sided more strongly with MIT on the issue of managerial freedom than did ARDC or the Army or the Navy. Though at first the Army preferred more rigid controls, it eventually came to terms. The Navy appears to have been least concerned about either the mission statement or the terms of program control.

Until the end of 1958, when the air defense net was being erected and the remainder of the task was being handed over to The MITRE Corporation, Lincoln Laboratory continued to administer with a relatively free hand the air defense R&D project that ADSEC and Project Charles had defined.

By September of 1951, MIT administrators had organized Project Lincoln around an explicit divisional structure and had appointed the principal staff members. Loomis had enough confidence in his new organization to invite the Tri-Service Joint Advisory Committee to meet in Cambridge on September 24 to confirm with Project Lincoln's leaders, with representatives of the Air Force Cambridge Research Center, and with Air Force and Navy liaison officers not only the research objectives of Project Lincoln but also the basic terms of the relationship between the Joint Advisory Committee and Project Lincoln's administrators.[12]

Project Lincoln was initially organized into five major divisions; two more were added before the end of the first year. Of these seven, the two most important to this story of the SAGE computer were Division 2 and Division 6. Division 2, "Aircraft Control and Warning," directed by George Valley, would tend to both short-term and long-range engineering improvements. Division 6, "Digital Computer," with Jay Forrester as its chief and Robert Everett as its associate chief, would design and develop the air defense computer. The two divisions would collaborate to continue the intercept field tests that Valley and Forrester had undertaken jointly in 1950.[13]

MIT, as the contractor possessing controlling authority over Project Lincoln, delegated immediate direction to Director F. Wheeler Loomis, who had exerted a positive influence in directing the Project Charles study, organizing Project Lincoln, and formulating Project Lincoln's

Figure 11.1
Jay Forrester, Secretary of the Air Force Thomas K. Finletter, Project Charles chairman and first director of Lincoln Laboratory F. Wheeler Loomis, and Major General James F. Phillips, November 1951.

administrative procedures. He was advised and assisted by the Lincoln Steering Committee, which included leaders of the project's divisions and other MIT personnel.

During Project Lincoln's first year, the Lincoln Steering Committee met regularly with Loomis. Its twelve members were experienced administrators, engineers, and scientists, each of whom had a record of long association with MIT and with national defense problems. Most of them were members of either the faculty or the administration of MIT, and most had worked with the Radiation Laboratory or the Servomechanisms Laboratory during World War II. Loomis had been associate director of the former, Brown director of the latter. Each had been associated with Project Charles, as either a member or a consultant, or had participated in its formation. There was, in consequence, an internal operating environment of ease, familiarity, mutual respect, and confidence.

One indication of Lincoln Lab's importance to MIT and of MIT's importance to the Air Force was the appointment of MIT's President James Killian and Provost Jay Stratton to the Lincoln Steering Committee. Killian and Stratton remained members of the LSC until May of 1952, when the relatively smooth tenor of Lincoln's organizational affairs and the press of other duties caused Killian and Stratton to designate as their replacement the experienced and creative Nat Sage.

Professors Jerome Wiesner and Gordon Brown were appointed to the Lincoln Steering Committee because of their experience in MIT and national defense affairs over the years. Indeed, it came as no surprise to observers to see Stratton later succeed Killian as MIT's president and, still later, to see Wiesner succeed Killian as science adviser to the president of the United States before becoming president of MIT in his turn.

Lincoln staff members on the original Lincoln Steering Committee included five division heads: Malcolm Hubbard of Division 1, George Valley of Division 2, Albert Hill of Division 3, William Pease of Division 4, and Carl Overhage of Division 5. Other key staff members included H. M. Mott-Smith, Ragnar Rollefson, and Jerrold Zacharias. Because of the importance of the Digital Computer Laboratory and its Whirlwind computer to the work of Project Lincoln, Loomis soon asked Jay Forrester and his associate Robert Everett to join the Steering Committee.

At the end of Project Lincoln's first year, Loomis returned to the University of Illinois and Overhage to the Eastman Kodak Company. In anticipation of his moving into the Director's office upon Loomis's departure, Hill was appointed Deputy Director in March of 1952. The summer of 1952 saw changes at the top staff level, but from then until 1958 the basic divisional and administrative structure remained intact, while the major administrative issues that would confront the Director and the Lincoln Steering Committee were more apt to be external, concerned with budgetary matters and general relations with the Air Force.[14]

Although Project Lincoln was to be operated by MIT, it was never intended to become integrated with the Institute's primary educational function. It was located in Cambridge only until the Air Force could build a suitable physical plant. While the project was in Cambridge, it became convenient for MIT to exercise its contractual authority temporarily through its Division of Industrial Cooperation, which Nat Sage headed. But in January of 1953, anticipating the move to Lexington, MIT formed a new Division of Defense Laboratories to take over the DIC's functions of legally and formally integrating the lab into MIT's overall corporate structure. In this way Killian and his advisers maintained their policy of emphasizing, as he put it, "the sharp distinction between defense research of the kind represented by Lincoln and the normal academic interests of the Institute."[15] Three years later, in the spring of 1956, Lincoln Laboratory began reporting formally to MIT's Vice-President for Industrial and Government Relations, Edward L. Cochrane, a retired admiral.[16]

While MIT was making internal upper-level administrative adjustments, Lincoln was growing rapidly. By January of 1952, only 6 months after Project Lincoln was organized, the number of employees reached 501, of whom 241 were clerical and other non-staff employees and 260 were directly involved in the technical engineering and scientific problems associated with the air defense project. By the middle of June the total had grown to 1020, of whom 610 were non-staff and 410 were staff. Six years later, on the eve of the transfer of many of Lincoln Lab's employees to the new MITRE Corporation, the number of non-staff employees stood at 1305 and the number of staff employees at 720. In the autumn of 1955, the total number of Lincoln Lab employees was set at around 2100 for planning purposes.[17]

Under the terms of the agreement reached between the three services and MIT regarding how to tackle the air defense problem, no exact definition or locus of military authority was established over either the budgeting or the programming of the R&D activities that Lincoln would carry on. Such a loose arrangement presumably would expedite the creation of a continent-wide radar net reporting aircraft sightings to a central computer with backup computers.

Actually, three unorthodox arrangements were piled on top of one another here: the AC&W (aircraft control and warning) system itself, control of the R&D programs that would produce it, and control of the budgeting of these programs. To design and build the system itself was a sophisticated and unprecedented technical task that all parties agreed lay beyond the contemporary competence of the military service that had primary responsibility for air defense: the Air Force. But such was not the case where budgeting, funding, and program control were concerned, for these were traditional managerial practices with well-known traditional procedures. It was precisely to nullify conventional interference from these directions that MIT had imposed first its special terms and then the provisions of the Lincoln charter. On the ground that there was too little time to follow conventional weapons-development procedures while Soviet and Chinese foreign policy was so militant, the Air Force policy makers had agreed to these terms.

Assent (formally called "concurrence"), not independent judgment, was stipulated. The agencies involved would include, on the one hand, MIT and Lincoln, and on the other, in descending order of responsibility, the Joint Services Advisory Committee, Air Force Headquarters, the Air Research and Development Command, and the Air Force Cambridge Research Center. However, the concurrence that was stipulated in principle did not happen so smoothly in practice. The Air Force did not follow the streamlined, sped-up, short-cut managerial procedures that some felt the nation's vulnerability to atomic attack called for; it followed established, routine procedures. In consequence, by the end of 1951 Project Lincoln was encountering a funding crisis.

The particulars are instructive. In February of 1951, while Project Charles was making its study and before the Lincoln charter was issued, Air Force Headquarters directed ARDC to assume responsibility for "administration of . . . long-range research and development laboratories." In August, after the Basic Agreement had been negotiated and the charter agreed upon, Air Force Headquarters directed ARDC to "assume full responsibility for the administration" of Project Lincoln and to provide "required Air Force support including that required for systems-wide testing."[18] To implement this directive, ARDC was further instructed to study the fiscal year 1952 budget and recommend the adjustments or additions needed to establish the new Lincoln program while incorporating the existing ADSEC program into it. For the following fiscal year and thereafter, ARDC should prepare and submit Lincoln's budgets as "part of the regular Research and Development budget."[19] In turn, ARDC Headquarters delegated these responsibilities and duties to its Cambridge Research Center.[20]

Such directives always required interpretation, and it was a long-established practice that funds approved by higher echelons should, more often than not, be smaller than the funds requested by lower echelons. The Lincoln Steering Committee reviewed the project's first operational budget during its meeting of September 12, 1951, two days before submitting it to AFCRC. Project Lincoln was asking for $11.85 million to be added to $3.7 million for expenses already incurred by ADSEC, which would bring the estimated costs for fiscal year 1952 to a total of $15.55 million.

Project Lincoln had available and in sight $12.1 million of the $15.55 million total; the smaller sum would be made up of a combination of funds left over from Project Charles, funds transferred along with classified programs from MIT's Research Laboratory of Electronics, and funds allocated by the armed services. The $3.45 million difference would be made up, the Lincoln Steering Committee anticipated, by funds from the Research and Development Board, which appeared to have available as much as $6 million.[21] But a month later the Navy reduced its contribution by $2 million, and if the RDB were willing to take the Navy's place its original $6 million reserve would be reduced to $500,000.[22]

The rules and criteria for handling and allocating funds were independent of the rules and criteria for framing R&D programs and estimating their funding costs, as Project Lincoln's administrators well knew. Although they had been reasonably confident that the RDB would provide the funds they needed, their confidence ebbed as weeks and then months passed. By December of 1951, LSC members were asking how soon the project should reduce its expenditure rate below that originally planned, for unless the funds were immediately forthcoming the program would have to be curtailed before monies at hand were exhausted.[23] It was the custom at MIT that personnel on projects under contract be given 3 months' notice of termination, and for this purpose MIT's Division of Industrial Cooperation maintained a termination reserve of $1 million.

Steering Committee discussions reflected growing concern and dismay. There was concern lest the RDB allocate no more than $3 million, and there was discontent with the established procedures that required Project Lincoln's budget to be processed through the intermediate echelon of the Air Research and Development Command, from there to Air Force Headquarters, and in this instance from there to the Research and Development Board at the still higher-echelon of the Office of Secretary of Defense. Killian and Stratton "advocated immediate discussions with the Air Force" to explain Project Lincoln's inability to "continue operation at full scale" if the RDB's contribution did not exceed $3 million. Both men also vigorously endorsed the recommendation made by Valley and Hubbard that a single channel be established between Cambridge and Washington for matters pertaining to the budget. As a first course of action they proposed that a meeting be called the next day to "clarify [the] DIC-Lincoln relationship, and to formulate written agreements on procedure."[24]

Meeting with Killian the next day, December 11, 1951, were Stratton and Malcolm Kispert from MIT and Loomis and Hubbard of Project Lincoln. They agreed that the project's annual budget would be prepared in the office of Project Lincoln's director "with the advice of the Steering Committee." From there the budget would go to the office of

MIT's president for review; it would then be submitted through the Division of Industrial Cooperation to the Air Force. Any further actions that would be needed to ensure funding of the budget would be the primary responsibility of Project Lincoln's director, who would act with the assistance of MIT's president and provost, "particularly where contact at high level necessitates their intercession."[25]

Once the men had agreed on future internal budget procedures, they turned their attention to the budget problem that had precipitated the meeting. The circumstances of funding for the budget, they agreed, must be clarified immediately. Project Lincoln would have to receive $4 million from the emergency funds of the RDB. Anything less, they concluded, would compel Loomis, assisted by Killian and Stratton, to bring the situation officially to the attention of the Air Force through discussions with General James F. Phillips, Commander of the AFCRC. The Air Force would have to be advised that if the funds approved by the RDB were less than the total requested the Lincoln program would be, as they expressed it diplomatically, "fiscally limited."

Loomis was advised that, in his talk with Phillips about funding for FY1952 funding, he should bring up the FY1953 budget (which was in preparation), since it was "essential that Project Lincoln arrive as soon as possible upon a basis of funding ten to twelve months in advance."[26] This was a matter of special urgency for MIT, which simply did not have the resources to obligate funds it did not have available. Project Lincoln would have to be brought into accord with funding practices applied to other MIT government research projects and be funded for the fiscal year and "a minimum of six months thereafter, thereby alleviating the use of Institute funds occasioned by the delay in government appropriation and allocation."[27]

On December 17, 1951, the Lincoln Steering Committee heard that action by the RDB was imminent, but that only $4 million would be allocated immediately—not enough to cover the deficit incurred by the Navy's withdrawal of $2 million. The possibility was voiced that the deficit generated by the Navy's withdrawal might be met by the RDB later. This was not good enough. Project Lincoln's immediate reaction was to stop all recruiting and to halt the purchase of machine tools.[28] Consistent with the plan of action laid out in Killian's office on December 11, Loomis reviewed the situation in a letter to Killian dated December 21. He questioned the reliability of the support of the services for Project Lincoln, and he advised that the whole situation be brought to the attention of "the authoritative level in the services, especially in the Air Force." Killian, in a letter dated the same day, called the problem to the attention of Air Force Secretary Thomas Finletter and enclosed Loomis's letter with his own. Loomis had pointed out that the time had come for the air defense project to be "carefully reconsidered" to determine if it was worth the cost and, if it was, whether "sufficient assurance of support, both current and future" would be forthcoming from the services to justify MIT's making the commitments necessary to continue the project beyond its present stage. He felt that the time was appropriate to bring the situation to a head, for enough understanding of the technical problems had been acquired to enable them to propose a realistic budget covering at least 2 years. He questioned the reliability of support from the services, in light of the $2 million reduction in funds available from the RDB, and he pointed to the "reported intention of the services to propose $13,000,000 for fiscal year 53 instead of the $18,200,000 needed." (Neither action was yet final.) Five years would be required, Loomis continued, to complete the principal part of Project Lincoln's task: the development of the centralized digital air defense system. He considered most of Project Lincoln's other programs to be essential to accomplishing this task. If they were substantially reduced, it would impair the main program. MIT, he argued, could not commit itself to a task of this magnitude without assurance of "adequate and continuous support despite the to-be-expected fluctuations in the international situation and in the size of the over-all military budget." Without such assurance, MIT would risk serious damage to its "reputation by a large and awkward instability in its employment of scientists and by having incurred the odium of a major technical failure." Adequate support, on the other hand, would lead to the success of a program essential to the national security, a program offering "the *only* prospect . . . for coping with the threatened air situation after a few years." Loomis repeated the warning voiced in the ADSEC report of October 1950: without such a system, "the much more expensive radar net, interceptors, etc. [were] likely to be almost totally impotent." The effort on which MIT was embarked, he concluded, was "wise national policy," but only if "undertaken with the firm intention of carrying the development through to completion in minimum time."[29]

Killian put three questions to Air Force Secretary Finletter while endorsing Loomis's arguments: Should Project Lincoln be continued? If so, how was it to be adequately financed? Should MIT continue to be the contractor? He saw emerging "a growing uncertainty in the services in regard to the funds which can be made available to the project," and he recognized the obstacles that confronted both the services and the laboratory. He was familiar, he explained, with both the budgetary problems the services faced and the amount of time it took the Lincoln Steering Committee "to reach conclusions as to the funds . . . required to do the task essential to carry through the mission of the project." He agreed with Loomis: the project had to be completed within the shortest possible time if it was to have any value at all. Budgets, he emphasized, must be "controlled by this significant fact." Consequently, both Project Lincoln and the services were "faced with the virtually inflexible conditions" of going all the way or proceeding no further. This was a decision only the services could make.

Killian pointed out that MIT had reluctantly accepted the air defense project at the insistence of the Air Force and as a national duty. It would willingly relinquish this duty if the services were of the opinion that another contractor was preferable. As an educational institution, MIT had limited working capital. If it was to continue with the project, budgetary arrangements would have to be made that would permit "the earmarking of funds to insure a continuation of the project beyond the period of a single year." The Institute could not carry the project, with its attendant "financial burden of salaries and other expenses," beyond the end of the fiscal year, unless the services advanced funds that would allow it "to work within its fiscal and financial limitations."[30]

Finletter made no written reply until early February, when he had his pins lined up in a row. But General Yates, Director of R&D in the office of the Deputy Chief of Staff for Development at Air Force Headquarters, replied quickly. On December 26 he wrote to Loomis that, during a meeting of the Joint Services Advisory Committee in his office on December 17 (before they had received Killian's and Loomis's letters), both the Army and the Navy had supported Lincoln Lab's request for $6 million to be forwarded through Air Force channels to the Research and Development Board. However, the RDB, at its meeting two days later, had responded with reservations to the services' endorsement of the request; it had "tentatively approved for allocation to Lincoln" only $4 million—"tentatively" because some of the board's members wanted additional explanation from the Joint Services Advisory Committee before giving final approval. That explanation was to be presented later that day, Yates told Loomis, and he was confident that full and formal approval for $4 million would "follow immediately."

Then Yates took up the subject of the budget for fiscal year 1953—a "what we ought to do next time" approach—and requested on behalf of the Advisory Committee "a more detailed and definitive portrayal of what your Directorate wishes to accomplish." The material that had been submitted through the Air Force Cambridge Research Center was too general "for the Committee to use as a basis for intelligent discussion of their respective interests." As soon as possible, then, "a detailed breakdown of . . . planned activities and the estimated supporting funds required" should be prepared and forwarded with a total commitment from the services of $14 million used as a guideline. The Advisory Committee, Yates offered, would provide "service representation" to assist in the preparation of the more detailed budget."[31]

Loomis accepted the offer of assistance with diplomatic alacrity, admitting courteously that it had become "all too apparent that we do not, ourselves, have a clear enough understanding of the nature of presentations and the kinds of breakdown that are likely to be wanted by the several budget review groups."[32] Loomis's diplomatic candor underscores the significance of the actions taken at the meeting of December 11 in Killian's office. Furthermore, his response, as well as that of Yates, reflects the adroitness that both sides were prepared to exercise in dealing with the issues at hand. Neither Yates nor Loomis was addressing the three fundamental policy questions that Killian had raised with Finletter. The answers could wait while the two sides exchanged pragmatic information.

In a letter to Killian dated February 5, 1952, Finletter emphasized the Air Force's desire to have MIT continue as contractor. There was, he wrote, reiterating an earlier argument, "no other source which affords an air defense development potential comparable to that of MIT." He sought to refute any erroneous impression held in Cambridge that the Air Force was undecided "as to the extent to which it is willing and able to support Project Lincoln." The Air Force, he insisted, placed "air defense matters in a priority second to none," and it "heartily" endorsed and supported "the planned activities of Project Lincoln."

Then, to pour oil on troubled waters, Finletter mentioned how favorably impressed he had been by the program and the staff during his recent visit to Project Lincoln. He recognized the extent of the project's task and the need for the Joint Services Advisory Committee to "stand ready to assist in the development of mutually satisfactory policies and programs." He acknowledged that it was difficult "for an educational institution . . . to operate on this scale without receipt of full and timely funding guarantees." At the same time, it was difficult for the armed services, "who must secure funds through justification at several levels of review," to secure those funds "without substantive information in considerable detail." He was confident, nevertheless, that such problems were well on their way to being resolved as a consequence of a meeting held in Cambridge on January 9. That meeting had been called to "clear the air on these fundamentals" and to open the way to "a better understanding of the administrative responsibilities incumbent upon both sides for the successful prosecution of the project."[33]

Finletter's criticism was not unreasonable, for it wasn't always easy to tell how carefully Project Lincoln's detailed budgeting of its FY1952 program of action had been thought out. Although administrators at Finletter's and Vandenberg's levels of authority and responsibility were expected to find good men and back them up, as they did Killian on one level and Valley at another, the accepted function of lower levels of command at ARDC and AFCRC was to monitor the actions taken by the "good men" and their subordinates, especially when it came to authorizing or endorsing the spending of money.

The "quick fix" program offers an example of the complexities of funding and programming that confronted both Project Lincoln and its funding sponsors. When Project Lincoln had budgeted $7.05 million for research and development in fiscal year 1952, it had proposed as one part of its Aircraft Control and Warning program a "quick fix" approach to provide "short-time improvement in traffic capacity" of the existing system at a beginning cost of $1.5 million. The second part of its AC&W program would be committed to developing the centralized air defense system recommended by both ADSEC and Project Charles, at a beginning cost of $2.8 million for fiscal year 1952. "Quick fix" would use a test facility to be constructed at North Truro, where an AN/CPS-6B radar was already located.[34]

The $2.8 million sum budgeted would take care of "the continuous operation and improvement of the Whirlwind computer" that ADSEC earlier had brought into the air defense project, would underwrite "the development by the Whirlwind group of a new computer expressly designed" for the future centralized system, and would set up a small warning system in New England. This "Cape Cod" test system would consist of some ten to fifteen radar sites "equipped with modified versions of standard search and height-finding radars." These would be connected to the Whirlwind computer, which would process and display the radar data received and give automatic direction to intercepting aircraft.[35] The balance of the $7.05 million in R&D money would be devoted to costs associated with communications and components, with weapons, with special problems, and with equipping Project Lincoln.

Secretary Finletter, in his letter to Killian, then turned to what had brought about the exchange of letters: the immediate funding of the FY1952 budget and the limits imposed by the services on the budget for FY1953. He assured Killian that the Air Force was taking steps to meet the part of the deficit incurred by the actions of the Research and Development Board. Furthermore, he promised that the Air Force would provide, over and above the Army's and the Navy's contributions, the funds needed to support the FY1953 program agreed upon by Project Lincoln's Directorate and the Joint Services Advisory Committee.[36]

All parties were in agreement that this first budget, for fiscal year 1952, had been a particularly difficult one to prepare. It had been formulated under pressure of shortness of time, and it had attempted to estimate unknown costs for unexplored fields of research and development. Furthermore, it was being put into effect under new and untried administrative procedures. What was required was a general clearing of the air.

By early March of 1952 it appeared that Finletter's reassurances were solid, for word came down that the missing $2 million he had promised had been reprogrammed and would soon be available. The Lincoln Steering Committee was willing to conclude that the funding for fiscal year 1952 was "guaranteed."[37] But although this was true for actions at the level of the Secretary of the Air Force, it was not yet true for actions involving the Air Research and Development Command, and there was a final spasm of frustration over the RDB's promised $4 million. The proposal that Nat Sage submitted to ARDC on March 21 was rejected "on the grounds that the detail given was inadequate for . . . a proper analysis of the reasonableness of the 'price.'" The Lincoln Steering Committee promptly found this latest bureaucratic interference "intolerable," and Jay Stratton suggested that the matter be taken directly to ARDC Command Headquarters in Baltimore.[38]

Sage went to Baltimore and succeeded in persuading his listeners that the RDB's contribution was really a special case and should not be subject to routine procedures. His proposal was then "hand-carried through the red tape," permitting "speedy arrival of the money," the members of the LSC were told, and they heard with relief on April 7 that the latest troubles had ended. FY1952 funds were either in hand or en route.[39]

As often happens in such situations, after calm had been restored and the final figures for the fiscal year were in, the LSC learned that some $3 million in FY1952 funds remained uncommitted. These were added to the $13 million anticipated in FY1953 funds, and the LSC contemplated the possibility that carryover funds from one fiscal year to the next had materialized.[40] Indeed, funding issues of comparable severity did not threaten the project's continuation until the FY1958 budget was being developed.

"We had all the money we wanted for SAGE until Sputnik," Joe Giordano recalled as he looked back over three decades of personal association with Lincoln Lab's Division 6 and The MITRE Corporation. "We had no problems with cost. The emphasis was on getting the most practical solutions, to make SAGE work."[41]

Giordano's recollections were confirmed by Jay Forrester, who added that as a member of the LSC and the head of Division 6 he had encountered no funding crisis as serious as those that earlier had beset Project Whirlwind.[42] Lincoln did not, of course, have unrestricted access to Air Force R&D funds, but continental air defense was such a high-priority item of federal policy that Lincoln's program was not even affected by the budget cuts that occurred during the Truman and Eisenhower administrations.

Lincoln, for all its special status, was not immune to traditional federal practices of budgeting, allocating, committing, and expending funds. Actual, ultimate budget figures allocated, committed, and spent are difficult to establish in retrospect, partly because amounts were constantly fluctuating in the time between the beginning of preparation of any specific fiscal year's budget and the end of the fiscal year. In practice, final budget amounts were smaller than the sums originally requested by Lincoln's divisions. Nevertheless, the picture that emerges is one in which the total sum finally agreed upon came to rest somewhere in between the maximums requested and the minimums allocated, and usually in Lincoln's favor. Funding problems were chronic, usually minor, and never allowed to exert crippling pressure on the technical development of the air defense program, even though the budgets for fiscal years 1952 and 1958 caused serious concern at the time.

One chronic source of lesser friction was the reaction of some Air Force members to MIT's insistence on having a Joint Services Advisory Committee. Since the greatest part of Lincoln's work was for the Air Force, they reasoned, and since this service contributed the greatest amount to the laboratory's budget, Lincoln should be sponsored solely by the Air Force. Besides, the monies involved represented "a large fraction of Air Force research and development funds."[43]

Nevertheless, Lincoln Lab maintained its special status, because it was considered an emergency "crash program" in the highest national interest. From fiscal year 1953 through fiscal year 1955, its annual budget hovered around $20 million. For fiscal year 1956 the budget went up to about $30 million, and it stayed at that level for fiscal year 1957. A major cause of the increase was the addition of manufacturing and production costs to the R&D costs of bringing the new air defense computer into existence. The Air Force advised Lincoln Lab to begin to identify these costs separately, and Lincoln complied, requesting $21 million for R&D and $8.2 million for production in 1955. Lincoln Lab was told that the manufacturing and production funds would come "through ARDC" and "could be handled just like R&D money."[44] This arrangement, because it avoided involving the Air Materiel Command (which customarily was responsible for production expenditures), was intended to provide more efficient funding-management procedures than ordinarily prevailed.

Consistently, the largest single share of Lincoln Lab's working budget continued to be devoted to the air defense system, which accounted for $11.29 million out of $30.22 million for fiscal year 1957. On no occasion was the air defense project ever placed in jeopardy by a severe reduction in development funds.[45]

In practice, final budget amounts were smaller than the sums originally requested by Lincoln Lab's divisions.

When the FY1958 budget was being prepared, in 1957, the Eisenhower administration was particularly insistent on balancing the national budget, and this resurgence of a policy ideal that the administration had sought to follow since Eisenhower had taken office (in January of 1953) generated a budget crisis for the Department of Defense.

In June of 1957, the Air Research and Development Command received instructions from Washington and notified its centers around the nation that the Pentagon was imposing an austerity program effective immediately.[46]

In February of 1957, Lincoln Lab had submitted its proposed budget for fiscal year 1958. Its total request exceeded $37 million, of which the SAGE air defense budget was nearly $15 million. In May, ARDC had approved 99.4 percent of the request: $36.759 million.[47] But in September, Air Force Headquarters reduced Lincoln Lab's total budget to $28.609 million. This was a 23 percent reduction, and its effect on the SAGE air defense portion of Lincoln Lab's budget was to reduce it to $10.855 million. Robert Everett, then chief of Division 6 and responsible for bringing the air defense computer into operation, responded that this amount would not even meet "current SAGE commitments."[48] Early in October, Air Force Headquarters notified Lincoln Lab that it was making one exception to the summer cutback: SAGE funds were increased to

$14 million, only $750,000 less than the initial request.[49] Sputnik (put in orbit on October 4) was, presumably, not responsible, since it took the better part of a month for the federal government to begin to reverse its military austerity program. In any event, SAGE's funding priority continued, for in the preliminary discussions on the budget for fiscal year 1959 SAGE was to be funded as requested, although Lincoln Lab's total budget faced further reductions.[50]

During the R&D period that began for continental air defense with ADSEC's appearance on the scene in 1950 and reached a plateau when the Air Force began to install the SAGE system in 1958, the program that Lincoln Laboratory carried on to harness a computer to radars never was held up or "stretched out" by lack of funds. Project Lincoln came into existence to do the job 3 months after the Bedford tests showed that aerial interception could be accomplished under carefully controlled experimental conditions, using a computer harnessed to a radar. But Lincoln would have been established regardless of the outcome of the Bedford tests, for it was a creature of high-level executive planning, not the product of Valley's and Forrester's temporary R&D collaboration.

ADSEC, after all, was never regarded at the time as more than a key to open a door. It simply was reapplying the practical wisdom, acquired during World War II, of forming some kind of flexible alliance of civilian technical experts with military programming and funding managers. Lincoln Lab was to be the long-term administrative solution to the problem of air defense, but it took a year to get organized and establish its working relations with the rest of MIT, with the Air Force, and with the ad hoc Joint Services Advisory Committee. Whatever its funding and administrative fortunes in dealing with higher echelons year after year, its own character and style of carrying on its internal activities were determined by other forces, other issues, and other men.

12

Organizing the Tasks That Faced Division 6

Lincoln's internal organization and its approach to carrying on its technical affairs were improvised, but they were improvised with much thought and care. This was not enough to guarantee success, of course; that depended on thousands of personal acts and hundreds of team decisions. It was these acts and decisions that wove the historical fabric of Lincoln's large successes and small failures.

As F. Wheeler Loomis first set up Project Lincoln, none of its major divisions was organized to work on the key element of the proposed air defense system: the computer. Loomis, before the end of September 1951, made no bones about acknowledging to Jay Forrester that Project Lincoln was "heavily dependent upon digital computers in its future program." In his view they had no choice but to turn to Forrester's Project Whirlwind team for the command-and-control centerpiece of the air defense system. Just that month, the team had been reorganized as the Digital Computer Laboratory. Now attached to MIT's Department of Engineering, it was, of course, directed by Forrester. The only real question was whether Forrester's group would work as a part of Project Lincoln or as an independent organization under contract to the project.[1]

Two weeks earlier, at the first meeting of the Lincoln Steering Committee, George Valley had raised the question of Forrester's participating in the LSC's meetings and had spoken briefly of Forrester's views on how to design the air defense computer. But the LSC was new at its job, it was preoccupied with all sorts of organizational matters, and its members would not be rushed—particularly when Valley said that Forrester was estimating it would take "between 50 and 100 engineers" to design the computer they would need. Forrester's thinking was too far ahead of the LSC members' thinking for them to find it credible. Valley diplomatically suggested that they "support a study group and a small amount of components research and decide after perhaps three months whether or not to go ahead with the design of a new machine."

In the meantime, the LSC was willing to ask Forrester to prepare "interim plans together with . . . financial requirements between now and 30 June 52," the eve of the next fiscal year. After all, scarcely 10 years earlier Forrester had been a mere graduate student. The funding arguments he had had with the Whirlwind computer's earlier sponsor, the Office of Naval Research, appeared to indicate that he was young, still unseasoned, and administratively impetuous.[2]

Valley had a faster timetable in mind, and so did Loomis. Before the end of the month, Valley was bringing Loomis and Forrester together at lunch—"principally for Loomis and me to become better acquainted," Forrester noted in his administrative logbook.[3] After lunch the men discussed the kind of administrative relationship that might be established between Project Lincoln and the Digital Computer Laboratory. Forrester, always keen for maximum operating freedom, expressed firmly his preference for "separate laboratories," at least during the period of time the two organizations would remain "physically separated." An immediate problem requiring prompt action by both Lincoln and DCL was the availability of useful space in which to carry on their operations. Both were using MIT facilities and needed to expand, and Loomis was facing a threatened delay in the construction of the promised Air Force air defense laboratory building that Lincoln Lab would occupy. Forrester was encountering resistance in obtaining permission from MIT to make use of the Whittemore Building near the campus. Loomis agreed to press their common cause in a forthcoming meeting with Provost Jay Stratton.[4]

On October 2, Forrester and the Associate Director of the Digital Computer Laboratory, Robert Everett, were attending a meeting of the Lincoln Steering Committee as guests. Loomis led the LSC into "a comprehensive review of current plans for Whirlwind II (a transistorized modern version of the present Whirlwind computer)." He discussed the need for more space and called the committee's attention to two issues that he felt needed to be settled before MIT's reluctance to provide more space would be overcome: where the new "Whirlwind II" computer should be located and what kind of operating relationship should be set up between the new Whirlwind II project and Lincoln. The computer's location "would surely be the Lincoln Laboratory," he felt, although "preliminary work" could be performed elsewhere.[5]

After being introduced by Loomis, Forrester outlined a three-phase program of action that his Digital Computer Laboratory was currently facing: conducting the Cape Cod air intercept tests, designing the air defense computer, and building "interim machines to do computing jobs other than Air Defense." Loomis broke in to mention another problem the LSC had to deal with: influential higher-level advisers were recommending "the immediate construction of approximately ten Air Defense computers."[6] Forrester's technically informed response was to categorically reject the proposal: there was no present *design*, even, for a computer that could do the job.

Then, in reply to a question from LSC member Jerrold Zacharias, Forrester launched into a description of the existing Whirlwind machine and pinpointed three major tasks that he saw lying ahead if they were to build Whirlwind II: further research on components would be necessary, basic circuits for the new computer would have to be designed, and the functional organization of the computer would have to be simplified and consolidated. These tasks would require, he estimated, approximately 286 man-years of engineering staff time spread over 3 years. According to the rule of thumb of allowing 150 square feet per person, this translated into 20,000 square feet of new space needed by March of 1952 and a total of 35,000 square feet by July. The new space ought to be not far from the Barta Building, where the Whirlwind machine was located. The additional space was essential, for not until the middle of 1953 could the engineers designing Whirlwind II move to Lincoln Lab. Until that move could be accomplished, Forrester added, the existing administrative arrangement should be continued.

Forrester had on hand at the moment a staff of "about 85 . . . of whom perhaps 15 are working on general applications under an ONR contract." The other 70 were working on the Cape Cod System, and the first order of business called for a number of them to train replacements so that they might be freed to move on to Whirlwind II design work. Leadership of the new program, Forrester explained, would "come from the group of people who have been through the Whirlwind I program." Forrester's confidence was tempered by Loomis's cautionary observation that the enterprise "would represent a very large fraction of the total Lincoln effort" and "would have to be executed at very high speed."[7]

Three days after his oral presentation, Forrester released a two-page, three-graph study that he and Everett had put together at Loomis's suggestion. It was a concise analysis of the "time and personnel required for the work leading up to production of digital computers suitable for air defense information centers." The two electrical engineers specified no starting date, since the timing of space and budgetary provisions was beyond their control, but they set 3 years as the time needed to provide the experimental prototype.

The experience the two young men had gained by working on Whirlwind I and on command-and-control problems since 1947 had convinced them that any existing computer was no better than a "laboratory model" for the job ahead. The air defense task would require a machine with greater reliability and less maintenance—one "tailored to the air defense application" and able to employ the special input and output equipment that air defense operations would demand.

The apparent state of the art suggested that transistors could be used instead of vacuum tubes in the computing circuits, and that bi-stable magnetic cores could be used in the internal storage, although transistors and magnetic cores were still untried in such applications. The other elements of the computer either had already been put to practical use elsewhere or had been at least "proven out in the research and development phase." The overall speed of the machine could not be permitted to delay execution of its programs if it were to keep up with events in the sky as they occurred, and this consideration left Forrester and Everett quite unconvinced that the ten-machine solution to which Griggs and Ridenour were attracted was the way to go. Rather, the computing capacity on which such heavy demands would be placed should be "much less expensive to build or maintain in a single high speed machine than in several of slower speed."[8]

In comparison with the estimated 286 man-years of engineering effort spread over at least 3 years (to accommodate a carefully planned, overlapping progression of design, construction, assembly, and test activities), Whirlwind I would have consumed about 370 engineering man-years over the 6 years that would have elapsed between December 1946 and December 1952, "at which time the Cape Cod terminal equipment should have been installed." Everett and Forrester conceded that their estimate for building Whirlwind II might well be low, since developing the air defense computer entailed at least as much effort as had been required to bring the digital computer to its current state of the art. And, of course, more building space for the project and its expanding personnel would have to be provided.[9]

At the start, three major activities would require the engineers' full attention: laying out the basic logical design of the machine, beginning research on components, and selecting a manufacturer to simultaneously construct the prototype and prepare to build the production model. After 6 months, according to their idealized schedule, they should be launching development work on basic circuits and beginning to translate the information coming from their studies of logical design and of components into specific designs and hardware plans. All these preliminary activities would be essential before they could begin constructing the prototype, but they should enable actual construction to begin at the end of the first year. It would continue for the next 18 months. While all this was going on, they should also be training the chosen manufacturer to produce the prototype and the production models.

A computer could not be built as one builds a wall, piling brick on brick. It was too complex. Construction would have to be overlapped with assembly operations. Assembly of the prototype would begin in earnest 6 months after construction of its components had begun, and this would require the same kind of tight, crowded schedule to which the Whirlwind engineers had already become committed. But 12 months of assembly work, begun a year and a half after the project had started, would bring them simultaneously to the end of both the construction period and the assembly period.

There was a standard rhythm underlying the startup of these major activities, a new element of the project beginning every 6 months. Designing the machine's undergirding logic and undertaking initial component research, the initial activities, would both begin at the same time. After 6 months, work would begin on developing basic circuits, identifying and selecting details of the modular circuits, and initiating formal design work. Six months later, construction would start; it would take a year and a half. Six months after assembly had begun, the engineers could begin a year of testing the performance of the emerging computer. Three years after the beginning of the project, they should have a tested, "well engineered prototype machine" operating.

Meanwhile, several activities essential to production of the model that the Air Force would use would have been begun. The first of these entailed choosing the commercial manufacturer. The engineers would have entered this phase at the same time they set to work on the prototype: at the very beginning. A year later (at the same time they began construction work on the prototype), they would begin training and acquainting the manufacturer with the computer that they were developing, and they would spend the next 2 years in this sort of collaborative association. It would not be enough that the manufacturer learn to build a pre-designed computer; the manufacturer would have to be in on the engineering development of the prototype. During the second of these two years (the third year of the project), production plans would be made, and during the last 6 months construction would begin on the first of the production-line machines. A year after construction had begun on the production-line units, the first should be ready for testing over a period that would take 6 months. Delivery of the production models would begin 4 years after the entire project had started.[10]

Forrester and Everett then turned their attention to the manpower needs that would develop. The engineering staff should peak during the second year at 110, drop back to around 90, and level off for the remainder of the project. Non-staff support personnel might be expected to peak at 200 to 225 at the end of the second year, then slide back to 150 by the middle of the fourth year. None of these estimates included Cape Cod System personnel.[11] The understanding of the job ahead that Forrester and Everett demonstrated to the LSC during the first month of its existence confirmed the judgments of Loomis and Valley that Project Lincoln should promptly commit itself to designing and building a wholly new computer. October was spent working out how the Digital Computer Laboratory should collaborate with Project Lincoln. Placing the DCL under contract with Lincoln made less sense than making arrangements so that "many of the activities of the Digital Computer Laboratory should become part of Project Lincoln," mainly the Computer Development Group and the Air Defense Group. Valley urged that he and Forrester be made co-chairmen of an "inter-Division Group" that would be formed from his Group 22 and DCL's air defense team. This approach, Valley argued, would have the advantage of creating closer ties between Group 22's and DCL's activities "than would normally exist between two separate Divisions of Lincoln." But DCL had both MIT academic and Industrial Liaison functions, Gordon Brown pointed out. If some of the DCL groups joined Project Lincoln and some did not, Forrester would have to wear two hats: one inside Lincoln and the other outside. Forrester suggested separating from DCL both the Cape Cod intercept-testing program and the project assembled to design Whirlwind II, while Loomis thought Whirlwind I engineering work might be added to these, leaving behind only the "unclassified," non-national-security applications of Whirlwind, which had their historical origins in the Office of Naval Research's continuing interest in (and partial subsidy of) Whirlwind.

But Forrester thought Loomis's suggestion appeared "less workable." Although they did not say so, neither he nor Everett was inclined to give up DCL's efficient freedom to control its own destiny, a freedom that was rooted in the years of experience that Project Whirlwind had amassed. As for dollar costs, Forrester estimated that to use Whirlwind I half-time on air defense problems (the Cape Cod tests) would cost Project Lincoln and its Air Force funding sponsor half a million dollars a year.[12] No one had to remind any of the participants that whatever work was done on Whirlwind I to tie it in with the Cape Cod tests would also be an essential investment in the designing and the engineering development of the new air defense computer.

Eight months would elapse before the director of Project Lincoln, amid the press of budgetary and administrative affairs, would begin to look for a manufacturer. Meanwhile, by the end of October 1951 Loomis and Forrester had worked out the details of an agreement linking the Digital Computer Laboratory to Project Lincoln.

Loomis set forth the essentials in a "draft letter" to Forrester that was formally approved by the LSC at its meeting on October 29. Project Lincoln would create "Division 6: Digital Computer," which would consist of Forrester as its Division Head, Everett as its Associate Division Head, and "those members of the Whirlwind staff who work on Whirlwind II and related developments." Project Lincoln pledged to contribute $500,000 annually toward maintenance and operating costs of Whirlwind I in return for using it for the Cape Cod tests. Assistance would also be rendered in financing capital expenditures for improvements and expansion. In accord with Valley's suggestion of October 8, the new computer division's air defense group would work as a team with Group 22 of Division 2 on the Cape Cod project, and Valley and Forrester would be the "co-leaders." In daily practice, Bob Wieser of Division 6 and Howard Boehmer of Division 2 continued to supervise the joint project.[13]

The internal organization of Division 6 followed the pattern already being used in Project Lincoln, and by the end of June 1952 six groups were in place.[14] Of these, two were immediately crucial to Lincoln's mission: Group 61, "Cape Cod," and Group 62, "Whirlwind II." Led by Wieser, the engineers of Group 61 had been collaborating with the engineers of Group 22 in Valley's division since 1950 and the days of the Bedford tests, before MIT and the Air Force agreed to create Project Charles. When the success of the Bedford tests opened the way to the Cape Cod tests, the collaboration continued. The common aim of these two groups was to see to it that the Cape Cod System was "built, operated, and evaluated as a model of a complete air defense system."

Group 61 would continue to give particular attention to air defense center planning, to automatic information processing (including data screening and automatic tracking), to computing the orders controlling interdict weapons, and to providing the digital equipment the Cape Cod air defense center would need. Since Project Charles had recommended that Division 2 be given the overall responsibility for Aircraft Control and Warning, its Group 22 was responsible for "the system activities of the Cape Cod air defense system" and for "procurement, installation, and maintenance of the radar and data links." Working under these two sets of overlapping responsibilities, the two groups continued to collaborate as closely on engineering problems as they had before the administrators created Division 6.

Group 60 in Division 6, "Administration and Services," was a special case. It had long been an integral feature of the Digital Computer Laboratory on the MIT campus, providing the routine clerical services on which all staff engineers depended in the day-to-day operations of the laboratory. The fact that Division 6 became a part of Lincoln Lab's organization did not alter the circumstance that it remained physically remote, along with the rest of DCL, from Lincoln Lab's ultimate Lexington location. Any reorganization that would have made the new division logistically dependent on Lincoln Lab support services, Forrester pointed out in the first Quarterly Report of Division 6, would have "seriously delayed" work in progress.[15] He and Everett both realized that the need to get the air defense job done quickly justified retaining the efficient but not inexpensive divisional administrative and service self-sufficiency. That self-sufficiency set Division 6 apart from the rest of Lincoln Lab, and Forrester and Everett wanted it that way. It was a continuation of the operational autonomy that Project Whirlwind had enjoyed under the benign, informed tutelage of Gordon Brown, and it had been informally endorsed fiscally by Nat Sage, Director of MIT's Division of Industrial Cooperation (which was responsible for overseeing contractual and funding relationships with military and industrial project sponsors). Both these men gave those they supervised a relatively free rein, convinced of the desirability of "turning a good man loose" in order to get a good project, and Forrester and Everett had learned to subscribe to the same managerial philosophy.

But that managerial philosophy was not acceptable to all at MIT or to all at Lincoln Lab. In 1950, before ADSEC was formed, funding relations between Forrester and Project Whirlwind's sponsor, the Office of Naval Research, had become difficult, and the situation had caused Provost Stratton to wonder if firmer guidance should not have been given the project's young leader. Others within MIT were even harsher in their judgments. Some insisted that another Project Whirlwind not be permitted; others resisted the creation of Project Lincoln under MIT's aegis.[16]

Within Lincoln Lab there were some who resented the operating autonomy of Division 6, seeing little reason why it should be permitted an "irregular practice." And some members of Division 6 were given to criticism of Lincoln Lab. John C. Proctor, an old Whirlwind hand and a leader of Group 60, recalled in 1964 that cooperation and coordination had prevailed over friction, dispute, and controversy, both within Division 6 and between that division and the rest of Lincoln Lab.[17] Nevertheless, when the Digital Computer Laboratory joined Lincoln Lab, it had a different managerial and operating philosophy. When it left Lincoln, 7 years later, its philosophy was still different.

Four months before the crucial Bedford test of April 1951 (and 9 months before Project Whirlwind joined Project Lincoln), Jay Forrester had begun to clear the decks administratively so that some of the DCL engineers could give serious thought to the design of Whirlwind II. Here too the objective he had in view at the end of 1950 remained unchanged after Group 62 of Division 6 was formed, with Norman H. Taylor as its Group Leader. Furthermore, the tasks of Taylor's group and of the Cape Cod test engineers were interrelated, because they expected the unique design characteristics of the air defense computer to be strongly influenced by information obtained from operation of the experimental radar net.

Of all the elements of the proposed air defense system, the computer presented the greatest design challenge. There was little or no engineering and military experience on which to draw in selecting the precise design details of such a computer. Both the computer and its "associated equipment having the characteristics required for an operational air defense system" would be more highly integrated than had ever been attempted. Whirlwind I had been conceived, 5 years earlier, as a more general-purpose machine. Although Whirlwind I was proving to be an informative and invaluable experimental device for the Bedford and Cape Cod tests, Whirlwind II and its ancillary equipment would have to be much more closely and efficiently tailored to the air defense functions they would perform. Sensors and weapons systems were already in the military inventory, and their strengths and limitations were familiar to their users. Not so the computer that would use and control them.[18]

Forrester had begun the engineering prelude to laying out the design parameters of a new computer as early as December of 1950, when he had relieved Norm Taylor of his daily responsibilities for the operation of the Whirlwind computer and put him in charge of new engineering design.[19] During the second week of January 1951, Taylor, Everett, and Forrester had visited the University of Illinois "version of the IAS computer" and had then spent two days in intensive discussion with engineers at Engineering Research Associates in St. Paul, Minnesota. They were interested in the possibility of using magnetic drums to store the surveillance information that would be coming in to Whirlwind I from the Cape Cod radars. They reported: "ERA has made about 40 of these magnetic drums including their experimental models and they feel that this system is well understood and ready for use as a packaged piece of equipment. A demonstration which they made for us seemed to bear this out both in performance and in the general commercial appearance of the equipment which they are prepared to ship."[20] Later in the report, they offered these details: "A tour of the ERA facilities indicates considerable manufacturing space for both machine work and assembly and wiring. Of their 550 employees 140 are engineers and the remaining 400 are divided between shop and clerical help." Although Cape Cod problems involving Whirlwind I were weighing more heavily on their minds at that moment, they realized that this information would be useful in designing Whirlwind II and selecting a manufacturer.

George Valley had asked Jay Forrester for his thoughts on the air defense computer that ADSEC was interested in, and later in January of 1951 Forrester offered his considered reply. The computer could be no off-the-shelf design. It would ambitiously yet pragmatically force the state of the art, and the five aspects Forrester chose to call to Valley's reflective attention illustrated what he had in mind. Drawing on his knowledge of current engineering problems and prospects in order to transform design possibilities into working actualities, he dwelt first on the computer's internal storage, then on a handful of desirable general characteristics, next on the size and packaging of the machine, then on a half-dozen ways to improve and simplify certain of Whirlwind's operations that could be applied to the next-generation machine, and finally on longer-range research possibilities that ought to be explored, depending partly on new knowledge that was uncovered and partly on the delivery date that would be set. The first four, he said, "can be done quickly and in a straightforward manner with ordinary engineering personnel," but the fifth would be seeking "more radical changes" that required more sophisticated R&D work.[21]

Internal storage was uppermost in Forrester's mind, particularly speed of access and capacity. "In the near future this means that the machine will continue to use electrostatic storage tubes," he remarked. There was the theoretical choice of either the British Williams type or the RCA Selectron or the electrostatic tube that Whirlwind had been developing for several years. While he felt that "all three of these can be made to work," Forrester was not prepared to say which one would be "more practical and producible" or how soon. A machine that would store electrostatically as many as the 2000 16-digit "words" Whirlwind was designed to hold should be able to handle its air defense tasks with the assistance of, say, Engineering Research Associates' magnetic drum as a slower and cheaper form of auxiliary storage.

Although the projected air defense computer would benefit from higher overall speed than Whirlwind possessed, the route to such speed, Forrester argued, lay not in increasing circuit speeds throughout the machine but in speeding up storage access. Until storage time (then 20–30 microseconds) could be reduced below 6 microseconds, Forrester saw little wisdom in raising circuit speeds beyond those of Whirlwind, particularly since Whirlwind's speeds were "not unusually difficult to achieve." It would be more useful to increase the register length to accommodate about 20 binary digits. Also, the operators could incorporate drum storage to use, perhaps as buffer storage. This course of action would enable the computer to acquire information from the radar sets and hold it on call for processing as needed. Indeed, combining electrostatic storage and drum storage in the same machine "shows promise of being a powerful design." Perhaps a special hardware supplement might be inserted to extract square roots faster, instead of employing standard order codes to obtain the answers from the arithmetic element of the computer. "I'm not at all sure of this," Forrester said, "but it should be explored."[22]

Turning to size and packaging problems, Forrester saw no serious problems ahead. They should be able to shrink the circuitry of the next machine 15 percent by eliminating test storage of toggle switches and vacuum tube flip-flops, which Whirlwind, as a pioneer design, had required. Special checking equipment to make sure that transfer of signals did not incidentally change their values had accounted for 10 percent of the equipment in Whirlwind. Now that circuit performance characteristics were measured and known, transfer checking could be dispensed with in Whirlwind's successor. Similarly, the maximum pulse rate of 2 megacycles per second designed into Whirlwind could be cut in half without serious sacrifice of overall machine speed, thus accomplishing additional small simplifications of the equipment. High-speed video pulses did not require special cables as extensively as the engineers had anticipated in Whirlwind, and experience with marginal checking indicated how it could be accomplished more efficiently. Significant simplifications in hardware design could be achieved in both these cases. Finally, further simplification might be achieved at the small cost of slowing the operations of some of the circuits. Such measures might include avoiding "overlap of certain simultaneous operations in several parts of the machine," perhaps eliminating the carry register and reducing the complexity of the arithmetic element.[23]

In the realm of more advanced research, the properties of crystal-diode gate circuits should be investigated, Forrester suggested. Whirlwind relied "almost entirely on 7AK7 vacuum tube gates," but crystal gates could permit important reductions in machine size in addition to simplifying wiring needed for filaments and power. Even more attractive were magnetic flip-flops and magnetic gates; even though they would not be immediately available, Forrester felt they would become important contributions to basic computer design. But more important than any of these was the promise down the road of "a high-speed, rugged and reliable substitute for electrostatic storage tubes," perhaps taking the form of "three-dimensional ferro-magnetic or ferro-electric storage, such as we are now working on and with which you are familiar."[24]

In April of 1951, Valley received a reply to a request he had made of John von Neumann at the Institute for Advanced Study. He promptly passed it on to Forrester and Everett. It concerned the tasks the computer would be asked to carry out: using radars to scan the skies, identifying and keeping track of targets, and plotting interception courses.[25] In May, Everett used von Neumann's technical analysis as a guide in the preparation of an analysis his own, in which he would concentrate on how a Whirlwind-type computer might handle the relevant radar data and develop intelligible patterns of information fast enough to control and defend the skies. Von Neumann had sketched a scenario in which 70 radars would report, in the course of 15 seconds, on the progress of 1000 targets advancing across a segment of the sky. Everett approached it as "the problem of many targets, many radars, 1 computer." Presorting of the data, he realized, would be essential to making efficient use of the computer's immense calculating capacity.[26]

Everett's eleven-page in-house Engineering Report,[27] which he characterized as "a rough analysis," concluded that a Whirlwind-type computer theoretically ought to be able to handle the problem and calculate the interception vectors if auxiliary drum storage were added and if special orders and operations were programmed to accomplish the necessary sorting of targets and convert the radar reports to map coordinates precisely and thus to locate target positions.

To Everett's way of thinking, a "brute force" effort to update the progress of approaching aircraft by directly comparing hundreds of radar reports with hundreds of already identified targets made little sense when presorting was possible. Presorting could be done area by area, since incoming radar data were being acquired in that fashion as a consequence of the geographical disposition of the radars. Everett began with the hard-headed calculation that the computer might receive as many as 5000 reports in any 15-second radar sweep, rather than the 1000 needed to match von Neumann's imposed quantity of targets. These conditions would allow only 3 milliseconds to carry out a number of tasks for each report. Every report would have to be compared with any one of 1000 targets, and the information would have to be converted to *x* and *y* coordinates with a fixed reference. Further, one should expect that about half the incoming reports would require a smoothing prediction, and resorting might be necessary. New targets and missed targets would have to be identified and taken care of. Various displays should be generated for the human operators, and for one-fifth of the reports an interception prediction should be made.[28]

One reason for the large number of reports that Everett had invoked was the likelihood of overlapping radar coverage. Another was "possibly doubling [the number] for multiple reporting." If these complications could be taken care of, the number might be halved to 2500. Since it was the nature of radar to pick up more reflected signals than were sought, and since there were well-developed techniques for weeding out some of the clutter, these should be employed at the radar site before the radar reports were sent to the computer.

The remaining clutter would have to be tracked and its identity established, of course, for such clutter would be part of the target data and would have to be reckoned with. Taking care of clutter and multiple coding could increase the time available to the machine, doubling to 6 the number of milliseconds available per report. The current operating speeds of Whirlwind would allow 120 machine operations in 6 milliseconds—too few for the job. But if Whirlwind could reach its targeted speeds, 300 operations could be accomplished in 6 milliseconds, and if the number of reports could be whittled down to match von Neumann's assumption of 1000 targets—not 5000 or 2500, as in the cases Everett had invoked—then 800 operations should be possible in roughly 16 seconds.

In that event, one now had values to play with "within the realm of reason, especially if special operations are allowed (which they certainly should be)."[29]

Everett next turned his attention to the areas under radar surveillance. If one were to divide each of these into an appropriate number of north-south strips, one should realize that the fewer the strips the more comparisons would have to be made, and the more the strips the more presorting would be required. He supposed 50 strips, and he assumed that drum storage was handling the data in units of 50 targets per strip. In that event, a standard Engineering Research Associates drum could provide 160 bits per target in 2500 drum addresses—once again, reasonable. Now suppose further that 70 radars were supplying reports containing 20 bits each. Then there would be room for "about 300 reports/radar/computer reference, which seems more than adequate."[30]

In addition, target altitude information would be coming in, presumably from a second set of radars located at the same sites. Since height-finder radars tended to scan more slowly, they would not double the quantity and density of incoming data. In any event, information would be arriving from two different radar sets, and information from any one set ought to be recordable in one revolution of the storage drum. Though it was likely that the data usually would not come close to filling the assigned drum section (with the result that a part—even a large part—of the drum's revolution would unavoidably be wasting machine time), Everett speculated that the waste would be negligible and that it might be made more acceptable by the opportunity to call for overlapping operation of the computer while the drum was revolving if that should prove necessary.[31]

The problem of matching radar reports with identified targets evoked further discussion of area coverage and strip widths. Since the area involved was "roughly 500 miles square," what would be an appropriate strip width? And should sorting take advantage of the geometry of the drum addresses, or should electrostatic storage be used? A number of alternatives presented themselves, each with its own advantages and disadvantages, and Everett discussed these in detail.

Then Everett turned to the problem of computing a collision course and to the associated tasks of keeping the sorting of targets up to date and keeping track of each target as it progressed from strip to strip.[32] After reviewing in detail the analyses he had made of the tasks of handling target data, radar data, and interception calculations, he proceeded to make summarizing estimates of the lapses of time involved. All things considered, he could make only a rough estimate of the time required, since the extent of necessary manipulation of the data in detail was still unknown. But if one allocated 500 microseconds for taking target information from the drum, another 500 microseconds to return it to the drum for future use, another 2000 microseconds to correlate data, and another 3000 or more microseconds to predict, then the total time elapsed would include 6 milliseconds per target and 6 seconds per thousand targets. Doable!

Everett calculated that radar information would occupy about half a second of machine processing time, broken down into 50 microseconds per report (or 125 milliseconds for all) to transfer and convert the data, plus another 300 milliseconds "for drum waiting."

Everett summed up the interception calculations in several steps. First, "scan the drum for interceptors, bringing out as many as ES [electrostatic storage tube] capacity dictates." Assuming 3 registers per interceptor and 100 interceptors, then 100 interceptor transfers of information would require 50 milliseconds. Second, scan the tabulation record in order to convert target numbers to the appropriate storage drum numbers. This would consume another 50 milliseconds. Third, in another 100 milliseconds, scan the drum in order to extract 100 targets. Then to the resulting subtotal of 200 milliseconds add another 300 for computing 100 interceptions at 3 milliseconds each, and increase the number of interceptions to 500. The total time for calculating 500 interceptions would then come to 2½ seconds.

For all three classes of machine operations involving target data, radar data, and interception calculations, a total of 9 seconds would be consumed out of the 15 seconds available in von Neumann's problem. "The above estimates," Everett concluded, "may be correct within a binary order of magnitude either way although they are unlikely to be high."[33]

Everett's "thought piece" was circulated in May of 1951 to Forrester, Taylor, Wieser, and engineers in the air defense group. It is significant less for any specific indication of what hardware form the new computer would take (there was none) than for its detailed examination of feasible ways and means by which essential information-processing operations of a Whirlwind-type computer might be exploited to cope with events as they occurred in the world of physical experience populated by electronic circuits, radar pulses, and jet-powered fighters and bombers.

Early in July of 1951, Everett and Taylor made a one-day visit to Engineering Research Associates. In August, Forrester ordered from that firm two magnetic drums to be specially built for use in the Cape Cod

System intercept tests.[34] A year and a half had passed since ADSEC had discovered Whirlwind, and 3 months had passed since the eventful Bedford intercepts. It would be another year before Project Lincoln, having worked through its administrative shakedown period, would begin looking actively for an industrial manufacturer to build the air defense computer. But already Forrester and his engineers were giving specific engineering attention to simultaneously readying Whirlwind I for the Cape Cod tests and examining the options and problems they faced as they prepared to establish the design parameters of the air defense computer. The level of internal communication within Division 6 was sufficiently high that these two enterprises were being melded while maintaining their separate identities. It all spelled progress, but the pudding had yet to be eaten.

13

Planning the Air Defense Computer

To Jay Forrester and George Valley in January of 1950, the need for a specially designed air defense computer was as obvious as the need to set up an experiment to test the feasibility of the idea that a computer could guide a defending aircraft to intercept approaching enemy bombers. Hence, while MIT and Air Force administrators were reaching a policy consensus and organizing administratively to carry forward both tasks, Valley, Forrester, and Bob Everett were looking into preliminary problems associated with designing the needed computer, and their engineers were undertaking the Bedford tests.

Both tasks were pursued assiduously. By the end of 1950, Forrester had concluded that preliminary studies warranted pressing ahead with the task of designing the air defense computer.[1] Four months later, Bob Wieser and Jack Harrington, in charge of their groups working together on Valley's ADSEC problem, achieved their first successful Bedford flight interceptions, using Whirlwind. Half a year after that April success, Forrester gathered a handful of his engineers for a week of intensive discussions, to develop "a feeling for the nature of the machine" that would one day become the air defense computer.[2] The group, which ranged from seven to ten in number, met daily from November 5 through November 9, 1951. The members drew on their personal experiences with the pulsed circuitry of Whirlwind and radar to appraise the task that lay ahead: to design a computer that would keep track of the movements of as many as 1000 aircraft at all altitudes in an area of 300 square miles.

Bob Everett was there for obvious reasons. Norm Taylor was there because he had been in charge of new engineering design ever since the first of the year. Charlie Adams was in charge of the mathematics group that had prepared the programs which ran the Whirlwind computer. Ben Morriss, who served as secretary and issued the minutes of the meetings, had acquired valuable experience in the Block Diagrams group, coordinating the major functioning portions of the computer, especially the in-out units.[3] Bill Papian had devoted the past 2 years to developing and testing the tiny iron rings that would possess the switchable polarity and remanence characteristics suitable for magnetic-core storage. R. L. Walquist had been involved in the Air Traffic Control Project since July of 1950,[4] later helping plan the "Cape Cod" multiple-radar experiment that followed the Bedford tests.[5] David Israel's programming experience went back to the Air Traffic Control Project before it had been absorbed by ADSEC.[6] David Brown was the leader of Project Whirlwind's Electrical Engineering Division.

The meetings were intentionally unstructured at the start, to encourage the "ventilating" of problems and the exploration of promising and apparently blind alleys. Topics and questions could be raised by anyone. Among the things discussed were the relative merits of vacuum tubes, transistors, storage drums, and magnetic cores, features of Whirlwind that might be used or should be avoided, the logical operations and design that should determine the selection of appropriate circuits, and hardware characteristics that limited or opened up design possibilities.

"The objective of this series of meetings," Ben Morriss reported after the first meeting, "was stated to be the development of a feeling for the nature of the machine as it affects the work of the next year." Present work could be modified, or a complete change could be made. One technical option was to substitute the new transistor for the traditional vacuum tube. Another was to think of replacing electrostatic storage with magnetic cores. Either alternative raised the question of what types of basic circuits should be investigated.

The components they selected and the circuits they put together would have to result in operation as fast as Whirlwind's, and preferably faster. Three years to produce a working prototype would be a reasonable length of time to move from theory and past practice to improved practice. Since it would be unwise to freeze specific design requirements in advance, ignoring essential, valuable knowledge that would emerge along the way, a flexible "general purpose" machine should be the design target. The speed of the computer could not be overemphasized when selecting and designing components. Neither could reliability, since the principles of operation of a digital machine imposed an elaborate combination of repetitive, modular construction patterns on circuits that would operate in sequence and also simultaneously in parallel.[7]

Bearing these considerations in mind and turning their attention to major circuit-control devices, the engineers concluded that the three most attractive devices were vacuum tubes, transistors, and magnetic cores (which Papian had been working on since 1949). Cores appeared to be the most reliable of the three, and transistors next, but both were new. They did not have records of reliability such as tubes had built up over the years, nor were they available in plentiful number and standardized quality. It was obvious that the particular virtues and defects of each of the three should influence their selection for different modular circuits. Cores should be tried as the information cells of the machine's internal storage ("memory"), because the physical properties of their electrically imposed magnetic remanence indicated they should be the most stable and reliable. Transistors, employing different electronic principles, could also function as "on-off, yes-no" switches; however, so far they appeared to be less reliable, in view of the nascent state of the art of the transistor. They had other downsides too: a sufficient supply of transistors to meet projected design needs could not yet be expected, nor would the project engineers be able to exercise the kind of quality control over their manufacture that they were already exercising in their efforts to develop suitable magnetic cores. Vacuum tubes were the only circuit-control devices with which the engineers had had long experience, and as a result of Project Whirlwind's efforts to extend tube life they believed that tubes were about as reliable as they were going to get. Perhaps circuits or subsystems employing cores, transistors, *and* vacuum tubes would make the best practical sense.[8]

Taylor began the first meeting by calling attention to three basic hardware features of the computer that presented both problems and opportunities: components, building blocks, and registers. The components included vacuum tubes; storage tubes; crystals; transistors; and magnetic cores, tapes, and drums. The building blocks included flip-flop circuits; amplifiers; magnetic gates, gate tubes, and generators; and destructive-read circuits. The registers, containing groups of binary numbers, were of different kinds, depending on their function—adders, counters, decoders, selection switches, circulating registers, and memory systems. The systematic integration of all these items would make or break the design of the computer. As Taylor reflected on the experience that had been acquired "after about 7000 hours of systems operation on Whirlwind I," he could single out several classes of difficulties that had been encountered in dealing with what might be called the logic of the system.

All the improvements that the difficulties called for centered around the performance of internal machine functions, the most crucial of which was the slow access to the "memory" contained in the electrostatic storage tubes. Flip-flop vacuum-tube circuits had been the weakest link in Whirlwind. Should they be replaced by transistors? The input-output equipment had proved to be an unexpected bottleneck. On the other hand, there was the demonstrated reliability of vacuum tubes in the 7AK7 gate-tube circuit module, which had had a failure rate of only 1 percent per 1000 hours. Germanium diodes had achieved a commendable 0.25 percent failure rate, yet that would be "rather high for the military computers which we are considering," said Taylor. They should be aiming for 0.1 percent.[9] And they should allow between 10 and 20 microseconds, no more, to execute an average order, including access time to the memory, and between 6 and 12 microseconds when storage access was unnecessary. Multiplication should take about 2 microseconds per digit, and part of the computer could start on the next order before another part had finished with the preceding order.[10]

Everett, looking for ways to increase machine speed by keeping the internal design simple, explored aloud the fundamentals of the logical organization of an idealized, simplified computer that he had been concocting with some care, examining the "machine language" that would accomplish such basic orders as "Clear and Add (ca)" and "Transfer to Storage (ts)" and pursuing step by step the execution of the orders and laying out a simple schematic block diagram.[11] Dave Israel subsequently calculated for the benefit of the group that 1000 aircraft aloft and under surveillance would generate as many as 260,000 machine orders within a single radar scan of 15 seconds. [12]

After an extended discussion of the state of the art of magnetic cores and transistors at their third meeting, the men turned their attention to how big and ambitious a machine the computer should be. The answer, in general terms, was that it should be as big as its function required—the function to be defined by the size of the air space being monitored, by the total number of aircraft the computer might have to handle, by the sharing of information among different radar-computer arrays to provide a complete picture of the sky, by the number of computers that would be necessary to span the continent, and by stipulating a computer of "reasonable" size.[13]

But the engineers found themselves unable to agree on what should be the balance among speed, simplicity, and cost.[14] It would be very nice to save on the number of components so long as they would not have to pay the price of increasing the number and variety of building blocks being used in the computer and consequently complicating the logic of its operations. Nevertheless, it was true that "the fewer the number of different types of building blocks which are used, the simpler the logic" and the better a job could be done of developing each block. But what *was* "simple" or "complex" logic? These definitions appeared to depend on "the particular approach being used by the individual and his ability or inability to explain the process easily." They were up against several truisms that were none the less stubborn for being truisms: ". . . a quite simple event may be very difficult to explain or define . . . new things usually appear more complex than things with which we are familiar . . . new ideas should not be immediately discarded just because they appear somewhat more complex than the familiar." The dynamic flip-flop was a case in point: although it was a complex circuit, it did not place undue strain on any of its components, and it might be an excellent application for the transistor.[15]

Everett's simplified core-computer design should theoretically carry out as many operations per second as Whirlwind while using only 200 vacuum tubes, and since cores, not tubes, were to carry on most of the operations, the number of tubes per operation per second would drop to 0.01. Everett agreed to write a memorandum on usefulness, complexity, and cost, and Forrester reminded everyone that the fundamental question at issue was not "whether to build a machine or not, but rather to build the best machine possible considering speed, cost, capacity and complexity. For the problem under consideration, all of the reasonable speed and capacity will be used, and more will be desired by the users."[16]

Forrester announced that weekly meetings would be scheduled in the future, to hear from smaller groups that were working on transistor problems, core problems, "and other problems discussed but not answered." Taylor was supervising these operations and would plan the meetings.[17]

Dave Brown opened the fifth and final meeting with a summary of where the engineers stood. If first things were to be done first, then the coming year must be devoted to "the investigation and development of the logical structure and the building blocks for the computer." Speed and reliability would dominate the design of the building blocks that would control component selection and hardware design. Indexing of information by the computer would, perhaps, take a considerable portion of the operating time.

Magnetic cores would be used for the central storage of information, and the arithmetic element and the control section of the computer would most probably include tubes, transistors, and cores as circumstances warranted. Although transistors appeared to be almost reliable enough for high-speed use, cores offered the prospect of greater reliability even though the techniques for using them remained to be developed. The engineers should proceed with the design and development of computing and control circuits employing tubes, transistors, and cores. They should think of mixing and blending all three of these in their circuits. And they should set up core-manufacturing facilities that would be under their quality control.[18]

Brown's information on transistors was mixed. Current data from the Bell Telephone Laboratories, where the transistor had been invented and where primary development was going on, indicated that transistors ought to become about as reliable as the vacuum tubes and germanium-crystal rectifiers used in Whirlwind: The tubes had a failure rate of 0.5 percent, the rectifiers a failure rate of 0.25 percent. The success of Bell Labs with transistors, in "a 16-digit serial multiplier . . . operated at a one-megacycle prf," and subsequently at 3 megacycles, led Brown to conclude that transistors would be fast enough for use in arithmetic elements, although their speed in control circuits remained to be investigated. But procurement information was bad on two counts:

> For some time to come, certainly for the next six months, very few transistors will be available. In fact, the supply may be so limited that circuit development will be handicapped. We are evidently quite low on the priority list. Of the 100 Type 1734 transistors produced each month at the present time, the number allocated to us is zero.
>
> The rate of production of transistors in the future is uncertain, and the demand will probably be large. Quality control will be exercised by the manufacturer and will be beyond our direct influence.[19]

There were no organized "bull sessions" on the design of the next computer in December of 1951 or in January of 1952; they resumed in February, when weekly meetings of the "Whirlwind II Planning Group" began. The number of participants grew from 13 to between 40 and 50 before the meetings came to an end. Seven of the original members of the November group were involved, and Taylor led these meetings too. The meetings tapered off in June and July as Taylor, Everett, and Forrester became more deeply involved in the next phase: bringing the manufacturer of the production model into their operation.

Like the five meetings in November, the February meetings served an educational function at the start and were initiated "to stimulate ideas" about the air defense computer. Their expressed aim was to solicit general comments from the engineers about what the new machine should be like.

At the first meeting, Jay Forrester set the tone by emphasizing "the importance of striving toward simplicity in a future machine." He called the attention of the 13 men in attendance to the concept of a single-register computer, which Everett had discussed at the second November meeting. It was an appropriate way to approach the initial problem of structuring the logical design of the new machine, and its virtue as a starting point for discussion was that it focused the engineers' attention on a subtle but critical design policy: balancing increased efficiency, capacity, and speed.[20]

At the second meeting, Taylor moved away from theoretical considerations and injected a pragmatic engineering approach by presenting a "plan to build a prototype system . . . to test out new ideas in circuitry, using both transistors and iron cores." The nine engineers present proceeded to discuss "a first approximation of a prototype system," one that would make a virtue of simplicity and include no more than a single register, a 16-bit word length, and a 256-word storage unit "made from the Magnetic Ribbon with an over-all access of approximately 20 microseconds." Taylor pointed out that such a prototype system was "quite an ambitious one and perhaps beyond the scope necessary to provide a testing ground for new componentry and new ideas." Yet it might be possible to carry forward "if we are willing to freeze the specifications at an early date and allow a small integrated group of engineers" to proceed to build it. Furthermore, such a course of action would "help us realize a Whirlwind II system with a minimum of delay." Subsequent discussion would enable them to "decide just how ambitious this prototype should be."[21]

At the third weekly meeting, Taylor got more specific. He reviewed the concept of a simplified computer that Everett had discussed in November, and he offered the systems logic of its design as a basis for developing a "Whirlwind I½" to serve as a prototype for the air defense computer. It would be more than a miscellaneous collection of critical circuits they would want to test, and more than "just a modern version of the five-digit multiplier" that had been constructed several years earlier to demonstrate the validity of the calculating circuits that then went into Whirlwind. It would be a functioning computer, but one of quite limited capacity.[22]

If they allowed themselves no more than a year to build Whirlwind I½, and if they built into it sufficient capacity to test the proposed circuitry and system logic of the air defense computer (Whirlwind II), they would be able to make realistic progress toward firming up the design of the production model. Since to do its job any such computer would have to have at least 12 binary digits, Everett suggested that the added effort of going to 16 digits would be well worth the advantage of matching the design of Whirlwind I, not to mention that it would make the new machine more useful after its original testing function had been served. It would be nice to be able to add even more digits, but doing so might complicate the design of the computer's control unit. They would be most interested in the performance of the internal-storage unit of magnetic cores, so perhaps it would be worthwhile to provide for parity checking of the storage unit so that they would have harder information on its performance.[23]

How general-purpose need such a computer be? To this question Everett responded that, no matter how flexible the control matrix might be made,

> it is even more important to have the remainder of the computer designed so that the possibly desirable *commands* can be performed (i.e., shifting, cyclic shifting, cyclic shifting among several registers, etc.). Once these various commands are built into the computer, it would be possible to design all the possible orders making use of these commands and if this does not result in an overwhelming number of orders, it might be desirable to build all these orders into the computer so that they will always be available instead of making it necessary for them to be read into storage each time they are desired.[24]

The 14 men at this meeting informally endorsed Taylor's earlier suggestion that the design of the Whirlwind I½ they were speculatively considering should be "frozen in two or three months so that an independent group could go to work. . . . Otherwise, with people getting new ideas, Whirlwind I½ would drag on and on and not provide any of the information that is desired concerning how WW II should be built."[25]

The next two meetings were devoted to examining and proposing various features that their WW I½ ought to have. In the middle of March 1952, Taylor took stock of the progress that had been made. The machine should be a 16-bit parallel computer (since handling the 16 bits serially would make it unacceptably slow), it should have a storage capacity of at least 256 16-bit words, and it should employ between 8 and 16 machine-control orders. Two registers should be external to Storage, and there should be suitable (still undefined) input-output equipment. Inevitably these characteristics raised a number of questions. Was the word length too ambitious for the tasks they had in mind? Should the magnetic cores be composed of Mo-Permalloy ribbon, or a ferrite composite? Was a magnetic switch better for their purposes than vacuum-tube drivers? What did the number of registers do to operating speed, and was this a serious factor? The more orders that were built into the hardware, the easier would be the programming, but did this tradeoff really tell them how many orders to build in? To select the appropriate in-out equipment, they had better be sure what the prototype was for and what sort of paces they needed to put it through.[26]

Regardless of the specific answers to these questions, the machine would have at least three reasons for existing. Most important, it would "test techniques and components which are to be used for the building blocks in WWII." It would also "try out new ideas (such as a single-register computer) which might be used in WWII." Finally, it would "provide incentive." And once its primary functions had been carried out, the computer could then be submitted to reliability and life tests of its components, as the prototype Five Digit Multiplier had been for Whirlwind I, or it could be made a functioning adjunct to Whirlwind I (perhaps as a terminal-equipment computer that could get around the speed bottlenecks Whirlwind I had encountered). When Wieser suggested that such dual-use considerations might slow the design work on the prototype, it was pointed out that the machine could always be turned over to the operators of Whirlwind I, who could then modify as they pleased.[27]

With regard to register length, at least ten digits were required to accommodate an instruction and its address, and "it would be a shame to build WWIA with almost, but not quite, enough digits to work efficiently with WWI." Sixteen digits would not require that much more design effort, and the prototype's word length would then match that of Whirlwind I Though a 256-word internal-storage capacity appeared optimum for testing needs, Everett urged the engineers to seize the opportunity to go to a larger module of 1024 registers, as it "might show up various electronic problems more rapidly." Since ferrite cores and a magnetic storage switch were being considered for the air defense computer, that should be reason enough to try them out. Meanwhile, the men would acquire a better idea of the in-out equipment needs as work on the projected Cape Cod air intercept tests proceeded.[28]

The thorniest question was how many control orders to establish. Groups of 4, 8, and 16 were considered without reaching a firm decision. Although it would be simpler to provide an 8-position control switch, Everett suggested a 16-position switch would leave room for growth, unless incorporating it would prove "electronically inconvenient." It was the simplicity-versus-complexity problem again:

> The problem boils down to a question of being ambitious enough to check the components and logic for WWII and yet not make WWIA so complicated that it can not be built within a year. We should not worry about packaging and compactness except where it is electronically desirable, such as in the memory. We should not hesitate to use a vacuum tube to substitute for an element which we feel can be built after a little more development. The first step in building memory will be to get a complete single digit operating reasonably well before continuing with the remaining digits. The goal will be to have WWIA, with a 1024-bit word memory, operating in less than a year or else find out why we can not do so.[29]

Taylor then returned to the theme of an overall Central Control. It was time to consider it "intensely," he argued. Discussion then moved on to current work on circuits employing transistors and magnetic cores, to the merits and defects of each (transistors could fail if overloaded for too long, and non-destructive read-out of a core might be an impossibility). Brown reminded the participants that they should not expect new hardware, such as cores and transistors, to fit into old diode- and tube-oriented block diagrams. After all, "a core is not exclusively a flip-flop or a gate in the WWI sense." New block diagrams should be constructed around the special capabilities of new components. Papian urged that special attention be given to "the basic uses and abilities of cores."[30] Meanwhile, it appeared—at least, as of mid March 1952—that the arithmetic element of the prototype should employ transistors and diode gates unless cores became a more attractive alternative.[31]

The idea of building a small prototype, which had dominated the group's discussion from mid February to mid March, did not lead immediately to concerted action. Instead, it gave way during the next 8 or 9 weeks to a variety of R&D discoveries, problems, issues, and investigations centering around the still undefined air defense computer itself. It would be easy to conclude that confusion and indecision had set in during the spring of 1952, except for the fact that the men were exploring and mapping unexplored R&D territory, uncovering possibilities, identifying alternatives, and surveying options. They were wandering in order to probe, questioning and doubting as a procedure for acquiring knowledge.

As weeks passed during the spring of 1952, Taylor became more and more aware that the engineers were running out of precious time.

Although the meetings were informative and constructive, with the men often pooling enough information to generate bases for informed decisions (and, on occasion, a consensus), they were contributing to a common objective in too diffuse a manner to be continued for more than a few weeks. The hardware components of Whirlwind II had to be selected and a focused course of design activity launched.[32]

Early in April, Taylor addressed privately to Forrester a technical memorandum proposing a schedule for building the computer. If they were to meet schedules set by Lincoln Laboratory and the Air Force, the prototype ought to be operating by the beginning of 1955, then 2 years and 8 months away. Such a machine, Taylor suggested, should incorporate a magnetic-core memory, a vacuum-tube arithmetic element, a central control, probably of "magnetic shifting-register type," and terminal equipment that would include one or more magnetic drums for buffer, and perhaps auxiliary storage of the sort they had already ordered from Engineering Research Associates for the Cape Cod tests.[33]

By the end of summer at the latest, Taylor said, the engineers should have at least provisional and reasoned answers to such basic logical design questions as how large the internal storage would be, how complex the arithmetic element should be and what kinds of operations it should perform, whether a 64-position central control switch would be large enough, how the elements of the computer should communicate with one another in an organized fashion, how much buffer memory should be incorporated, what system of program timing should be employed, and whether all arithmetic operations should be concentrated in a central element or whether the terminal equipment should be outfitted with its own arithmetic element, particularly since it would incorporate buffer and perhaps auxiliary storage. As he looked ahead, Taylor realized that various decisions "will have to be rather arbitrary in nature." "Optimum or not," he continued, "June, 1953 is a deadline for all decisions, with the preliminary which affect the actual circuitry, made by June or July, 1952."[34] To Taylor, it was time to fish or cut bait.

14

The Memory Test Computer

While Norm Taylor was considering the overall schedule for building the air defense computer, he and Everett and Forrester were narrowing the focus of their engineers' attention on the basic logical planning needs associated with framing the "block diagrams" from which all the hardware designs would flow. To this end, they initiated a series of twice-a-week meetings. In all, 16 meetings were held between April 8 and June 3, 1952. The number of attendees ranged from nine to thirteen, depending on the demands of various laboratory tasks. Forrester, Everett, and Taylor urged their colleagues to make "some fundamental preliminary decisions . . . on the logical plan of the machine" soon—preferably by July 1, so they would have a year to assemble and test the machine. The goal was to put it into operation by the end of 1954.

In addition to their knowledge of Whirlwind I, four of the men were personally acquainted with the designs of the CADAC, CALDIC, ERA 1101, and IBM 604 computers. Their intention was to make "as complete a survey as possible for the logical designs open to us, and of components whose performance is now known or at least predictable," and to proceed from there to devise "an acceptable quantitative means of evaluating and comparing proposed design—some sort of 'figure of merit', as it were, to aid us in coming to a decision."[1]

Like the Whirlwind II Planning Group meetings, the Block Diagrams meetings began with a discussion of Everett's theoretical single-register magnetic memory computer. The group promptly turned to the question of whether one register was enough. They concluded that for practical purposes at least two registers would be required. After raising a number of other technical questions, they concluded that first day's session with a reminder by Forrester that a block diagram on paper was needed "so that concrete proposals could be dissected and modified or rejected in the course of future meetings."[2]

During April of 1952, the smaller group of engineers in the Block Diagrams meetings and the larger group (including many of the same men) in the Whirlwind II meetings raised a variety of questions. In the Block Diagrams meetings, they examined and re-examined such subjects as magnetic-core access time, "read" access time, and "write" access time, the efficiency offered by Ken Olsen's use of magnetic cores in a storage-selection matrix switch Olsen had designed, and the value of applying Boolean symbolic logic to computer design.[3]

Early in the third meeting, Forrester felt impelled to back off for a moment and look at what they were doing. W. A. Hosier's report of the meeting mentions Forrester's suggesting that "there is perhaps a hierarchy of components and decisions applicable to the machine—that is to say, an order in which its aspects ought to be considered so that later decisions will have a minimum of undesired retroactive effects on previous ones" and that "if tentative systems are elaborated into enough detail, we may have a better notion of what this order is, and be able to impose a pattern of growth on WWII that fulfills itself somewhat more easily and consistently than did that of WWI."[4]

At the fifth Block Diagrams meeting, held April 22, the July 1 deadline was discussed, and Forrester explained what it meant. The engineers should know by the end of June, if not before, what kinds of physical components they would be needing after committing themselves to, say, three-dimensional memory or a basic type of arithmetic element. This done, they could spend the next 9 months "deciding the number of components and how they should be interconnected." If they established both the components and the major details of the block diagrams in 1952, then in 1953 they could be busy developing "the full design from the complete block diagrams."[5]

At a meeting of the Whirlwind II planning group late in April of 1952, Bill Papian led the engineers in a detailed analysis of "some of the problems which are arising in connection with the magnetic memory and the magnetic switch."[6] Papian had been experimenting with magnetic toroids since the autumn of 1950, and a review here of what he and others had been doing between 1950 and 1952 will serve as an example of the rising level of experimental and theoretical understanding in the laboratory during the years immediately preceding the concerted effort (begun in 1952) to jell the design of the prospective air defense computer.

Papian and his fellow engineers were accumulating their technical understanding of core storage's promise and its problems from their own investigations, from daily shop talk, and from reading the Biweekly

Reports and other technical memoranda that were circulating in the laboratory and available in its library. During the autumn of 1950, Papian had set to work on the simplest form of a "breadboard" bench-test model of a magnetic-core memory grid composed of a 2 × 2 square of metallic-ribbon toroids. These particular magnetic cores were chosen for the demonstrated rectangularity and bi-stable polarity of their hysteresis-loop properties. Their response times were longer (20 microseconds) than Papian and Forrester had desired, but they were the best available in 1950. Multi-turn coils of fine wire were applied to each core in order to detect and alter its magnetic polarity. These coils would be powered by type 6AS7 vacuum tubes.[7]

By the middle of October 1950, successful preliminary testing had opened the way to systematic trials. Since the wiring scheme electrically interconnected the coils, background electrical "noise" from these, in addition to selected signals, was expected to interfere with the clarity of the signals received from any coil that was being read. Indeed, if enough coils were energized, however slightly, what was to keep their combined effect from triggering a particular coil's polarity unintentionally? Although the noise proved insignificant where a matrix of only four cores was concerned, it was likely to be problematic in any array large enough to be useful as a "memory."

Perhaps if Papian altered the polarity of each core in every column and every row, in a checkerboard pattern, the noise would be self-canceling. He rearranged the circuits in October, and this did tend to cancel the unwanted noise. By early December of 1950, several weeks of cycling of information around all four cores had established their acceptable performance of the simple experimental task they were given: to retain or change their polarity reliably, over and over and over again.

Papian then set up a more elaborate test to measure the behavior of two cores adjacent to two active cores. While a given polarity was stored in the two passive cores, a pattern of alternating polarity was cycled a number of times through the other two. Then the signals retained by the first two cores were "read" and "rewritten" so it could be determined how much the original polarity had deteriorated. The deterioration was found to be "limited and reasonable," so Papian proceeded to record output-pulse shapes and calculate signal and information-retention ratios in order to get a clear picture of how the cores were performing. By the following April, he could point to qualitative tests which showed that "a core holds its information to a reasonable degree in spite of disturbing selections in neighboring cores."[8]

As Forrester reviewed these technical events 10 years later, he concluded they were a milestone. Not only had cores operated successfully together for a first time under coincident-current excitation, but the basic operating mode of a three-dimensional system had been achieved by including in the November tests a "z-axis inhibit winding" at right angles to the other two. Especially encouraging was the fact that the 20-microseconds switching time of the cores represented a 500-fold improvement in speed since Forrester had tested the original Deltamax magnetic cores in the summer of 1949.[9]

Response time was as important as signal retention and identity, and it was plain that the metallic cores were still too slow. In October of 1950, Forrester and Papian had initiated research efforts to improve the quality of the cores. Since additional funds would be needed, Forrester had turned to George Valley. He followed up a phone call by sending copies of Papian's thesis and of his own report R-187 (the substance of which would appear in the January 1951 issue of the *Journal of Applied Physics*) to Valley and to John Marchetti. "The idea of using magnetic cores for high-speed, high-density internal storage," Forrester had explained, "is in the research stage." In view of its promising outlook, it should be pursued. Though the development cost should be no greater than that for electrostatic storage tubes, reliability and service life would likely be much better.

Turning to the factual details, Forrester had admitted that the present metallic-core response times of 20–50 microseconds were "5 to 10 times slower than one would wish as an ultimate goal." Nonmetallic cores offered "entirely adequate speeds under a microsecond," but the shape of their hysteresis loops indicated submarginal polarity retention. It was to remedy these deficiencies that Forrester was seeking funds to underwrite the needed industrial research, which would benefit both the work on Whirlwind and Arthur von Hippel's scientific inquiries into the phenomenon of magnetic remanence in MIT's Instrumentation Lab. Forrester proposed to continue working closely with von Hippel, who was, he thought, "in a position to make the greatest contribution."[10]

During 1951 a three-pronged attack was leveled at the problem of achieving truly functioning cores. Should they be composed of metallic, ferroelectric, or ferrite materials, and of what chemical composition? Forrester picked up Harold Hazen's suggestion that ferroelectric slabs might have the desired electrical properties. Papian put a new graduate student, Dudley A. Buck, to work on this problem early in 1951. By June, the hysteresis curves that Buck was getting from barium titanate slabs were rectangular enough to merit further study.[11]

As various manufacturing firms began to develop core materials, Papian and his assistants began to test cores from the Glenco Corporation, from the General Ceramics and Steatite Corporation, from the Alder Products Company, from the Armco Steel Corporation, from Magnetics, Inc., and from other manufacturers. Because there were a number of prospective electronic applications of magnetizable materials beyond Forrester's core-memory scheme, more and more firms had become interested in iron cores. The MIT engineers gradually began to receive better cores for testing at the same time that they were exploring operational options. Everett pointed out late in the spring of 1951 that changing the ratios of switching to nonswitching magnetizing forces from 2:1 to 3:1 or higher should improve both the response times and the signal ratios, and tests confirmed his calculations. Slow but thorough progress was being made.[12]

Of course, there was the usual engineering tradeoff: additional external equipment was required, and for a while it appeared that the three-coordinate arrangement might have to be sacrificed. But since the engineers were primarily interested in gathering information on core performance under various conditions, as they moved from the research phase that Forrester had initiated into the development phase, the progress in understanding on the theoretical level quite compensated for the immediate practical drawbacks encountered, especially since this understanding might well enable them to surmount the practical drawbacks. They realized that the more detailed an understanding they acquired of the phenomena involved, the better were their chances of achieving practical results. They were committed to interweaving pure and applied research.

By the middle of 1951, though Everett and the others were aware that neither the steel nor the ceramic cores offered an adequate solution to their storage problem, the encouraging test results indicated that it was time to write up an Engineering Report on the four-core bench tests and to proceed to a single-plane 16 × 16 core array.[13] Olsen would undertake a master's thesis in connection with his work on the 256-core project. This ambitious new memory unit would probably use ¼-inch-diameter metallic Mo-Permalloy cores supplied by Magnetics, Inc. Coordinate currents of about 300 milliamperes would allow single-turn (i.e., pass-through) circuits employing non-selecting amplitudes in 3:1 ratios and driven by "either large driver tubes (6AS7s) or some rather tricky transformer coupling designs," with "flexibility rather than packaging . . . emphasized."[14] Much work had been done. Much more lay ahead.

In August of 1951, Olsen suggested a design for a new type of "driver breadboard for the 16 × 16 array," and before the end of the month he was at work on a proposal that he carry out master's degree research to put to experimental test the more efficient technique he had devised for selecting any particular core in the storage matrix in order to read or alter its magnetic polarity.[15] A solution already in practical use—the one that would be incorporated in the Memory Test Computer a year hence—was the reliable diode-crystal matrix switch. Originally developed by David Brown in 1947 for use with Whirlwind's electrostatic storage, it employed crystals and vacuum tubes to selectively power, or "drive," the functioning units of internal storage and "read" or "write" their positive or negative charges as directed. This switch could readily be adapted for use with magnetic cores if its use as internal "memory" should prove as feasible as it appeared.[16]

By mid September of 1951, Olsen had found that a single core would respond to his new matrix-switch control. He moved promptly to construct a 16 × 16 array, using ceramic cores. Though these did not switch their polarity as readily as metallic-ribbon cores—"high coercivity (relative to the best metals)"—their polarity was less readily disturbed and would be less likely to give ambiguous information on how well Olsen's matrix switch would work. Olsen explained it as follows:

> A new scheme has been worked out for selecting and driving with cores instead of crystal matrices and hard tubes. A 1 × 1 array works well and a 16 × 16 ceramic array will be built in the next week or two to evaluate the scheme more completely.
>
> In the original scheme the memory cores had to discriminate between two critically adjusted values of current, but with the new scheme a non-selected core will be kept from switching by an effective shorted turn, so the currents are not critical and can be increased to speed up the array.
>
> The magnetic switch may have other uses. It might select heads on a magnetic drum or tape recorder, or decode flexowriter tape and drive the keys of the typewriter directly.[17]

Olsen would undertake careful experimental study and testing of his device, investigating its reliability and the effects of its operation on the performance of single cores. He knew that rapid switching of polarity in the cores would generate heat, and he knew that their magnetic properties were adversely affected by increasing temperatures. He would build apparatus to "switch several test cores for a long period of time to investigate this problem."[18] As was so often the case, speed and reliability were fundamentally important. As Olsen expected, his investigations took longer than the couple of weeks it would take to assemble his device.

So it was that two 16 × 16 matrices of cores came under study in the laboratory in the autumn of 1951. One was composed of the steel-ribbon cores whose performance was most familiar to the investigators. The other consisted of ceramic-ferrite cores, which were slower but perhaps more reliable under the test conditions Olsen had in mind and about which more performance data had to be gathered. Papian had two reasons to be optimistic about ceramic cores. One was the receipt of a small test batch of "some tiny (0.22 inch O. D.) Ferramic cores along with assurances that sizes and coercivities will continue to improve in the downward direction." The other was Olsen's switch, "which may well be the answer to the ceramic array driving and selecting problem." The memory array of steel cores, which was "mechanically quite complicated," would take several weeks longer to assemble than the ceramic array.[19]

Olsen spent the autumn assembling, debugging, modifying, and testing. His work was punctuated by events such as the following, reported in November of 1951: "Difficulty has been encountered in producing switch transformers (driver cores) with the necessary consistency. This does not seem to be a permanent problem." In December, Olsen was "reorganizing" and "simplifying" his testing gear and calling attention to a new core, "Ferramic H," whose properties suggested that it might be a good switching core. In January he began testing a 2 × 4 section of the 16 × 16 array, using high permeability, low-energy-loss Ferramic H switching cores, only to discover that, because their hysteresis loop was far from rectangular in shape, "there is an output from the switch while it is being set up that may cause difficulty in a large array."[20]

To obtain adequate performance of the cores, it was necessary to understand a variety of their properties in order to select an appropriate core size and appropriate winding designs. The performance of ceramic cores over a wide range of temperatures helped to give Olsen and his assistant, E. A. Guditz, information on which to base realistic expectations. They found that, as temperatures went up, the bi-stable properties of the cores became less stable, and resistivity decreased. As resistivity decreased, the cores took longer to switch. It was the old story of tradeoffs. While the rectangularity of the loops was better at the temperature of liquid nitrogen, the coercive force then needed to change polarity was greater.[21]

Another tradeoff was encountered in selecting the optimum power conditions and the number of wire windings that were required in order to "read" and "write" electromagnetically on the memory cores. If more powerful switching cores were required, they would have to be larger; and larger switching cores would require more windings. Olsen and Guditz realized that the number of windings could be reduced if they were willing to use "slightly more complex electronics." Alternatively they could keep the circuitry simpler by increasing the windings, but this raised the question of whether special core-winding machines should be obtained, modified, or built. Investigating the "slightly more complex electronics" alternative and recognizing that "it is necessary to have drivers which can be gated off and on by flip-flops," they built a "breadboard" driver to help determine what kind of a design compromise they should seek. If they used the same winding to drive the switch and to serve as a selecting winding, more complex circuitry would be necessary. Their small "breadboard" might give them suitable data.[22]

Meanwhile, ceramic cores of MF 1118, a new material with an impressively discriminating signal-to-noise ratio, became available during the winter of 1951–52, and Olsen and Guditz assembled a 16 × 16 grid of the new cores to identify their properties and test their performance.[23] To avoid delaying testing of this grid in the spring, Olsen assembled a second one to use in his master's thesis research.[24]

Magnetic-core storage was under intensive development study. Though it looked promising, there was no way that those on the inside of the development cycle could convince those on the outside that it was a sure thing. The Whirlwind computer went on using electrostatic storage tubes, but Taylor, Everett, and Forrester realized that the experimental results obtained from testing the cores opened up new R&D options for them to explore, even while they were attending to the larger scheduling problems their laboratory faced.

Such was the status of research into memory and magnetic-switch applications, and such were the prospects for magnetic storage, by the spring of 1952, when Taylor asked Papian to acquaint the engineers with some of the challenges affecting the upcoming design of "Whirlwind II." All realized that the design of the air defense computer's memory would affect the computer's basic logical design.[25]

Taylor and the others also realized that the technical development of magnetic memory was only one of a number of engineering problems that had to be called to everyone's attention. Forrester, acting on this impetus and on a suggestion made at the April 25 meeting, drew up, with the help of Wieser and Israel, "a set of minimum boundary conditions and relative values of machine parameters . . . to guide logical planning."[26] Forrester selected five basic characteristics of the Whirlwind I machine to form the skeleton of the boundary conditions he laid out. He then added flesh by identifying four levels of improved performance beyond what Whirlwind had demonstrated. He called the first level of improved performance to be identified with the air defense computer's design "poorest acceptable performance." Then he added the other levels in "steps of equal value to the air defense application." Speed went from 55 microseconds per order for Whirlwind I to 40, 30, 20, then 10 microseconds. Storage size went from 1000 registers for Whirlwind I to 2000 at the "poorest acceptable performance" level, then to 4000, 8000, and 16,000 registers.

The very first characteristic on Forrester's list was "Reliability." Experience with Whirlwind and its troublesome electrostatic storage tubes impelled Forrester to distinguish between scheduled and unpredicted "down time" for maintenance per 24-hour period. Steve Dodd's data on Whirlwind I called for 2 hours of scheduled maintenance per day and 1 hour of unpredicted "down time." For Whirlwind II, Forrester reduced the latter successively by increments of one-half, from half an hour at the "poorest acceptable performance" level for a new machine, to a quarter of an hour, an eighth, and a sixteenth of an hour. With respect to scheduled maintenance, his demands were even more rigorous, cutting "down time" from an hour at the "poorest acceptable performance" level to a quarter of an hour, then an eighth, and finally zero.

How long should it take to build Whirlwind II? Whirlwind had taken about 6 years. If the longest acceptable time for its successor were to be set at 3½ years, then equal steps of improvement would be marked in 6-month increments, decreasing from 42 months to 36, then 30, then 24.

The final characteristic on Forrester's list of design parameters was the order-code program that would operate the machine. After designating for the "poorest acceptable performance" of a new machine the same order code that Whirlwind used, Forrester offered as indicators of a more powerful code "the following three quantities in any sequence: (a) Trigonometric orders plus square root (b) Provision of a B-box (c) Automatic orders for correlation."[27]

Forrester suggested that the "boundary conditions" be applied to Whirlwind I, then to a very-least-acceptable Whirlwind II, and then not just to three more machines corresponding to his last three improvement steps of equal value but to any number of machines combining different characteristics and different levels of performance. However, to arrive at the final characteristics of whatever machine they would build, he urged that the engineers not select values more than two improvement levels apart. To avoid "a gross unbalance in machine characteristics," they should be sure that all performance levels were at, say, the second Whirlwind II level before attempting to incorporate any from the fourth and final level.

Forrester matter-of-factly dismissed his "poorest acceptable performance" machine. Though it would be a useful computer, it "would not handle the job we desire to handle nor be available as soon as we need it." Offering further guidance to the engineers of the project he headed, he pointed out that his analysis tied the factors of reliability and the time it would take to build an operating prototype more closely to speed and storage than had the standards which the Planning Group had been employing in their discussions. Now it was time to consider first "a machine meeting the poorest acceptable performance requirements and to make improvements . . . in a way which gets the greatest accomplishment for the least effort. The figures will be reviewed by the Air Defense group as they get new information and they will be revised as necessary in the future. They provide an adequate set of specifications for present planning work."[28]

Consonant with this line of policy thinking and with the heartening experimental results of Papian's and Olsen's investigations, Forrester agreed in May to revive the "Whirlwind I½" idea, which Taylor had proposed in February and had allowed to lapse in late March. This course of events reflects the fact that the shift in design thinking during these months was actually more subtle, complex, and graduated than is suggested by the names the engineers attached to the designs they were contemplating. From the outset, they were in agreement on the wisdom of "starting simple" and using magnetic cores, whose reversible polarity implied a variety of uses. Indeed, cores were mentioned or discussed at all but two of the nineteen "WWII Planning Group" meetings between February and December 1952 of which we have records.

At the sixth Block Diagrams meeting, held in late April (just before Forrester issued his "boundary conditions" analysis), Forrester reiterated the importance of performance reliability, pointing out that, if necessary, the air defense coverage and the machine operations performed would be reduced in order to make the coverage reliable. Others would be emphasizing speed, versatility, and other characteristics, but it would be either the Digital Computer Lab engineers or no one who would defend reliability. They should be on guard against the fallacy that reliability of the sort Forrester had in mind was prohibitively expensive. They should remember that "initial cost of the computer is a minor item in the total expense of an air defense network: that maintenance and operation costs far surpass it, not to mention telephone line rentals."[29]

Then the discussion turned to the familiar question of speed and complexity versus reliability and simplicity. All realized that the design price of higher speed was more complex circuitry. Forrester observed that, once a machine could keep up with events as fast as they occurred, there was the question of when to be satisfied with the compromise between simplicity and speed. Since greater speed and greater reliability of machine operation would require more design effort, it was helpful to think of a family of curves generated by plotting speed against reliability, using design effort as a parameter. Once one had passed the minimal speed and minimal reliability required, one's selection of a design parameter could rest on other criteria, so long, said Forrester, as "our design effort has brought us to a curve in the 'green pastures' where we exceed these minima."[30]

Forrester's chart "Boundary Conditions for Whirlwind II Design," which he had worked up with the help of Israel and Wieser, became available at the beginning of May, and the conditions it stipulated provoked a query from the Block Diagrams group to which there was no easy answer: Why should a computer that was more difficult to design be given *less* time to complete? Whatever answer was vouchsafed at the time, the effect presumably was salutary: in the end, the engineers did indeed develop a computer that was enormously superior in speed and performance to any then on the drawing boards.[31]

By the middle of May, the Block Diagrams meetings were taking a close look at the terminal equipment and its influence on the precise design of the computer. As Taylor had noted earlier, the best information on what the design configuration of the terminal equipment might be like would be forthcoming from the past Bedford radar flight tests and from the Cape Cod testing program then in progress. Such information should enable the designers of Whirlwind II to sidestep the problems that had been encountered when the issues of how information should enter and leave Whirlwind I were tackled after the computer had been designed.

At least Whirlwind II should not have to cope with the climate of improvisation that dominated the Cape Cod program. Since Whirlwind II would be advancing the state of the art, it would encounter its own problems, such as how best to deal with the radar "noise" (unwanted information) generated by "the presence of stationary radar returns (ground clutter and, relatively speaking, rainstorms)." It would be most economical to eliminate clutter from the information before it reached the computer, especially since "it has been conservatively estimated that 5% of the clutter would equal total aircraft returns on a typical radar. This means that unless 96% of the computer's correlation time is to be poured down the drain, the great bulk of clutter must be stopped before it reaches the computer." Placing an opaque mask over the portion of an observer's CRT screen that was presenting stationary ground clutter would remove it from the observer's attention, but there was still the issue of how much incoming information from interleaved radars would have to be deposited in buffer-drum storage. Slow-moving rainstorms posed another clutter problem that might better be dealt with at the radar end than at the computer end of the transmissions. When it was suggested that "our approach to the clutter problem will probably be more sure-footed after the Cape Cod system has been put into operation and we have confronted it 'in the flesh,'" Taylor cautioned the engineers to recognize that the system linking Whirlwind I to the Cape Cod radar system was committed to "an excessive (though unavoidable under the circumstances) amount of unproductive data-manipulation." And when Ed Rich indicated that "in-out control, not the in-out register, is the present bottleneck," Ben Morriss remarked that it was a mistake to "make the computer stop and wait for terminal equipment to complete operations."[32]

By early May, tests on an "experimental breadboard" arrangement of ferrite cores in a 16 × 16 array were giving such excellent results that "now it should be possible to write the required specifications for a final array." Read-write time, when carried on at a low repetition rate, was down to about 6 microseconds, with an access time of about 2 microseconds.[33] Perhaps it was this information, as much as any, that got Forrester, Everett, and Taylor back to considering a rudimentary computer less ambitious in design but not unlike Whirlwind IA.

Taylor opened the May 16 meeting of the Whirlwind II Planning Group by reintroducing "the question of building a prototype WWIA computer, which in the recent past has been regarded as impossible because of pressing time schedules." There would be perhaps three or four computers at each air defense site, Taylor pointed out, as few as 10 or as many as 50 sites scattered across the continent. Ultimately there would be from

30 to 200 Whirlwind II machines. By the time they finished building the experimental prototype of Whirlwind II, they would have to know what the production model should be like. "If our experience of the techniques involved is obtained solely from this WWII prototype," Taylor said, "it is likely to be inadequate, for production will follow too closely on the heels of the prototype." All the circumstances pointed to "the desirability of going back to our original intention and building a WWIA which will test memory and other components under something approaching operating conditions."[34]

Taylor then proceeded to tie Whirlwind IA closely to Forrester's "boundary conditions" characteristics. He saw the best completion time, 24 months, as so short as to allow little better than a duplicate of Whirlwind to be built, "since no time would be available for development of new components." Automatic marginal checking routines might reduce scheduled maintenance to 15 minutes, but he did not see how to reduce unpredicted "down time" to 15 minutes unless standby machines were available. Three and a half years should give them time to construct a Whirlwind IA that would straddle evenly Forrester's first two performance levels, starting with "poorest acceptable performance." The machine would then take 42 months to build, and it would be necessary to provide for as much as an hour per day of scheduled maintenance and half an hour of unscheduled maintenance on the finished machine. A machine meeting Forrester's second level would have a 4000-word internal storage capacity, employ an order code one performance level better than that of Whirlwind I, and operate at a speed of 30 microseconds per order.

Forrester demurred at the time schedule that Taylor proposed, then temporized. Forty-two months was "absolutely an outside figure," partly because there were people on the outside who regarded 30 months as long enough. If the Digital Computer Laboratory was going to want more time than that, it would probably have to provide special justification. On the other hand, as Taylor had pointed out, there was always "some question as to just when the machine is to be regarded as completed." Forrester felt the computer ought to be usefully operating with its terminal equipment. Certainly an advantage the project engineers should gain from developing Whirlwind IA "would be a better estimate of how much improvement we could realize in WWII by taking longer to build it."

Taylor noted that little thought had been given to auxiliary, external storage for the air defense computer. If they had no time to develop anything better, magnetic-drum storage could be readily adapted, but if they were given some time they should consider an external core memory as well. An early decision on, say, drum storage would provide time to improve associated circuits, perhaps incorporating transistors, but significant development work would take a couple of years and ought to be started forthwith if at all.[35]

Referring to the Whirlwind IA specifications he had invoked at the March 14 meeting, Taylor considered the possibility of working concurrently on both a 16 × 16 planar array of cores for the IA machine and an experimental 64 × 64 storage array ("approximately what is planned for WWII"). Testing the latter would give the engineers a better idea of the problems they would run into when operating a more massive array. And they should plan on "at least 2000 hours' running time" for IA—6 months of elapsed time—if they wanted any information on the components' performance reliability. Recalling that Whirlwind I's electrostatic storage tubes "ran 1700 hours before developing significant faults," Taylor concluded they should allow themselves no more than 9 months to get Whirlwind IA up and running.

The men had been acquiring an understanding of their overall design challenge that was at once more profound and more detailed. The performance of magnetic cores was a telling example:

> W. Ogden remarked at this point that building a large memory, or even one plane of it, may be difficult at the moment for want of sufficient uniform cores. P. Baltzer gave figures on the wide dispersion of output voltage for one thousand metallic cores recently tested. With a uniform input pulse of 2.6 ampere-turns, the output voltage on a single-turn sensing coil varied from .24 volts to .80 volts, with a mode of .56 volts. Of the original one thousand cores, 250 were within 5% of the mode and 500 were within 10%.[36]

After further discussion, Forrester turned the group's attention momentarily to the larger picture, suggesting that "research effort in the nation as a whole is not well-balanced from the standpoint of achieving maximum reliability in electronic components." He went on to compare transistor research efforts with vacuum tube quality control, and then to consider circuits. "Flip-flop service, it was pointed out, has been harder on tubes than most other applications." Everett's rejoinder was to wonder if redesign of flip-flop circuits wasn't called for. Perhaps inserting more tubes might lengthen the service life of each tube. "He stated further that if we are, in fact, going to have to use vacuum tubes in considerable numbers in WWII, we should be putting more effort into testing and improving them to make them more reliable. Transistors, he believes, are at best somewhat of a gamble, whereas there is little doubt that satisfactory reliability could be achieved with vacuum tubes if the necessary effort were put into improving them."[37]

Forrester concluded the meeting with the philosophical observation "that we may have to choose in WWII between achieving reliability through sound, conservative engineering on the one hand and introducing revolutionary new components on the other. Trying to do both things might well spread our facilities and talents too thin."[38]

During the rest of May, through June, and into July, the basic character of Whirlwind IA continued to crystallize as the project gathered momentum. By the middle of July, the machine and the project had been given the name by which they were to be known thenceforth: Memory Test Computer (MTC). A "WWIA Planning Group" of about a dozen engineers came into existence and shortly became the "MTC Planning Group." Kenneth Olsen was placed in charge of the project, and Taylor began to shift his attention to Whirlwind II, but the gestation of MTC continued to be closely watched over by Forrester, Everett, and Taylor that summer and fall.[39] In early July the design parameters were firm enough so that preliminary specifications could be spelled out to guide the men who would "proceed with the Block Diagrams of WWIA" preparatory to moving on to obtaining the specific hardware:

> Since the main purpose of WWIA is to test memory and other components, reliability and conservative design will be emphasized throughout. Also, since time is short and since it is essential that the great majority of circuits be circuits of proven and unquestioned reliability, it will be built as far as possible to begin with out of standard plug-in units or test equipment.[40]

By mid July, Rollin Mayer had drawn up a block diagram to serve as the basis for the logical design of the computer, and Bill Papian had enough test information on experimental cores to indicate how the machine's core storage would be structured.[41] In April, Papian's group had tested three arrays of cores, one metallic and two ceramic, and he had reported to the Whirlwind II Planning Group on work with one of these: a 4-inch-square plane of 256 "Ferramic 1118" cores (nicknamed "Cheerios") in a 16×16 array, with the switching cores arranged around the perimeter. The cores were "single-turn," i.e., energized by straight wires on which the cores were strung. So far, the engineers had tested the performance of an 8×8 corner of the 16×16 plane, observing how crucial it was to have nearly uniform magnetic properties in all cores. Since sending pulses through the arrays released energy in the form of heat, temperature was another factor they were watching closely. It appeared that differences in core temperatures in the array might be more critical than the absolute value of their temperatures.[42]

The 8 × 8 section had, by one test procedure, taken a total of 4 microseconds to read and then do the rewriting needed to restore the information. An alternative procedure had required a total of 6 microseconds, but speed was not always the significant criterion. A metallic-core array using vacuum-tube drivers to supply the pulses, instead of a magnetic switch, took 8 microseconds to complete a read-write operation, and since the performance and reliability of vacuum tubes were better understood than those of magnetic-switch cores the metallic core circuit was seen to be a fully competitive alternative.

Papian, Taylor, and Brown agreed that "we should try to design the logic of the computer with plenty of time for components to function" before trying to speed up the operation of the components. Furthermore, actual performance values should be used, not those that looked reachable. Papian had taken care to point out that the ferrite cores he was referring to "have never been tested with another operation starting immediately after the six microseconds, but investigation of the waveforms indicates that it could be done."[43]

Whatever the tests revealed, it was of overriding importance to develop MTC's core storage as soon as possible. "We now feel," said Papian toward the end of April, "that we are ready to work toward a 16 × 16 array. The switch cores to be used in the array will be tested individually, either on the array itself or on a single core tester. . . . After the 16 × 16 array is running, we should try the 'Z axis inhibit' method of writing." This would shorten the read-write time while providing experience in powering 256 cores with one driver.[44]

By the middle of June, Papian was able to report on preliminary success with "Z axis inhibit" tests, which had demonstrated operation times of 5–6 microseconds and which promised times as low as 4 microseconds. Likewise, the outlook was attractive for core-driver vacuum tubes. Papian was sharply aware that the cores they were receiving from different manufacturers, custom-produced to the Digital Computer Lab's specifications, were not at all uniform, even within the same batch. This greatly complicated the problem of distinguishing signals from background "noise" in circuits as complex and densely spaced as core arrays, particularly when they went to larger arrays in three dimensions.[45]

By mid July, Papian could describe the storage element that was planned for the Memory Test Computer. It would contain metallic cores "wound of ⅛-mil Mo-Permalloy ribbon, ⅛" wide, wrapped about 10 times around a hollow ceramic core of ⅛" outer diameter." There were a number of reasons for settling on metallic cores. Foremost was their uniformity of properties, "requiring less time and effort for core testing and development." Second, the alternative ferrite cores would require larger driver tubes of higher power, "such as 715's, instead of 7AD7's." Third, the rectangularity of the hysteresis loops exhibited by ferrite cores left something to be desired and did not offer "quite such good selection ratios as those of the metallic cores." Finally, though Brown's and Papian's groups would continue strenuously trying to upgrade ferrite performance, in Papian's view it would take twice as long to develop ferrite storage as good as metallic storage.

Metallic storage would take 9 months to develop. They could expect a read-write operation to take 20 microseconds, half of which would be taken up by switching. "However, it has been found necessary to switch the core completely in order to get a satisfactory read pulse. This means that, for the read and re-write cycle, the core need be only partially switched, traversing a smaller hysteresis loop and taking only 12 micro seconds or so for the whole cycle." One difficulty they would have to cope with by switching from the "Cheerio" ferrites to the metallics (with their long, narrow holes) was the problem of passing straight wires through the plane of a core at the proper angles while still providing effective physical support for each core. Should they embed the cores in a plastic sheet with their planes at right angles to the sheet? They were working on a number of proposals.[46]

The simple design of the Memory Test Computer would probably preclude a central power bus, would provide for marginal checking routines, and would allow for interchange of some components (so that transistors and new circuits could be tested). The magnetic-core memory would contain "32 × 32 = 1024 registers, 17 bits (16 carrying information and one parity check). Selection by 3 superimposed currents of relative magnitude. . . . Vacuum tube drivers on each of the 3 'coordinates' selected by crystal matrix driver. . . ." Ten of the bits would contain the number to be operated on, and six of the bits would carry the order code dictating the operation. The Arithmetic Element of the computer would include an accumulator consisting of 16 flip-flops, an A-Register consisting of 17 flip-flops (to serve as a buffer between storage and the Accumulator), and probably other flip-flop registers to serve as a program counter, a storage switch, and a control switch. The order code would not include multiplication or division, "since these orders contribute little or nothing to the testing of other components." Later in July, Taylor reported formally and for the record to Forrester that the order code selected would carry out some 23 operations (clear and add, subtract, transfer to storage, cycle right, cycle and punch, etc.). The terminal would use punched paper tape, a photoelectric read-in, a typewriter, and a display screen.[47]

During July and August of 1952, work progressed on circuit design and the physical layout of the Memory Test Computer. Special pains were taken to ensure compatibility and interchange of components, where feasible, with the Whirlwind machine. Before the end of July, the eight-foot-high racks of the machine were being erected "on the third floor of Whittemore 2 . . . under the direction of K. H. Olsen and R. von Buelow." Standard test equipment would form the backbone of its structure. "Except for the magnetic memory, no circuits fundamentally different from those of WWI are likely to appear in the machine at first." The engineers had moved away from the enthusiasm for transistors and cores that had dominated their wintertime thinking and had become cautious in the face of all the problems and uncontrolled, independent variables they had encountered as they shifted their consideration from the logical theory of the machine to the nitty-gritty level of known and unknown performance of components and circuits, especially if these were novel.

They had not abandoned their quest for the new and the improved, however. It was a part of their plan for WWIA, as it became "MTC" in their minds, that "new circuits now being worked on in other groups will be incorporated into the set-up in the course of time." Taylor had emphasized the wisdom of minimizing the use of new circuits at the start. "In this way, any difficulty occurring in testing a new circuit, such as the magnetic memory, can more easily be localized and assessed than it could if unpredictable failures were occurring in various circuits throughout the machine."[48]

It was already well established how MTC would function as an intermediate development between Whirlwind I and the air defense computer. The engineers understood that they could not wait to complete a battery of tests on MTC before proceeding with Whirlwind II. MTC would probe the reliability and the durability of various circuits and components. Although this could be done by bench testing isolated components, circuits, and subsystems, what was imperative was to investigate new stability and reliability problems likely to arise when new components and circuits were integrated into the system:

> It is possible to interconnect equipment as it becomes available and perform a running life test on the growing assembly. However, it is difficult to evaluate the history of components in such a varying system. Instead, we should build a functioning machine, set it off in a corner, let it run, and see what happens. Only in this way can we feel a little more confident that all of the pieces will fit and work together as a system without the necessity of locating and correcting major difficulties which could arise only in an interconnected system.[49]

It was under this philosophy that the Memory Test Computer was built.

But Whirlwind II required a different course of action. Everett already knew a number of the basic characteristics of the projected air defense machine. It would be a specialized computer, but it would be a design descendant of Whirlwind I in many respects. It would be a parallel binary

Figure 14.1
The Memory Test Computer.

machine. It would employ a single-address code, and it would contain one order in each word. Word length was not yet established, but it should be long enough to contain location data on any target detected by radar. If both *x* and *y* coordinates were to be contained in a single word, word length should fall somewhere between the 22 bits needed for a machine order and 30 bits. The arithmetic element would accommodate the word and would be designed to handle the bits simultaneously and discretely, employing this "parallel" electronic design in preference to a serial, sequential, one-bit-at-a-time arrangement, because of the speed at which it would be expected to operate. The parallel design would entail a provision for high-speed carry, but a number of design alternatives would have to be explored before the arithmetic element's precise design could be fixed.

Magnetic cores appeared to offer the best opportunity for the rapid, reliable storage of the data and for the machine-order program that would be required. Preliminary performance indications provided by cores arrayed in 16 × 16 and 32 × 32 planes suggested that a 64 × 64 array, providing 4096 registers, would be feasible, but the details of driving (powering) and controlling each core's bi-stable electromagnetic state would have to be experimentally coordinated with appropriate logical design options. Although the general design of the internal "memory" of Whirlwind II was known, the details remained to be worked out. Everett recognized that the simpler the design the likelier was reliable operation, and he recognized the appeal of a "simplified machine making use of some of the core storage registers to perform some of the storage functions." But recent investigation of this sort of design indicated that "this technique results in excessive loss of performance" and would not provide sufficient efficiency.[50]

The overall operating speed of the air defense computer would have to be at least 25,000 single-address operations per second. In view of their experience with Whirlwind I and the engineers' acquaintance with the general state of the art, Everett set an upper limit that "probably cannot conveniently be made greater than 100,000 single-address operations per second." These and other preliminary considerations made it clear that the most complicated element and the most difficult to design would be the control element of the machine. Here again, various design opportunities presented themselves. Everett and several others were, on the one hand, looking into a redesign of the control used in Whirlwind I, "retaining the same basic technique," and, on the other hand, investigating the possibilities and hazards involved in developing new designs incorporating magnetic switches and stepping registers. Everett was under no illusions; the control, he cautioned, was the "most difficult to understand and troubleshoot."[51] The Bedford tests and the projected Cape Cod System tests would establish the practical design parameters for input and output equipment.

Two and a half years into the enterprise on which they had embarked when George Valley and his committee had appeared with their air defense problem, Forrester and Everett refused to be stampeded to an early design decision. They realized that their position on the "cutting edge" of computer engineering design required them to survey and understand the theoretical possibilities and to determine which of these to convert to workable, reliable hardware, software, and operating procedures. The Memory Test Computer would provide some of the answers; however, it was in no sense the kind of machine that the air defense computer would have to be.

Other approaches were being pursued by various groups and by single investigators in the Digital Computer Laboratory, and the complexity of their activities was reflected in the lab's internal organization even as it changed its name and became a functioning division (still separately situated on the MIT campus) of Project Lincoln. Division 6 was pursuing several interrelated research and development projects. The one with the most manpower (Group 61, under Bob Wieser) had nearly 40 investigators at work preparing the Cape Cod tests. In Group 62, under Norm Taylor, approximately 25 engineers were working toward preliminary designs of the still-unformulated "Whirlwind II" (as it was then called) that would be the electronic brain of SAGE.[52]

And these were not the only groups contributing. The 12 investigators in Pat Youtz's Group 65 ("Storage Tubes") were working on Whirlwind II R&D and on getting Whirlwind I ready for the Cape Cod tests. By the summer of 1952 it was quite evident to Forrester, Everett, and their senior engineers that the electrostatic storage tubes intended to serve as the computer's "memory" were complicated, expensive, and disappointingly unreliable despite all their efforts over a period of years to bring them to the level of performance that a reasonably reliable computer should possess. For the special task of an air defense computer, the electrostatic storage tubes would be not merely disappointing; they would be unacceptable. Either the reliability of storage tubes would have to be sufficiently improved or an alternative device would have to be found that would do the trick. The likeliest alternative appeared to be the magnetic-core approach that Forrester had begun investigating in 1949. A third alternative was a ferroelectric device.

For a number of reasons, the exact configuration of the air defense computer would not be worked out alone by DCL-become-Division 6. Perhaps the most important of these reasons was that MIT, DCL, and Lincoln never had any intention of going into the business of manufacturing in quantity the computing machines that the Air Force would need to monitor the skies of the continent. Consequently, while Forrester and Everett continued to lead their DCL engineers in probing the technical problems, by the middle of 1952 they were involved in the delicate task of finding a manufacturer that would agree to accept their design and to help build the prototype and the production model. Before they were through, they had enlisted Taylor and Wieser in that task.

15

IBM and the Air Defense Challenge

It took Project Lincoln a year after it received its charter to make sufficient internal administrative and technical progress to permit F. Wheeler Loomis to relinquish his post as Director of Lincoln Laboratory and return to the University of Illinois, as he had planned to do. In July of 1952, at Forrester's prompting, Wheeler's successor (charter member Albert G. Hill) and the Lincoln Steering Committee turned to the deferred task of selecting a manufacturer for the new computer.

In June, Norm Taylor had attended a professional committee meeting that was making arrangements for the forthcoming national Second Joint Computer Conference. While there, as he reported June 9 to a Division Group Leaders' meeting, he had gotten into an informal discussion with an engineering colleague, John C. McPherson, IBM's Director of Engineering. When Taylor had mentioned some of the problems he anticipated would be encountered in manufacturing the projected air defense computer,

> MacPherson [*sic*] was very interested. IBM now has twelve per cent of its personnel on government military work and would like to have this work in the digital field rather than the analog field as at present. MacPherson will contact Forrester if IBM management is interested. We should hear between one week and one month.[1]

Eleven days later, two representatives from IBM visited the Digital Computer Laboratory to "discuss production of Air Defense Computers." One was McPherson; the other was James W. Birkenstock, Executive Assistant to IBM's President Thomas J. Watson Jr. It was agreed that several engineers, led by Jay Forrester, would "pay a return visit to the IBM offices, plant and laboratories in New York, Endicott, and Poughkeepsie."[2]

Forrester's approach was to cast a wider net, however, and Hill agreed. At the Group Leaders' meeting of June 9, Forrester had pointed out that

> the trouble with IBM would be its traditional secretiveness. We should by all means look into the problem of finding a suitable manufacturer very soon. A number of companies need to be considered. Any suggestions should be brought to Forrester's attention. However, we must avoid indiscriminate discussion of this problem with different manufacturers who might create considerable disturbance in competition and pressure to receive the job.[3]

From a list of 15 or 20 possibilities, Hill and Forrester settled on five manufacturers that they felt were the "most promising" to approach without further delay. On July 21, Forrester told the Lincoln Steering Committee: "It is important that a manufacturer be brought into the picture as soon as possible and that the selected company be in a position to contribute to the engineering."[4] A crucial question was how many "good people" a manufacturer might be able to put on the job. The five firms Hill and Forrester had in mind were the Bell System (more specifically, Bell Telephone Laboratories and Western Electric), IBM, the Radio Corporation of America, Raytheon, and Remington Rand. Once the necessary technical information was collected, a decision would be made and the Air Force would be brought in to close a formal contract.[5]

In a preliminary meeting, Forrester and his engineers acquainted representatives from each of the five firms with the general program and with the objectives they had in view. Forrester outlined a "proposed basis for cooperative work" between Lincoln and the chosen manufacturer and described the general framework within which "cooperative research and development would be carried out." Finally, Forrester made it clear that "technical supervision would for some time reside with the Lincoln Laboratory."

The kind of technical information that Forrester and Everett were looking for from the manufacturers included the number of research engineers experienced in electronics R&D and the number of engineers involved in production engineering. Another indicator was how many vacuum tubes a firm used in digital computer production per month. Another was the size of a firm's customer-service engineering department. Yet another was the number of employees with "specific training in the operation and maintenance of . . . electronic machines."[6]

By the end of August, Bell Labs and RCA had withdrawn from further consideration, each explaining that its technical staff was too busy with prior commitments to undertake so demanding an assignment.[7] A team of four Division 6 engineers—Forrester, Everett, Taylor, and Wieser—visited each of the three companies remaining in contention. They spent one day at each of three Remington Rand "development groups": the Laboratory for Advanced Research in South Norwalk, Connecticut; Engineering Research Associates in St. Paul; and the Eckert-Mauchly Division in Philadelphia. They spent another day visiting Raytheon's plants in Waltham and Newton, Massachusetts. They spent a day inspecting IBM's laboratory and plant in Poughkeepsie, a day and a half at the lab and plant in Endicott, and a day inspecting the Watson Scientific Computing Laboratory and visiting the corporate headquarters in New York City.[8]

In response to a written query from Forrester, an assistant to President Tom Watson Jr., who had visited DCL in June, stated that IBM had more than 200 engineers qualified to carry on electronics research and another 71 "engaged in production engineering and release engineering work." In a month, the firm used about 60,000 tubes in digital equipment production and another 3000 tubes in field service. Of the 3942 customer-service engineers IBM had trained, 774 were qualified to handle digital computer maintenance problems. And there were 70 company mathematicians who knew how to operate the computers that IBM built.[9]

After their visits to the various companies, each member of the Division 6 team completed a "score sheet." When the scores were totaled, IBM was some 500 points ahead of Remington Rand; Raytheon was a distant third. There was "no doubt of proper order of preference," Forrester told Hill; IBM possessed "a much higher degree of purposefulness, integration, and esprit de corps" than its nearest rival. Remington Rand was weaker for a couple of reasons: its two best-qualified electronics groups had been recent corporate acquisitions, and they "had not yet worked with each other or with a Remington Rand factory." IBM possessed a "closer tie between research, factory and field maintenance."[10]

Taylor's assessment added substance to Forrester's report. He found IBM more "willing to place a significant number of key people on the job immediately." The firm had more experience on which to draw, and its "top management was sympathetic to the job and anxious to complement the MIT organization in any way they could." Furthermore, IBM backed up "its systems activity with more basic studies of components aimed at improving reliability than any other organization in the country with the exception of Bell Laboratories." Taylor judged IBM's field maintenance organization to be developed "to a much higher degree" than those of the other companies.[11]

Forrester reported the team's findings to the Lincoln Steering Committee on August 25, and they were informally endorsed by the committee. In the words of the minutes of the meeting: "IBM seemed the best choice by a substantial margin."[12] Later that fall, when requesting Financial Officer Paul V. Cusick of MIT's Division of Industrial Cooperation to draw up a formal subcontract with IBM, Forrester mentioned the visits to IBM's facilities that he, Everett, Taylor, and Wieser had made:

> We considered, among other things such items as top management interest; availability of competent technical staff personnel; experience in digital computer work; experience in placing into production machines of complex electronic equipment; size and experience of field service organization; accessibility to MIT, etc. . . .
>
> They have considerable background in the field of parallel digital computers, having over one thousand computer machines in service in the field. This means that working with IBM will not only greatly reduce the problem of familiarizing the technical personnel with this subject but we can expect to realize the benefits of many original contributions by their people. Furthermore, IBM has under way a comprehensive program covering research, development, and application of electronic and magnetic components of the type used in the circuitry of digital computers. This means that a long stride has already been taken toward our objective of reliability of operation for Whirlwind II.[13]

Remington Rand, disappointed at losing to a rival of long standing, challenged the decision to the point of attempting (as Birkenstock later told Emerson Pugh) to "use the political clout of General MacArthur [whom they had recently hired] to unhook the MIT-IBM agreement." But those efforts came to nothing.[14] Remington Rand, having purchased the Eckert-Mauchly Computer Corporation in 1950 and Engineering Research Associates in 1952, was presumably confident it had acquired an expertise in computer R&D that not only went back further but also was superior in quality and experience to IBM's.[15] However, Forrester, Everett, Taylor, and Wieser were interested in demonstrated performance. When the decision to select IBM was questioned by a visiting vice-president of Engineering Research Associates, Howard Engstrom, Forrester could only repeat the judgment that Remington Rand, with its two computer divisions, was "not yet a well integrated organization."

Forrester noted in his administrative log that "Engstrom agreed and seemed to imply that he felt from our standpoint that we had made a good selection in choosing IBM."[16]

Although the political implications of the selection of the manufacturer of the air defense computer did not generate a major crisis, more than a tempest in a teapot did result, and Forrester's habit of thoroughly documenting significant actions stood him in good stead. In November of 1952, when he was formally requesting that a suitable subcontract be drawn up between MIT and IBM, he once again explained in detail the reasons for the selection of IBM and the steps preceding it. In May of 1953 he filed a full report with Albert Hill.[17]

In the autumn of 1955, when an investigator for the House of Representatives Committee on Appropriations visited Lincoln Laboratory to investigate the contract with IBM, "he appeared to feel," Forrester related to George Valley, "that the selection of the company had been handled satisfactorily." Forrester's estimate was later confirmed in January of 1956 when the House Committee's Director for Surveys and Investigations filed a report stating that his staff concurred with the selection of IBM. The political issue appears to have died there.[18]

In September of 1952, engineers from MIT and IBM began to work together informally on the air defense computer without waiting for the formal contractual arrangements that permitted collaboration to begin in earnest late in October. Forrester and Everett were following their past practice of giving precedence to engineering policies, issues, and actions that, in their view, administrative, fiscal, and legal actions existed to implement. Either the practical wisdom of this approach was apparent to the policy makers in the Air Force, MIT, and IBM who recognized the urgent need of the air defense project or they saw no option. The course of events suggests that they were knowingly embarked on a "crash program" of the sort that had come into being during World War II.

It was clear from the start that the MIT engineers would control the design of the computer and that they would participate in determining how to manufacture it in the necessary quantities. Since it would be a joint venture, each side would be expected to defer to the other's judgment in the areas it knew best. MIT had no experience with multiple-item production, and IBM had no experience designing military or commercial computers of the high performance standards that Whirlwind II had to meet with respect to swiftness, reliability of operation, and complexity of hardware and software. Neither the business machines that IBM manufactured in quantity nor the custom-designed computers it built for university laboratories were required to perform virtually without error, and never before had swift, reliable, error-free operation been quite literally a matter of life and death. For the engineers and the administrators on both sides, this was a joint enterprise without precedents to guide it.

IBM, under the leadership of young Tom Watson, wanted to get into military R&D and the manufacturing contracts associated with it.[19] MIT, under James Killian, Jay Stratton, and its governing board, was willing to engage in ad hoc R&D for air defense, provided that MIT's primary educational activity could be geographically insulated from the project.

Both the educational corporation and the defense-industrial corporation were led by men acutely conscious that the United States was involved in a small war that might be a prelude to the World War III that none of them wanted. Here, unquestionably, was something that they could do to strengthen their country's position in a time of threatening international and ideological rivalries.

Forrester, Everett, and their engineers—now Division 6 of Lincoln Laboratory for organizational purposes, but still the Digital Computer Laboratory at MIT—wanted to develop the computer that they were confident they knew how to build even though all the design parameters were not yet set. They knew what type of machine they would build because of the practical experience they had gained from Whirlwind I and from the promising Bedford and Cape Cod tests and because of the theoretical grasp of computer design concepts and possibilities that they had acquired since 1947. The IBM engineers, in addition to being accustomed to following the policy instructions of their top management, were attracted by the creative prospect of moving into a new design realm.

When the Division 6 and IBM managers and engineers began to work together, in September, several weeks before a formal contract was signed, much of the R&D ground ahead of them was untraveled and unmapped. So, too, was the joint-management approach they were embarking on. Ahead of them and the engineers and mathematicians who would join them in the next 6 years lay thousands of man-hours of hard work. It proved to be fascinating work for most of them, as they felt and probed their way forward, but it was also demanding. At times it was exhilarating, at times exasperating, and on occasion it found them at the limits of their competence if not beyond. The outcome was neither inevitable nor predictable.

Since Division 6 was taking the initiative, Art Kromer, acting as Norm Taylor's chief assistant in September, surveyed the preliminary, detailed, technical actions that the Division 6 and IBM leaders had identified as necessary to get their cooperative venture rolling. A number of the MIT engineers should expect to spend "several consecutive days" examining and discussing with IBM engineers the "operation, reliability, maintenance, etc." of the new IBM 701 computer. Others should be comparing the programming features of the 701 with those of Whirlwind I and examining the option of using punched cards for input. Still others should undertake similar comparative studies of the logic and circuitry of the two machines, paying special attention to vacuum tube reliability, to transistors, to magnetic drums in particular, and to magnetic materials in general. Regarding magnetic materials and memory, MIT's investigators should satisfy themselves about IBM's research on techniques and components. They should examine the advantages and disadvantages of parallel versus serial transmission of digit signals in various circuits. They should look at the relative merits of flip-flop and delay-type circuits, and they should consider the possibility of using both types in the same machine.

MIT should obtain 701 technical literature from IBM, and that brought to mind the need to establish reliable lines of open communication between the two organizations, to identify specific contact points at the working level, and to make sure that appropriate offices at higher levels in the two corporations were kept informed, "such as IBM New York office (Mr. Zollinger), MIT Project Lincoln office (Mr. Cusick)."[20]

Beginning October 1, Division 6 and IBM should undertake a number of joint activities, with MIT and IBM personnel working side by side. These activities should include preparing "reports and block diagrams for the several sections of WWII, providing logical design and performance objectives," and setting up for the design engineers a standards book covering circuits, techniques, components, drafting standards, and physical standards to be applied to hardware, materials, finishes, and appearance.

The men would have to go to work also on plans for packaging the new machine and for producing the instruction books that Air Force personnel would be using. They would have to figure out how to handle the huge quantities of magnetic cores that would be used in the machine memory—how to store them, test them, and sort them in production quantities. They would have to determine how MIT laboratories and IBM custom engineering groups should be organized to collaboratively "conduct tests for reliability and evaluation of components and manufacturing techniques."[21]

Though Division 6 was committed to developing core storage and building the Memory Test Computer, Forrester accepted the necessity of hedging his bets. Early in September, he "indicated very strongly" to Pat Youtz and Steve Dodd that they should give careful attention to IBM's extensive experience with electrostatic storage using the Williams tube. And, of course, DCL's efforts to harness the MIT electrostatic-tube design to the forthcoming tests of the Cape Cod System would continue. Meanwhile, DCL should be ready to replace magnetic storage with electrostatic storage "in the event that magnetic storage was not ready for the first group of new computers."[22]

Before the end of September 1952, and a month before formal negotiations to collaborate had been concluded, the MIT engineers began to take a close, hard look at the IBM 701 computer. As far back as March of 1951, Bob Everett had relayed a description of the "Defense Calculator" that he had heard IBM representatives give at a meeting in the IBM office of Erich Bloch.[23] The 701 had been born after the outbreak of the Korean War in the summer of 1950, when the younger Watson had persuaded his father to permit the corporation to establish a military products division and expand operations into the field of "electronic stored computers." When approval was forthcoming, the corporation moved promptly to undertake design, development, and manufacture of the Defense Calculator, subsequently given the commercial designation IBM 701.[24]

The 701 program was well underway and had produced a prototype by the time the MIT and IBM engineers joined forces to produce Whirlwind II. IBM found the looming air defense project neatly implementing its corporate policy and plans, offering both an additional market and an attractive opportunity for rapid advance in computer technology.[25]

The possible use of the 701 in an air defense system was not overlooked, for the Air Force favored considering the use of available industrial equipment whenever it appeared feasible. During the autumn of 1952, while contractual arrangements were still under discussion, that possibility arose but quickly subsided.[26] The University of Michigan, working with the Air Force's Rome Air Development Center on a separate "Air Defense Electronic Environment" system, was considering using the 701 at the urging of General Donald Yates of Air Force Headquarters. MIT engineers, in preliminary discussions with IBM, had given some thought to using the 701 in the Cape Cod experimental system, for some of the 701's technical characteristics were similar to those of the existing

Whirlwind machine. But after studying the machine in operation and discussing various technical details with IBM engineers, they found the 701 inadequate to the task and dismissed the possibility.[27]

The Whirlwind engineers concluded that, although the 701 had many impressive features, it was, as Taylor said, between 10 and 50 times less reliable than Whirlwind. The frequency of undiagnosed random errors, averaging one every 15–25 minutes, revealed the 701 as unready to carry out the control functions that would be demanded of the air defense computer. Taylor was struck by the fact that administrative pressure to keep the prototype operating for study and demonstration runs permitted only 4–8 hours per week to be devoted to the analytical studies needed to improve the performance of the machine. To get it operating at peak reliability would require, he felt, 40–60 hours a week of engineering study. Until another 701 was up and running, that would not be possible. In view of the present design condition of the machine, it would take 9–12 months to "clean it up" so that it would be "useful in a control problem."

Taylor was impressed to find that the IBM engineers' preferences, as distinct from IBM management's philosophy, agreed with the R&D philosophy of the DCL engineers. The quality control simply had not been stringent enough, and the practice of replacing faulty units made it impossible to investigate the defects in circuits and components that made the 701 frail and unreliable. For example, for a while, the machine, after being shut down over the weekend, had suffered Monday-morning symptoms that the IBM engineers came to refer to as "morning sickness." But these symptoms disappeared as September weather set in, and engineers could only speculate that the humidity might have been responsible. Meanwhile, there had been considerable difficulty with poorly soldered joints in the removable plug-in components.[28]

Nine of the DCL engineers had spent 3 days at Poughkeepsie, and the gist of their 14-page report was communicated to the Group Leaders by Taylor:

> The 701 computer is now in the same state that WWI was in 18 to 24 months ago. Intermittent errors appear to occur every twenty minutes or so—errors in storage are perhaps even more frequent. The margins [of acceptable performance of components and circuits] appear to be low, and it is difficult to make the machine operate as a whole. The engineers doing [the needed] systems work have very little time on the machine, only about ten hours per week.[29]

The MIT engineers reached their informal but compelling technical verdict against the 701 in October, a month after Division 6 and IBM had begun to work together. Although they were concerned that the IBM engineers had not been given the opportunity to get to the bottom of the technical troubles, the technical information they had picked up confirmed their earlier impressions that IBM was their best choice as a manufacturer. Since they knew that DCL would be dictating the design and the quality control on the prototype of Whirlwind II, they saw no problems ahead that could not be overcome.

In the managerial and technical areas, IBM's mastery of the state of the art was not as advanced as DCL's. But the MIT engineers had many reasons to believe that IBM could be brought up to that level, and they were committed to doing just that. They realized, too, that they could learn from IBM.

Even before completing contractual arrangements with the Air Force, IBM had rented office space on the third floor of a former necktie factory building at 27 High Street in Poughkeepsie, about 3 miles north of the existing IBM Engineering Laboratories, for use by the engineers who would be working on the new project.[30] As organizational arrangements began to be worked out, it became clear that Jay Forrester (at the top of the DCL/Division 6 management structure) and Nat Sage (MIT's industrial liaison officer) had as their counterparts at IBM James Birkenstock and his assistant, J. E. Zollinger. Tentative joint engineering functions and assignments were set up between Division 6 and "Project High" (as IBM called it), and engineers from both organizations were given responsibility for a number of key functions. Norm Taylor's IBM counterpart as Design Engineer was to be John M. Coombs. Arthur Kromer's counterpart as Liaison Engineer was to be Thomas A. Burke.

Thirteen major tasks were identified at the start. Not all were undertaken immediately. Eight engineers from MIT were matched with ten from IBM in order to get things moving. The tasks assigned were dictated by the nature of the digital computer and the character of the forthcoming development-and-manufacturing enterprise, even though the precise design parameters of Whirlwind II remained to be settled in detail.

The details of the arithmetic element would have to be laid out. Although transistors were not yet available in the quantity and quality required, their potential use in an arithmetic element and in other circuits should be kept in view. Vacuum tube circuit designs would have to be arrived at, as would the designs of the equipment that would control and integrate the internal operations of the computer. The power supplies needed and the design of the terminal equipment would have to be determined. Research and identification of suitable magnetic materials for cores would be continued, and details of the overall mechanical design and packaging of the computer would have to be worked out. The modular character of the computer's circuitry, together with the multitude of component parts and the complexity of the whole, made it imperative that design and performance standards be imposed at every step in the construction of the computer. Careful attention would have to be given to the composition of instruction books and spare-parts lists.[31]

The scheduling of the collaborative enterprise was of paramount importance, and the men in charge lost no time in pooling their ideas. Informal planning discussions began in September of 1952 when Taylor and Kromer of MIT met with Coombs and Burke. Out of their discussions came a preliminary schedule of the design objectives that IBM and MIT anticipated committing themselves to if and when the formal contract to proceed was made. The highly detailed schedule that was drawn up spanned 27 months, from October 1952 through December 1954. It stipulated seven major concurrent R&D activities, five of which would center on the air defense computer itself and two of which would provide information essential to building the computer. If the program could be adhered to, a prototype of the air defense computer should be "available by the middle of 1954."[32]

It is, of course, not possible to know in advance how realistic any schedule will prove to be. This was the third schedule to have been generated within a year. The first had been prepared when Forrester and Everett had been asked to present a schedule to the Lincoln Steering Committee in the autumn of 1951. They had visualized a finished prototype in 3 years and a tested production unit a year later.[33] A second schedule had been prepared by Taylor 6 months later, in the spring of 1952, before the search for a manufacturer began. It had allowed 2¾ years.[34] The third schedule, generated by the four MIT and IBM engineers working together in September of 1952, allowed 2¼ years.[35] Each of these schedules had appeared to offer a reasonably hard-headed estimate at the time, nor were their deadlines very far apart. Besides, the primary function of schedules as planning tools lay not in their accuracy but in the array of R&D activities that they set forth.

Taylor's estimate of the preceding spring had been made when selection of a manufacturer was still a theoretical prospect. He had outlined to Forrester a "plan to obtain a WWII computer in the minimum amount of time—January 1, 1955," and he had built this estimate around several informed assumptions that the prototype should have a magnetic memory, an arithmetic element that would use vacuum tubes (because of their demonstrated reliability and availability) rather than transistors, magnetic-drum auxiliary terminal equipment, buffer storage for both input and output if necessary, and a control unit employing new magnetic shifting registers. Taylor had considered the engineering effort that would be required and had concluded that at least six groups would have to be organized: one to handle memory, one to handle the arithmetic element, one to handle control, one to handle terminal equipment, one to handle "logical design and systems," and one to handle mechanical design. Circuitry decisions ought to be firmed up by June or July of 1952 and hardware design ought to be frozen a year later if the January 1955 target date Taylor had in mind was to be met.[36]

The schedule worked up in the autumn of 1952 by the four MIT-IBM engineers set a deadline that was 6 months closer than Taylor's estimate of the preceding spring. It also accepted the notion of a novel relationship of junior and senior partners to carry the design effort forward.[37] It called for five coordinated R&D efforts that would focus primarily on the structure and major functions of the machine itself, and it stipulated two other categories of R&D effort to which manpower also would have to be devoted: liaison with the Cape Cod System field tests and R&D work on transistors, on magnetic circuits, and on the circuits for the video tubes (scopes) that would be used to display information to the operators.

The five machine-oriented R&D efforts would cover Mechanical Design and Manufacturing Specifications, the Arithmetic Element, Memory, Terminal Equipment, and Systems and Control (including wiring and power distribution to all parts of the computer). As was detailed in a two-page bar graph covering the next seven quarters (up to the middle of 1954), each of these R&D efforts except Systems and Control would progress from its own study and evaluation phase into a design phase. All five would go from design into a procurement and construction phase, then to final testing.[38]

This schedule that Taylor, Kromer, Coombs, and Burke drew up in September of 1952 also assumed that the Memory Test Computer would already be under construction at the beginning of the final quarter of that calendar year. Testing of it ought to begin around the middle of the first quarter of 1953, and the tests should prove useful in developing the Whirlwind II prototype. By November of 1952 (less than 2 months away when this schedule was made up), serious study and evaluation (the first phase) would have begun on the Arithmetic Element, on the Memory, on the Terminal Equipment, and on Mechanical Design and Manufacturing Specifications. During 1953 this study and evaluation effort would advance to actual design.

The first of the major efforts that the MIT-IBM schedule enumerated was design. Design work was to commence at the start of 1953 for Arithmetic Element and for Machine Design and Manufacturing Specifications. Design work on Systems and Control would begin halfway through the first quarter of 1953. Design work on Memory would begin at the start of the second quarter. Design work on Terminal Equipment would begin at the start of the third quarter. In the meantime, by the middle of the second quarter of 1953, information from tests of the MTC should be contributing generally to the design of the Whirlwind II prototype—particularly in the areas of Memory, Systems and Control, Arithmetic Element, and Mechanical Design and Manufacturing Specifications.

The Procurement and Construction phase would begin in the fourth quarter of 1953 and would affect Systems and Control, Arithmetic Element, and Memory. Mechanical Design and Manufacturing Specifications would not enter the Procurement and Construction phase until about 6 weeks later (the middle of the fourth quarter of 1953). Terminal Equipment would be the last to move out of the design phase, at the end of the fourth quarter. For all of the five aspects of the prototype machine, Procurement and Construction would come to an end in the middle of the 1954 calendar year.

Final Arithmetic Element and Memory testing of the "Mod. 1" prototype would begin during the second quarter of 1954, final Systems and Control testing in the middle of 1954, and final Terminal Equipment testing in the fourth quarter. No separate Mechanical Design and Manufacturing Specifications testing would be necessary, since it would be taken care of incidentally in the other final testing.[39]

The schedule was ambitious and detailed but not unrealistic. In October of 1952, when Norm Taylor sought the opinions of the DCL Group Leaders on the estimate that he, Kromer, Coombs, and Burke had developed a month earlier, they seemed to feel, Dave Brown reported, that this schedule was reasonable in calling for a prototype to be constructed by mid 1954, less than 2¼ years away, and extending final testing of the computer and its terminal equipment into the fourth quarter of 1954.[40] Predicated on knowledge that Forrester and Everett had not had at their disposal a year earlier and Taylor had not had available in April of

1952, the new schedule allowed 2¼ years for completion of the prototype. Forrester and Everett had predicted that it would take 3 years, Taylor that it would take 2¾ years.

Whether the latest schedule would prove practical would depend in part on the sort of information that might be garnered from a trip, planned for November, to vacuum tube manufacturing plants operated by General Electric, RCA, and Sylvania. Until the engineers had a more definite idea what types of reliable tubes were available in sufficient quantity, they could not confidently lay out hardware circuit plans in practical detail. MIT and IBM would each send a half dozen people on a joint mission to "make a study of the actual production lines and to have discussions with the tube engineers of the three major companies in the business." They were interested not only in the manufacturing facilities these plants possessed but also in their engineering and manufacturing capacities to respond to specific tube-design needs.[41]

To an outsider, all these details of the planning of the many R&D tasks that were ahead of the MIT and the IBM engineers in the autumn of 1952 were bewildering in their cumulative effect and quite difficult to appraise, as the foregoing account suggests. There was little that the higher-level administrators of MIT, IBM, and the Air Force could do but identify good men and allow them freedom.

16

Collaborating on the XD-1

When formal joint operations between the MIT engineers and the IBM engineers commenced, on October 27, 1952, a new arrangement was brought to the attention of all concerned. "The only practical way to carry out proper and logical administration of this effort," said a memo to all Engineering Staff members of the Digital Computer Lab (drafted by Arthur Kromer and issued by Harris Fahnestock in his capacity as Forrester's assistant in charge of administration), "is to establish a central point of contact through which all matters are channeled." The object was to "provide records of the interchange of information between the two organizations, make arrangements for personal visits and all other related matters."[1]

The engineers spent the next 6 weeks sizing up the challenge of the new joint operations. They began to exchange technical documents and to converse. Since Division 6 (still the Digital Computer Laboratory, on the MIT campus) was taking the design initiative and was more conversant with what was in the making, more documents came from MIT in the early weeks than from IBM. While the MIT engineers were acquainting the IBM engineers with the design details of Whirlwind I and the Memory Test Computer, IBM managers were introducing the Whirlwind engineers to the characteristics of IBM's "Defense Calculator," the 701.

Within this general framework of activity, the two parties then began to discuss the logic underlying the functioning of their machines and to examine the design details of "Programs, Basic Circuits, Internal Storage, Arithmetic Element, and in-out equipment (i.e., drums, tape units, scopes, etc.)." As weeks passed, they moved into increasingly detailed and technical study of basic circuits, of the logic underlying the systematic operation and integration of machine functions, and of programs (some already written, some yet to be written). After examining IBM's punched-card equipment, they decided to hook up a punched-card in-out system to the Memory Test Computer (which was under construction). It was clear to all parties that the complex arrangement of circuits and components—"diodes, cores, vacuum tubes, transistors, etc."—would require early standardization.

The MIT engineers devoted extensive effort to acquainting the IBM personnel with the state of the knowledge and the state of the ignorance that prevailed "regarding the use of a digital computer for data processing in an Air Defense System." In the course of this, they took IBM personnel to some radar sites and to the Lincoln Laboratory facilities then being built at Lexington. They also demonstrated taped and live interceptions controlled by Whirlwind. "Application of components such as diodes, cores, vacuum tubes, transistors, etc. have been discussed at considerable length," Kromer reported to his DCL superiors. IBM personnel spent 110 man-days at DCL that first month. An MIT contingent spent 16 man-days at IBM's Poughkeepsie facility.[2]

Within 10 days after the October 27 signing of the letter of intent to establish formal contractual relations among the Air Force, MIT and IBM, Art Kromer of DCL and John Coombs of IBM agreed that the exchange of technical information between the two groups would have to extend beyond face-to-face contacts and written reports to include "a more or less formal conference at reasonable intervals."[3] Art Kromer promptly suggested to Norm Taylor that they set up a schedule. The meetings should include "a relatively large number of people from each location." Each attendee would report on the progress of his work, and administrators at the level of Taylor and Coombs, or higher, would ask questions and make assignments as the situation required. They could meet at a midway point between Poughkeepsie and Cambridge—say, Hartford, Connecticut. Kromer proposed several target dates: "Tuesday, December 2; Tuesday, December 18; Thursday, January 8; and continue on about a three-week interval." Minutes of the conferences would provide formal record of the actions taken and could be provided to those who had been unable to attend.[4] "Art, this sounds fine," Taylor scrawled at the bottom of Kromer's memo, "but I think we need to get underway a little before we start—perhaps Dec. 18 would be a good beginning." But there were a number of preliminaries that had to be digested by the two organizations before they were ready for the sort of meeting that Kromer and Coombs were proposing, and Taylor's suggested date proved to be a month too early.

One of the first preliminaries was a December 5 meeting (held, by coincidence, on the day Subcontract No. 20 went into effect) at which the schedule for designing and constructing the Whirlwind II prototype

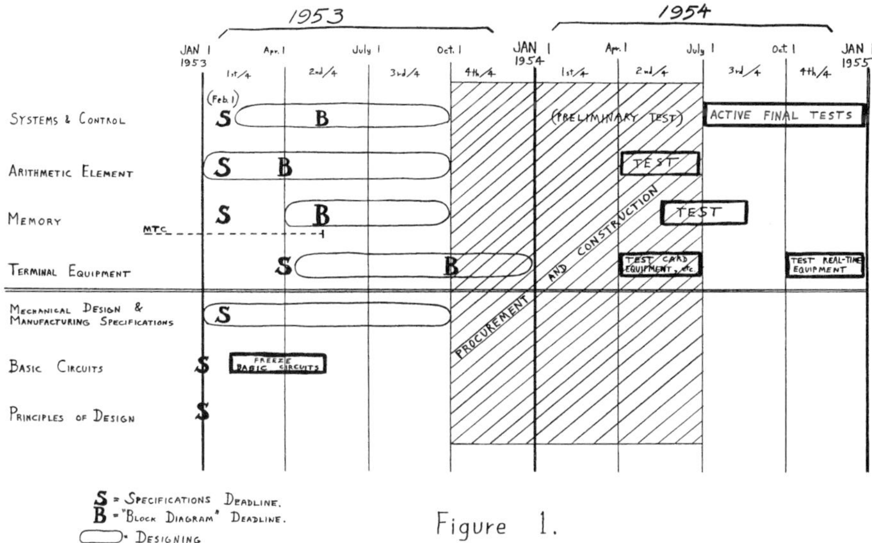

Figure 16.1
Kromer, Mayer, and Taylor to Whirlwind II Planning Group, "Design and Construction Schedule, WWII Prototype," M-1753, December 11, 1952.

was discussed. Norm Taylor, acting as spokesman for Division 6, described the schedule to the 42 MIT and 11 IBM engineers present, gave a broad overview of the work that the MIT engineers had been carrying on prior to IBM's involvement, outlined the scheduling decisions already made, and reviewed "for all parties concerned the projected future program and time schedule" under which their collaborative effort would be carried forward.[5]

The schedule would, if met, produce the Whirlwind II prototype by the middle of 1954, a year and a half away, and allow the remaining half of 1954 for final testing and "shakedown" of the computer. The schedule was much like the one that Taylor, Kromer, Coombs, and Burke had come up with in late September. In addition, it included a number of design decisions that already had been made, the most important of which were the conservative decision (made in June) to go with vacuum tubes rather than transistors or magnetic cores for the arithmetic element and the radical decision (made the preceding spring) to develop a magnetic-core memory. The prototype of such a memory would be installed in the bare-bones Memory Test Computer. A small staff would still continue "basic work" on transistors in order to keep up with the changing state of the art and its implications for later computer designs. Meanwhile, a small MIT group had already gone to work on the vacuum tube circuits that the arithmetic element would need.[6]

Taylor explained to the assembled engineers that the schedule had been developed "by beginning with the 1955 deadline and working backward to 1953." He admitted that it was a tighter schedule than any he knew of for such an ambitious machine, and remarked that it "presents a real engineering challenge to everyone." By 1955, Whirlwind II would have to be able to run programs continuously for 23½ of every 24 hours, with no more than one or two intermittent failures—and efforts should be made to eliminate even that rate of failure. It would have to be ready for prompt integration into a radar network similar to the Cape Cod System or larger.

The discussion and the questions raised at this early December meeting made it clear how crucial was the need to adhere to strategic target dates in the schedule. These would be "the latest dates beyond which we should not make a major change in the related decisions." Failure to observe these deadlines would leave "no leeway for the delays bound to be encountered in the later stages of procurement, construction and testing." Building on their experience with Whirlwind I, the MIT engineers expected to start the process with systems studies "showing what the system has to do and how it can be arranged to go about doing it." Out of these studies would emerge two major kinds of specifications: specifications providing "the broad numbers indicating the speed, memory size, desirable order code, etc." of the computer, and specifications establishing basic circuit criteria of stability, reliability, and other such characteristics. The latter would qualify selected circuits to function as basic building blocks. By the end of the second quarter of 1953, the basic design of the circuits should be frozen. In the meantime, these major specifications would provide a basis for laying out "block diagrams." (According to Taylor, they came to be called block diagrams "for want of a better name.") The block diagrams, which would "show in more detail how the parts of the computer must be interconnected," were the initial stage in the design of the machine itself, for they made it possible to generate in an intelligent way "circuit schematics, wiring diagrams, mechanical assembly drawings, bills of materials, and the methods of interconnecting various parts." To ensure that all parts of the computer would work together, the engineers involved in systems studies "should keep a watchful eye on the interconnections, timing sequences, starting and stopping of pulses, etc."[7]

Like an itinerary for a safari, the schedule would depend on circumstances encountered along the way. The engineers would have to be ready to go where the game they were tracking led them. Thus, while design was best begun after the block diagrams were complete, Taylor pointed out, "a number of block diagram decisions cannot be made until more is known about the Cape Cod system," because of the practical restrictions its operation would impose on the design of the terminal equipment. Thus, although most of the block diagrams should be frozen by October 1, 1953, the engineers should be prepared for the likelihood that some would not be frozen before design work would have to begin.

All these considerations, Taylor explained, indicated that the most critical period in the project would occur during the third quarter of 1953, since by that time the designers should know whether the decisions being made affecting the Cape Cod System and Whirlwind II were proper. Obviously, deadlines were created to be met ("we should feel that the urgency is NOW"), especially since Whirlwind II was to be the "important step in the transition of the air defense system from its state in 1952 to its state in 1958." But even more important was the reliable operation of the finished computer—"an unreliable one on time would probably be worse in the long run than a reliable one a little late." In consequence, they must ever be on guard, Taylor warned, against the lure of promising but untried ideas when reliable, tested, acceptable alternatives were known. "The overall problem is one of following an engineering program which is on a strict time schedule and which must be based on certainties and not on hopes, if at all possible."

Yet there was one glaring, hazardous exception to this rule of procedure: the internal "memory" of the machine. Taylor candidly admitted what they all realized: "We do not know about anyone who has to date built a very reliable memory system." This circumstance imposed on the MTC engineers a very special burden, for "MTC must provide us with the answers concerning the memory system by the middle of the second quarter of 1953, regardless of anything else that the MTC group may eventually want to build into their computer."[8]

On January 20, 1953, the engineers from IBM and MIT held their first "Hartford Conference." Attendance was lower than at the December 5 scheduling conference. There were 14 MIT engineers (led by Forrester, Everett, and Taylor) and 13 IBM representatives (12 engineers from Poughkeepsie, led by Coombs, plus G. Solomon from IBM's New York headquarters). The time had come to take a broad overview of what had been accomplished at MIT since the days of Project Charles and, in the light of the engineering information that had been exchanged so far, to begin moving toward the consensus on design details that would establish the technical character of the forthcoming machine.[9]

John Coombs opened the conference and briefly explained the purpose of the Hartford meetings, which he anticipated would be monthly. At each meeting, the work of the preceding month would be reviewed and the work of the month ahead would be outlined, in order that "better coordination of all work should result." In addition to an exchange of knowledge and technology, there would be "an opportunity for IBM and Lincoln to review each other's work."[10]

Jay Forrester was next. Before going into a discussion of the computer project itself, he outlined briefly its historical antecedents, emphasizing the sense of urgency that had accompanied its establishment. He explained that IBM had been brought into the program because Lincoln Laboratory was a research and development organization and needed an industrial associate that possessed the competence and experience to "convert the results" of the laboratory's work "into actual equipment or hardware." This was, Forrester believed, "the first instance of this sort of association." Lincoln Lab's role in the collaborative effort would be to "coordinate the engineering and design work so that the resultant system [would] be best suited to the overall Air Defense problem," Forrester explained. Division 6 would be responsible for "detailed design work on portions of the system where necessary," while IBM's tasks would be "construction of the prototype model and . . . production planning."

The project's urgency mandated the tight schedule with its early target of a constructed first prototype system by the middle of 1954. The system would comprise, Forrester pointed out, "not only the computer, but a great host of related equipment for data input, output and display purposes." Nor did this system mark the end of the air defense program, for after they had installed the "Transition System" described in Lincoln Technical Memorandum Number 20 (informally called TM-20), it would be reasonable, Forrester said, "to assume that . . . the system would be extended further to continental United States Defense and possibly other applications."[11]

A question that might well be asked at this time, Forrester continued, was "Why undertake the design of another computer?" He had two answers: To adapt any one of the existing general-purpose machines would be "in itself a big engineering job, in view of certain specialized problems and the large amount of additional electronic equipment required for input and output." Also, none met the essential operational requirements for ease of maintenance and reliability of operation. Recent military experience with field radar and radio equipment had disclosed "serious maintenance problems" in that direction. These would be compounded by maintaining under field conditions the even more complicated and sophisticated equipment being planned for the air defense system.

"No effort should be spared," Forrester insisted, "in having the design possess the greatest degree of reliability possible." By arranging for "extremely close monitoring of the functioning of the equipment at all times to observe possible errors and to permit manual intervention to change the operation of the system whenever it may be desired," they could make the system "highly flexible and thus more useful to the military."

Forrester concluded by commenting on the Cape Cod System and its projected successor, the Lincoln Transition System. Cape Cod would be "a small but complete Air Defense system where all functions can be carried out in a manner similar to that contemplated in the ultimate system." While Cape Cod was being operated and demonstrated in the months ahead, the Air Force would be evaluating the feasibility of Lincoln Lab's approach to the air defense challenge. At the same time, the performance of Cape Cod would be providing information on which the engineers should expect to base design decisions regarding the ultimate system. For these reasons, perhaps IBM should assign a representative to work with Wieser's Cape Cod group. In regard to the projected Lincoln Transition System, the IBM engineers should refer to the details of TM-20, bearing in mind that it visualized a prototype system that "could be placed into operation in a relatively short period of time" and which would be using the first of the new air defense computers that the engineers were about to design and build.[12]

After Forrester's background remarks, the engineers turned their attention to design details of the new computer. John F. Jacobs of MIT led off with a discussion of the circuitry of the arithmetic element that would be the machine's calculating heart. To date, three basic types of suitable circuits had been explored: existing vacuum tube and diode circuits, promising transistor circuits, and possible vacuum tube and magnetic core circuits. Jacobs promptly ruled out transistors, on the ground that transistors with the requisite reliability could not be procured in the quantities needed. He then ruled out existing vacuum-tube and diode-circuit designs: all those in use in Whirlwind I, in the ORDVAC built by Eckert and Mauchly, and in IBM and National Bureau of Standards machines, were too slow. Time that would be spent trying to increase their speed would better be spent on new designs. As Jacobs saw the design prospects, a useful measure of overall machine speed would be "the weighted average time for a single address order." The air defense prototype could allow no more than 20 microseconds per order.[13] For this reason, magnetic ferrite-core circuits looked extremely attractive. "It appears," Jacobs said, "that the read access time will be in the order of 2 microseconds with a total read and write time of 7½ microseconds."[14]

H. D. Ross of IBM followed Jacobs's analysis with a discussion of arithmetic-element block diagrams. Although it was too early to make final decisions regarding the circuits to be employed, both IBM's and MIT's engineers recognized that, in view of the state of the art, development of the arithmetic element would be "quite straightforward" compared to that of other parts of the computer. In consequence, a number of tentative decisions could be offered: delay units of Haven's design would be too unreliable; a flip-flop register would probably do nicely; the IBM 701 machine's use of diode gating for an adder appeared attractive; use of ones-complement arithmetic made excellent sense electronically; and use of direct electrical connections instead of a "general purpose buss" ought to reduce power requirements while simplifying the design of the machine. Ross expected a proposal to be "prepared soon with block diagrams and descriptions of how various operations will be carried out."[15]

Next, Richard Best of Division 6 spoke on the basic circuits that would have to be developed for Whirlwind II. He grouped them under six categories for discussion purposes: three groups of circuits associated with memory, power supply, or in-out equipment, and three groups to handle diode logic, pulse logic, and in-out logic. He rated current technical knowledge of these according to whether they were reasonably well known, whether only a little was known about them, or whether they were completely unknown.[16]

Ross of IBM then returned to the podium to discuss "Principles of Design Criteria to be used in WWII." Focusing his attention on standardized circuits and standardized components, he called the engineers' attention to two IBM reports on the subject and indicated the character of the unfinished business that lay still ahead. Later in the meeting, J. A. Goetz of IBM explored more deeply the issues of component reliability and standardization. Pointing out that it made the most sense at this juncture to identify reliable components before attempting to standardize their design, Goetz referred to several visits that a group of IBM and MIT engineers had made to "leading representatives of the vacuum tube industry to discuss the needs of WWII with these people." Following up on these visits, they should soon decide what tubes they would use so that the manufacturers could get started on "actual engineering and manufacturing of tubes of the quality desired." As a practical matter, they should move ahead to establish standards that could be "incorporated into Government JAN or MIL specifications, so as to insure continued availability of these high quality items as approved components for military computer applications." They would have to work with both industry and professional societies as well as with military agencies.[17]

Since October, IBM's logical design group had been developing some "initial thoughts . . . regarding a computer for the Air Defense problem," Morton M. Astrahan reported. It would be a machine basically like Whirlwind I. It would use ones-complement arithmetic, allow for a word size of 15 bits including the sign (plus or minus), and carry out "addition, subtraction, multiplication, division, transfer, sub-program, etc." in the way that Whirlwind I did. The group also had come up with three improvements to recommend: A "40 to 60% increase in number of operations performed in a given time" could be obtained if the new machine were to have two arithmetic elements operating simultaneously. Automatic indexing of program instructions could increase the computer's effective capacity by one-fourth. Finally, if all the in-out units were designed to look exactly alike to the computer, then the in-out equipment could operate while the machine was carrying on its programming operations.

These ideas were incorporated in a proposal that already was under discussion by Division 6 and IBM investigators, Astrahan said, "and further conference discussion is planned for the near future. These meetings should lead to certain conclusions which will serve as the basis for design of the machine."[18]

Magnetic-core memory was taken up by the next three speakers, all from MIT. William Papian reviewed the history of MIT's experience with metallic and ferrite cores, pointing out that it was too soon "to draw final conclusions regarding the application of this type of memory to a computer until it has actually been operated in one." To this end, the Memory Test Computer was under construction. Papian expected it to be operating during the coming spring. Its memory would consist of 17 planes of magnetic cores, each plane arranged in a 32 × 32 square of 1024 bits, similar to the sample plane Papian had brought to the conference. Papian wrote:

> The cores will be vacuum tube driven, using 6AS7 tubes as drivers. One of the contemplated problems in making a ferrite core memory is the matter of actual construction. The first plane which was exhibited took 5 to 6 man days time to wire. Since the operator

> was experienced in handling this type of item, it is felt that future units can not be made in less than 5 days time. Attention to automatic or semi-automatic arrangements for wiring these planes is important. . . .
>
> Not much effort has been given to the development of a memory for WWII since it is felt that this decision can be made only after MTC has been in operation. It is thought, however, that WWII will require approximately 30 digit planes, each being a 64 × 64 array. Vacuum tube driver circuits will probably be employed in this machine in the same manner as in MTC.[19]

J. H. McCusker followed Papian to discuss the separate problem of producing magnetic cores of the quality and quantity required. Specifications for the ferrite type MF 1326 B cores then being manufactured by General Ceramics Company called for "a 1 microsecond switching time and a 100 millivolt output signal," but Division 6 was finding that, even when a plus-or-minus 15 percent tolerance was allowed, 55 percent of the cores were unacceptable. Nevertheless, they expected to have enough cores for the Memory Test Computer by the middle of February.

Still ahead lay the problem of the vastly greater number of cores that would have to be obtained and wired together for the air defense computer, assuming the Memory Test Computer would perform as desired. A measure of Forrester's commitment to the project was afforded by his recent decision to install at MIT a laboratory to investigate "the conditions encountered in the manufacturing process of the cores and their effect upon switching time, coercive force, etc."[20]

Kenneth H. Olsen then reported on progress in the building of the Memory Test Computer. The basic logic of its operations was like that of Whirlwind I, as was the fabrication of the major portions of the machine, since they sought as reliable a machine as possible as soon as possible, in order to test the unknown variable of a core memory. As a result, preliminary operation of parts of the machine had yielded only 20 unaccountable errors in 750 hours of machine time. Sixteen of these were probably caused by normal voltage fluctuations in the laboratory's normal power supply to which MTC had been wired, and the other four likely were caused by shorts in vacuum tubes.[21]

N. P. Edwards of IBM then discussed other uses of magnetic cores that appeared attractive in computer circuits, and he was followed by IBM engineer E. H. Goldman, who reported on investigations recently begun on buffer storage and radar-data display based on requirements indicated in Lincoln Report TM-20.[22]

The first Hartford Conference closed with remarks by Taylor of MIT and Burke of IBM. Taylor summarized the scheduling situation that had been discussed at the December 5 joint meeting, emphasizing that if they were to have the Air Defense prototype installed and operating by the target date of January 1, 1955, then between January 20 and October 1, less than 8½ months away, the two organizations would have to complete the "engineering work in connection with the preparation of specifications, block diagram work, development of basic circuit units, actual equipment design and all the other things necessary to permit actual construction to begin." By the end of 1953, they should have built up their professional manpower to the estimated peak of "approximately 235 persons" required if they were to meet their goal. This might seem to be a large number, Taylor admitted. He concluded on a habitually pragmatic and optimistic note, turning on what his colleagues affectionately called his "sunshine pump": "When these people are divided among the many fields to be covered, the number remaining in any one activity no longer appears large."[23]

T. A. Burke's closing comments were more sober, businesslike, and institutional in flavor. IBM currently had 20 professional, two technical, and four nontechnical personnel assigned to the project and expected to assign six more to Project High within the next 2 weeks. Noting the dollar ceilings and the latest purchase order and requisition backlogs on test equipment and material then prevailing, halfway through the 6-month subcontract, Burke emphasized the need to negotiate forthwith an Air Force Prime Contract in order to permit work to continue beyond the end of April. On that official note, the first Hartford Conference ended.[24]

Technical collaboration had begun in earnest, and an essential formal step had been taken to establish on administrative and policy levels the consensus that already was building between the two groups of engineers. Both levels of action were needed. Without agreement at the working level regarding ends and means, the joint enterprise could not get fairly started. Without agreement at administrative levels, it could not continue. And as yet the success of this unorthodox collaborative effort was far from ensured.

17

Bringing IBM up to Speed

As the winter months gave way to the spring of 1953, there were signs of progress in the unorthodox collaborative efforts of MIT's Division 6 and IBM's Project High. The first Hartford Conference had indicated the wisdom of organizing a number of joint committees to investigate major problem areas and determine effective courses of action, and this had been initiated. A tentative decision had been made to incorporate the informational capacity that a 32-bit word length would provide, and IBM had undertaken the detailed work of developing the block diagrams and circuit designs that would enable dual arithmetic elements to operate together as one or as two separately and simultaneously. As the work progressed, it made sense for a member of Jacobs's arithmetic-element group to begin spending all his time at Poughkeepsie in order to maintain adequate liaison.

By the end of the fourth month of collaboration (February 1953), the number of IBM staff engineers on the project was approaching 40, by the end of the fifth month it was nearing 60, and by the end of May stood at approximately 85. (Table 17.1 shows the levels of effort reported by IBM and MIT engineers from November 1952 through December 1954.) Groups of engineers from MIT and IBM were discussing component standardization with manufacturers and military laboratories; other joint groups were exploring the national state of the art with respect to display-tube design. Work on in-out equipment and the associated buffer storage and information displays was, as expected, getting off to a slow start. "This phase of the development," Kromer reported early in April, "contains more unknown factors, both with respect to mode of operation and equipment design. Thus, it is important that a large amount of attention be devoted to this matter throughout the entire program."[1]

In March of 1953 the second Hartford Conference convened, attended by 18 IBM personnel and 16 Division 6 personnel. The attendees heard

Table 17.1

		M-Note no.	IBM man-days*	MIT man-days†
1952	November	1739	110 (14)	16
	December	1780	125 (14)	–
1953	January	1817	37 (26)	11
	February	1891	55 (40)	4
	March	1955	69 (60)	25
	April	2125	128 (70)	56
	May	2221	144 (85)	36
	June	2272	172 (108)	93
	July	2374	106 (144)	88
	August	2406	66 (164)	57
	September	2455	96 (180)	103
	October	2505	63 (179)	99
	November	2563	40 (200)‡	88
	December	2610	51	127
1954	January	2679	29	108
	February	2714	78	133
	March	2766	107	127
	April	2822	191	158
	May	2879	50	54
	June	2901	63	70
	July-Aug.	3033	137	213
	Sept.-Oct.	6M-3142	54	211
	Nov.-Dec.	3270	112	123

*IBM man-days spent at Division 6, as given in the M-Note. ( ) = number of IBM personnel assigned. Sometimes the numbers were listed as "approximate."
†MIT man-days spent at Poughkeepsie (sometimes listed as "approximate"). No data available for December 1952.
‡"The size of the development engineering group assigned to the project at IBM has stabilized at approximately 200 people of staff level."—M-2563, p. 3.

progress reports on the Memory Test Computer, on magnetic-core circuit and driver designs, on the willingness of vacuum tube manufacturers to provide the high-quality tubes required, on the continuing acceptability of the general operating specifications for Whirlwind II that had been presented at the first conference, on the progress of arithmetic element designs, on ways to achieve with appropriate circuits overall control of the computer's operations, and on the overall scheduling picture.[2]

Another general Hartford Conference was held in April of 1953, and many of the same people attended, including 17 from MIT and 22 from IBM.[3] The increased attendance at the March and the April meetings probably reflected the growth of Project High, which had increased its professional staff from 22 in January of 1953 to 71 in April.[4] At each of these meetings, the agenda consisted mainly of reports on work done and decisions reached during the intervening period.

In retrospect it can be seen that all of these three Hartford Conferences were serving an interim purpose in coordinating the early collaborative efforts of the MIT and IBM engineers until their accelerating joint design efforts rendered the conferences superfluous and generated a need for a different type of meeting. Collectively called Project Grind, the later meetings were intended to "grind out" solutions to specific preliminary design problems. IBM and MIT engineers met for 7 days of intensive discussion in late June and early July of 1953; there was also a single meeting in August and another in October.

Meanwhile, other technical meetings were held in Hartford. Nine days after the April 21 meeting, 13 people from the two organizations attended a seminar on basic circuits in order to determine "the circuit requirements of the input-output group" and those of other groups who needed "slower [and] less expensive circuits" than had been designed up to that time. These requirements incurred further consideration of "basic voltages and tube types" that had to be agreed upon.[5]

Of greater moment were two joint meetings held in Hartford in May and June of 1953 to consider how the modular circuits used in many of the registers of the computer should be packaged in convenient pluggable units containing (as seemed most likely) six vacuum tubes per unit, "with backup designs for four-tube and nine-tube units." These would be useful when either routine or unusual maintenance and repair actions were indicated.[6]

The various early Hartford meetings contributed significantly toward establishing an effective consensus on the design, and while neither definitive nor final, they were later judged by participants Astrahan of IBM and

Jacobs of MIT, when looking back from a perspective of three decades, instrumental as a "catalyst for initiating action." The meetings had provided a "means for identifying overlooked aspects of the machine," they had created a "forum in which people could interact on a personal level," and they had established "a modus operandi . . . between the IBM and MIT staffs" which enabled them to proceed toward basic agreement on the configuration of the computer they were about to build: "It would have a single-address order code in a 32-bit word. The memory would have a read-write cycle in the range of 5.5 to 7.5 microseconds for 8192 words of 33 bits, including a checking bit. Data words required only 16 bits, so each retrieval involved two data words."[7]

Such technical decisions were possible during these first 6 months of the MIT-IBM collaborative effort, because they were also the products of a larger pattern of collaborative activity that took shape during the winter and spring of 1952–53. It was a pattern that posed new administrative and technical challenges to Forrester, Everett, and their Group Leaders to gain and maintain control of accelerating and expanding activities, and Forrester and Everett saw to it during February of 1953 that enough time was spent on "the general organization and planning of the engineering work necessary" to carry out the R&D that lay ahead. Kromer explained:

> Detailed lists of items were prepared, and programs showing the time in which the work required for each of these is to be done are in the process of being prepared. This information will be collected at a centralized location, and control will be established so that the pace of the program can be carefully followed and attention given to any phase of it which seems to be losing ground. Through this means we hope to successfully complete the development work and construction of the prototype models in line with the extremely tight schedules that have been set as targets for this work.[8]

These policy actions were paralleled by daily implementing activities. For example, between January and June of 1953 the design characteristics of the arithmetic element were established. After first considering the equipment that would be required by several alternative designs, the MIT and IBM engineers settled provisionally on the use of a split accumulator and a multiplying technique "based on an adder which uses an asynchronous add and combined shift." IBM was then able to begin detailing the appropriate block diagrams and circuit designs that helped the two teams reach tentative design decisions in March. Arrangements were made at that time to have an engineer from Jack Jacobs's group spend full time with the IBM engineers for a while, and IBM was asked to "furnish schedule dates for the completion of the engineering and construction of models of the arithmetic element." Forrester at the beginning of June was consequently ready to propose formal transfer of "the principal responsibilities for engineering on this portion of the machine from MIT to IBM."[9]

Corresponding action had been initiated in January to design and develop the display scopes and consoles that the human monitors of the air defense computer would need, and in March this effort was expanded to include the other input-output equipment. "This phase of the development," reported Kromer, "contains more unknown factors, both with respect to mode of operation and equipment design. Thus, it is important that a large amount of attention be devoted to this matter throughout the entire program."[10]

In April of 1953 two input-output meetings were held at Hartford and one at Cambridge, and a two-day visit was paid to the IBM plants in Poughkeepsie and Endicott to "survey their work in the field of magnetic drums" and consider whether such drums might be used as buffer storage on the air defense computer. The men decided to spend until July 1 making a preliminary survey of all aspects of the problem. In July they would begin to "establish specifications for various pieces of equipment in order to permit detail engineering design to proceed and have construction of the models completed as required for operation during the latter half of 1954." Four more meetings in May, three at Hartford and one at MIT, explored the situation further and confirmed the need to go to work on July 1 framing explicit specifications of the equipment required. The input-output design challenge was just one of the reasons why Jacobs and Astrahan were to look back and conclude that designing "the central part of the machine turned out to be the easy part of the job."[11]

Another study group was created in May to "prepare recommendations for the general physical layout and arrangement of the arithmetic element and control section of the machine." After two joint meetings at Hartford to review the group's work, the decision was taken to follow the conservative, time-saving course of imitating the layout already being built into the Memory Test Computer, arranging digit-row equipment to run horizontally and registers to run vertically, while using standard plug-in units.[12]

In still another direction, after informal discussions Division 6 formally asked IBM to take on additional tasks related to developing and manufacturing the air defense computer. Further development work on the auxiliary drums and related circuits for the computer needed to be undertaken. Also, they agreed it was time IBM should set to work on a research and development program for ferrite cores, giving special attention to "the improvement of manufacturing know-how and quality control in the interests of increased uniformity of the product." In response to a May 5, 1953 proposal from Coombs to implement part of IBM's new prime contract with the Air Force and organize "pilot production facilities for the reliable manufacture of magnetic cores," Forrester sharpened the formal intent: "We hope that the primary objective will be *the production of satisfactory magnetic cores at a time and rate adequate for the air defense program.*"[13] This assessment by Forrester of the magnetic-core R&D situation at the beginning of the summer of 1953 is worth closer study for the way it reflects the extraordinary attention to detail that Division 6 maintained while relating such detail to larger goals and keeping track of what needed to be done next. First, Forrester ran over the delivery schedule that had been set up for prototype and production-model deliveries of the air defense computer. Then he described the three methods of attack on core R&D that he saw available to meet the schedule, explained why the second method was to be preferred, and closed with a detailed, courteous suggestion on how IBM and MIT should proceed in order to achieve their common goal of perfecting on time a magnetic memory for the computer. Was he meddling, trying to tell IBM how to do its job, or was he simply looking ahead helpfully? It did not simplify matters any at the time to realize, as all knew, that the practical value of magnetic-core memory had not yet been demonstrated.

Forrester and Everett were always looking ahead, and it was Forrester's habit to call attention to the most efficient course of action when he saw it lying so clearly ahead. He did so more conspicuously than Everett, for both his personal style and his position as head of the Digital Computer Lab since the group's wartime days, when it had been a subdivision of Gordon Brown's Servomechanisms Laboratory pursuing Project Whirlwind, caused him to maintain a higher profile. One of the consequences was that some regarded this sort of foresighted personal behavior as presumptuous meddling at the time, even though subsequent events frequently bore out his analyses.

Bearing these considerations in mind, one finds this instance of Forrester's management of magnetic-memory R&D illuminating. In his letter to G. R. Solomon at IBM's New York headquarters he opened his review of the overall delivery schedule of the air defense computer with the observation that since both the time schedule and the rate of production depended on the progress made in designing and building the equipment, they should be re-examined from time to time. Here was the schedule "we hope to meet" to put high-speed core storage in the first prototype machine planned for completion by the middle of 1954[14]:

Specifications on cores	July 1, 1953
Specifications on memory circuits	Aug. 1, 1953
Specs. on mechanical design	Oct. 1, 1953
All cores and other parts procured	Jan. 1, 1954
Construction complete	May 1, 1954
Testing of memory as independent unit, complete by	July 1, 1954
Combine and test with arithmetic element and control by	Aug. 1, 1954
Arrive in Lexington, Massachusetts	Aug. 15, 1954

Forrester expected the second prototype to trail the first by 6 months, and the production models might begin to arrive "as early as November 1, 1955" at two-month intervals for the first four, after which succeeding production models should arrive a month apart. Accepting this delivery schedule, Forrester turned to the magnetic-core R&D program and identified three alternative courses of action: a short-range, a middle-range, and a long-range approach.

The short-range approach was the one that Division 6/DCL had used to obtain cores for the Memory Test Computer and would likely be the means of obtaining cores for the first air defense prototype. It entailed the use of "high-capacity rapid automatic selection equipment for selecting high-quality cores from production with a wide dispersion in core characteristics" (i.e., with poor quality control).[15]

At the other extreme lay the long-range approach. Basic research and new knowledge lay at the heart of this approach; it was presently being pursued in Division 6.

For IBM Forrester recommended the middle-range attack because of the "tremendous improvements" he saw it bringing. It "calls for refinements of production techniques which are now obvious," he said, "on the basis of our knowledge at the present time." These included using a "precision molding machine and . . . a special tunnel kiln" to manufacture the cores. Such an approach was preferable to greater emphasis on the well-conceived basic research approach IBM was planning. On the basis of MIT's experience, "we would suggest that less work be planned for growing and studying single crystal ferrites and that considerable effort be made to study the micro-structure of polycrystalline ferrites."[16]

IBM should keep closely informed on developments at MIT, General Ceramics, and elsewhere in order "to become familiar with the problems and to benefit from the development which has been necessary for prototype-core production." "I suggest that we review the progress in this area in September," he concluded, adding that continuing liaison could be maintained between IBM and David Brown's magnetic materials group in Division 6. As far as Forrester was concerned, it was plain that IBM should begin by following MIT's lead.[17]

Every month of the first half-year of collaboration had seen more men go to work on more problems. Every month had seen a winnowing and sifting of technical options, an exploration of technical problems in greater depth and detail, and a sharpening and hardening of design requirements, as study replaced relative ignorance with feasible options, feasible options with specific designs, and specific designs with hardware subcomponents under preliminary test. Sprawling like an amoeba, the enterprise in its separate actions was neither random nor aimless in its movement.

But Forrester, Everett, Taylor and the others in charge of specific investigations sought tighter coordination in order to maintain their control of events and meet the research, development, and manufacturing deadlines they had set. As spring gave way to summer it was time to alter their technical decision-making strategy once more. It was time for Division 6 to initiate the "Project Grind" meetings, to assess what had been done and what needed doing after nearly 8 months of collaboration.

18

Tight Schedules, Slippage, and Project Grind

While the nation's leaders were hammering out the defense policy that in October of 1953 would give presidential authorization to the development of a continental air defense system, Lincoln Laboratory's Division 6 engineers and IBM's Project High engineers were continuing their collaborative effort, developing engineering plans and resolving engineering problems in which they had been immersed since the autumn of 1952.

In January of 1953, at the first Hartford Conference, the MIT and IBM engineers had begun to establish a preliminary consensus on what they needed to do. Five months later, it was time to assess their progress. Art Kromer, the liaison officer of Division 6, was increasingly apprehensive that the IBM engineers were not shouldering their share of the burden. The signs were small, but they formed a disturbing pattern suggesting that the IBM managers in Project High did not share the sense of urgency needed to reach the goal that Kromer described succinctly: a "prototype computer, with its associated equipment, installed and operating by January 1, 1955."[1]

Early in June of 1953, some 7 months after the collaboration effort had begun, Kromer voiced his concerns to Jay Forrester. Forrester, mindful of a visit he would be making to IBM on June 11 and 12, asked Kromer to draw up a bill of particulars.

Charging that Project High's administrative support of technical decisions was so weak that members of its own staff were informally complaining, Kromer cited specific examples, such as delays encountered during a recent three-week period in setting up a joint study group to develop specifications for power supplies. Also, work on the arithmetical element, on control circuits, and on machine memory was moving too slowly and appeared to Kromer to be "badly disjointed."[2]

Instead of taking any initiative in coordinating joint activities, Project High's managers were moving only at the insistence of Division 6, and even then further delays occurred. Project High administrators, said Kromer, were in many instances "unwilling or unable to act in any positive or definite manner and then follow through with supporting paperwork where necessary to back up their action." Nor were they hiring or organizing the staff to remedy the situation.

The picture was not unremittingly bleak. Kromer could point to a few exceptions: "The In-Out work is being guided by Bob Crago and is moving along better than any other phase of the work. Basic planning of logical design also is moving quite well. This is sparkplugged by Mort Astrahan who personally does most of the fundamental work."[3]

Nevertheless, design progress on basic circuits and mechanical design was moving at a perilously slow pace, threatening serious delay unless, said Kromer, "strong leadership provides incentive and coordination." Even programs being pushed within IBM, such as drum development, were threatened. Too much time was being taken to organize a program on standards for all phases of the project other than the phase concerned with electronic components. (IBM still had not empowered the Project High members of the joint Standards Group to screen electronic designs or to challenge the selection and application of components.[4])

As the months had passed, Kromer had become increasingly impatient with the slowness with which IBM was allowing its technical personnel to educate themselves to accept and put to work "well established principles known to our people at the very start of our joint work with IBM." Instead, what he had encountered, with increasing exasperation, was the circumstance that "many months were required to haggle over relatively minor details before IBM agreed to adopt the proposals of the MIT people as the basis for proceeding with design."

Kromer became convinced that IBM's traditional corporate conservatism was blocking progress. "It is my feeling," he declared, "that in contrast with the feeling here [at MIT], no one at IBM has the feeling of urgency and the honest desire to pitch in and do the kind of a job that is needed on the schedules that are contemplated." Thus, when he was discussing with them how to go about controlling procurement of items with which to construct the two prototypes, he had come to realize that so far they had been giving "nothing other than 'lip service.' . . . Mr. Coombs indicated that he had not realized procurement of material could begin almost at once."

And when Kromer reflected on actions taken at higher levels within IBM, he realized that "the New York office of IBM has not served in any fashion but to pass along information between here and Poughkeepsie.

If delays occur in getting action at Poughkeepsie, the New York office seems satisfied to accept any reason offered by Poughkeepsie."

When Kromer turned his attention from engineering matters to IBM's conduct of business relationships with MIT or with commercial firms that it might ask to supply needed items, and when he reflected on IBM's contractual relations with the Air Force, he found that "IBM's ultra-conservatism . . . frequently caused delays and misunderstandings of the real intent and goal of the project." He felt compelled to conclude that neither their attitude nor the way they conducted daily business offered any solutions. "They do not seem to have the proper frame of mind in this field. . . ."[5]

Kromer's was not the only appraisal, and his indictment—communicated privately to Jay Forrester and Norm Taylor (and consequently to Bob Everett)—was tempered within the next couple of weeks by his immediate superior, Taylor. Reporting in his capacity as head of Group 62 and as the man responsible for the technical progress of the defense computer's design, Taylor noted in the June 19 Biweekly Report that a 3-day visit he had just made to Poughkeepsie had left him with "the general feeling that activity at that location is proceeding at a better tempo than is indicated by the various reports which have been written." Though it was true that basic circuits, the magnetic drums, and the magnetic-core procurement testing "seem to need a little more attention," Taylor concluded that "other activities seem to be progressing at a fairly satisfactory rate." The situation was improving, he felt. "The IBM group is beginning to feel the urgency of the situation and is taking measures to overcome some of the obstacles which always lie in the road of rapid progress."[6]

Whether one responded to Kromer's privately communicated gloom or to Taylor's "sunshine pump" news (published in his report and circulated throughout the laboratory), there were other particulars upon which to base a judgment of the progress being made. Taylor's and Kromer's separate assessments of IBM's development of magnetic-drum storage, for example, reflect the quality and detail of the information to which the Division 6 engineers had access. In April of 1953, Taylor had joined Forrester, Everett, and four other MIT engineers in a two-day visit to Poughkeepsie and Endicott to observe in technical detail IBM's R&D work on magnetic storage drums. As Gus O'Brien and Ken McVicar reported, the team was impressed by the evidence "that IBM can produce drums and heads of excellent quality." But one item that had given them pause was the lack of work done on the electronic circuits the drum system would need.[7] There was substance to both Kromer's view and Taylor's.

Whether or not the true situation lay somewhere between Kromer's and Taylor's appraisals, Forrester and Everett's response appears to have been not to confront IBM's corporate management with the problem and ask them to fix it, but instead to redouble their efforts to establish a more intensive joint pursuit of the design goal that the two engineering groups held in common. Thus, by the end of the third week in June they were setting Project Grind in motion. A month and a half had passed since the Air Force had decided to withdraw its support of the University of Michigan air defense program.

Whereas the Hartford meetings had served as "an information exchange, a catalyst for initiating action" (as Jacobs and Astrahan recalled them), Project Grind was prompted by the awareness of the leaders of Division 6 that the preliminary design phase of the joint enterprise was already running out of time. Their solution was to hold as many meetings of MIT and IBM engineers as it took "to grind away at each topic until a decision was reached."[8]

In Project Grind, Forrester and Everett—characteristically—sought to bypass the sensitive administrative and personnel relations problems and go directly after the technical design problems, making it clear circuit by circuit and system by system to the engineers specifically concerned just what needed to be done, how they should attack each problem, what the IBM engineers would (not should) be doing, what the MIT engineers would be doing, and in which directions of design activity they would accomplish progress. Identifying in this way the unfinished technical business that lay ahead, and showing the way to proceed by engaging in doing it, the MIT engineering managers sought to enlist the informed and focused commitment of the IBM engineers pervasively at the working level (in Project Grind) and in this way to build a fire under their administrative superiors.

It was at this level of involvement and understanding that the leaders in Division 6 decided to undertake remedial action, making it clear to the IBM engineers taking part that if operational testing of the prototype was to begin at the start of 1955, as planned, this was the way they would have to participate promptly in joint action in order to carry out the detailed steps called for in the schedule. In a sense, the new course of action was an effort to overwhelm the IBM engineers at their vulnerable technical level of understanding. As Astrahan and Jacobs recalled years later, the time to make design decisions had arrived. "Choices had to be made primarily on the basis of the experience the individuals had with the subject area under consideration."[9]

They had run out of time for the kind of extended study and discussion of alternatives that characterized IBM's usual way of doing business. Three decades later, Forrester recalled that, when Thomas J. Watson Jr. had visited Forrester and Everett's Digital Computer Laboratory on the MIT campus, he had asked "How do you do so much with so little?"[10] Business as usual was a concept and a philosophy that had altogether different meanings for IBM and for the young MIT engineers. It should be remembered, however, that Project Grind itself was as extraordinary an enterprise for the MIT engineers as it was for their IBM counterparts.

Part of the answer to young Tom Watson's question lay in the anticipatory philosophy of management that Forrester and Everett had employed since they were graduate students in Gordon Brown's Servomechanisms Laboratory during World War II. Part of the answer lay also in their persisting philosophy of keeping engineers in the laboratory informed about all major technical phases of work in progress, both by word of mouth and by frequent written reports that devoted as much attention to problems as to solutions. This sharing of information and involvement contributed to the enthusiasm and the esprit de corps that characterized the group both as MIT's Digital Computer Lab and as Lincoln Lab's Division 6.

It was in this atmosphere of doing business that Taylor, as head of the computer design group, formally notified sixteen of the Division 6 engineers on June 22 that they would be participating, as needed, in "a concentrated session . . . in Building B in Lexington starting at 8:30 Wednesday morning, June 24." He explained in detail what would be expected of them. Each should be ready to lead discussion of the design investigation he was familiar with, as listed in the preliminary program that Taylor set forth. "As a particular subject comes up for discussion, those engineers whose name appears opposite this subject should be prepared to present the present thinking on the subject, the proposal most likely to be accepted, and the reasons supporting such decisions. Where such data is not available, a recommendation should be made for a procedure which would gather together enough data so that the decision can be made at an early date."[11]

Between June 24 and July 2, the Division 6 and IBM engineers joined in six days of intensive discussion, and in these formal joint meetings they reviewed critically the status of development and design work. The sessions were divided into two series of three days each, first at Cambridge and then at Poughkeepsie, with a three-day break in between. A seventh meeting was held on July 15 to gather up loose ends.[12] In these meetings the men focused on all design aspects of the newly named "AN/FSQ-7, Combat Information Central" air defense computer and its prototype model, the "XD-1." Before they were finished they had covered the computer from one end to the other.[13]

With the thoroughness that had long been a customary way to anticipate and avoid trouble in the MIT Digital Computer Laboratory, Rollin Mayer of Division 6 prepared and Kromer circulated detailed minutes of the daily discussions, not only "to put on record some of the decisions made and some of the reasons for these decisions," but also to solicit "comments from all parties concerned," particularly any errors or omissions. The aim of the meetings was both to ensure "continuation of development along existing lines in certain areas" and to outline "programs for additional work . . . in other areas."[14]

The participants from MIT numbered from eight to twenty during the seven meetings, depending on the technical subjects under discussion. The five MIT engineers who Taylor had announced would always attend did indeed participate in all seven meetings in their entirety: Everett (as the ranking Division 6 engineer), Taylor, J. F. Jacobs, R. P. Mayer, and K. H. Olsen.[15]

Those attending from IBM ranged from eight to eighteen. Only M. M. Astrahan attended all seven meetings in their entirety. R. P. Crago missed only the third meeting. Four other IBM engineers were in attendance much but not all of the time: Coombs (the chief engineer of Project High), H. D. Ross, N. P. Edwards, and D. J. Crawford.[16] Forrester, whose time was taken up with Air Force and Lincoln relations, attended none of the meetings.

The IBM and MIT engineers brought under intense joint scrutiny every major area that they felt required some degree of consensus on design issues which were still unfinished business. As the minutes of the seventh meeting put it, the leaders of Division 6 felt it was imperative that any and all problems "be brought into the open so that decisions can be made as soon as possible." Those attending agreed that "everyone should get even tentative plans for various parts of the system so long as everyone knows that they are tentative." The conclusions reached at the meetings, and some of the reasoning behind the conclusions, would be recorded and circulated in the minutes. Final decisions would later be "given in writing by Lincoln Lab., Division 6, acting on behalf of the Air Force."[17]

The engineers immediately identified thirteen major technical areas, and they hammered away at these in the six meetings held between June

24 and July 2, 1953. At the seventh Project Grind meeting, held on July 15, seven more areas were examined, some of which had been taken up but not finished at earlier sessions.[18]

In his Biweekly Report after the seventh Project Grind meeting, Taylor called "satisfactory" the progress made by IBM in four areas: "Crosstelling," "Power Supplies," "Manual Inputs," and "Subcontract of Video Mappers."[19]

The object of Project Grind was to put a cap on efforts of the weeks during which the engineers had been meeting informally since the preceding October. Whatever technical opinions they had been forming, swapping, or modifying, and whatever judgments they had been sharing, would now be explored and hardened into formal decisions. Thus, at the first meeting of Project Grind the fourteen MIT engineers and the eight IBM representatives wasted little time on preliminaries; they moved directly to settle design details of the slowed-down video (SDV) system, centering their attention on the video inputs, the video mappers, and the video input registers. Their object was to distinguish the items that remained unfinished from those that had been identified and taken care of. Group 24 would be responsible for design details and would aim at September 1 as the deadline for completing the "breadboard" design on the laboratory bench model, so that IBM might use it to determine the design of the production model that would follow.[20]

On the second day of Project Grind, twenty MIT engineers met with twelve IBM engineers to arrive at mutually informed decisions on the design of the marginal checking feature the computer would incorporate to increase its reliability. From there they proceeded to examine power supplies and magnetic memory.[21]

A glance at the problem of designing power supplies for the computer will indicate how Project Grind operated to push ahead the joint enterprise that MIT's Division 6 and IBM's Project High were carrying on. Kromer's earlier dismay at the lack of progress on power supplies, which he had detailed in his memo of June 9 to Forrester, was confirmed when this issue was spread out on the table at the second and the seventh Project Grind meetings. It was a problem that underscored the genuine need for Project Grind, for it became apparent when the subject was brought forward that insufficient work had been done so far, particularly since the power supplies for the computer and its accessory equipment would have to be among the first items to be installed.

As Taylor reported in the next Biweekly Report to the Division 6 engineers, there was "a rather wide difference of opinion as to just what should be used in the XD1 machine." More than selection of the power supplies was at issue; they must plan for and avert "worst case" emergency breakdowns in the supply of electric power to the computer itself, its drum buffer systems, its monitoring consoles, its air conditioning system, room illumination, and other accessory equipment. If electric power should fail, so would the computer at the heart of the air defense system. "Any standby power unit," they decided, "should be able to take over within one minute of power failure." For emergency backup the XD-1 prototype would use a diesel generator, the design requirements of which would be decided later. Experience with this approach should tell them how to proceed with the production models to be installed nationwide.

At the first of their two meetings on power supplies the men agreed on the quality "floor" they must impose, but when they turned to the basic items of "primary power, d-c power supplies, marginal-checking power, and standby power to carry the load in case of commercial-power failure" they found that, although they could draw up some tentative specifications, they didn't know enough yet to make the crucial design decisions. Concluding that "all of the quantitative information in this area must be reviewed again" before they could firm up their overall power-supply specifications and proceed to implement these with specific design work, the engineers agreed that they would have to "investigate the points listed above more thoroughly" and meet for another session.[22]

But there were some design considerations they could explore right away. For example, since power tolerances in normal electric power lines were too wide to be acceptable for reliable computer operation, the computer would use motor-powered generators for normal operation in order to isolate computer voltages from disastrous transient power-line surges. Furthermore, the motor and the generator sections would have to be electrically isolated from any destructive "capacitative coupling of large transients" occurring in commercial power lines.

Power-supply considerations also inevitably brought to mind the effect of the normal surging startup loads on vacuum-tube filaments, so the engineers agreed that power customarily should be cycled on and cycled off slowly enough to prevent premature aging of affected components. Further, such sequencing of power on and power off must not add any unwanted information onto the auxiliary and buffer storage drums hooked to the in-out system of the computer.

Still another consideration was dissipation of the heat generated by the thousands of vacuum tubes in the machine. Power to run the air conditioning system must be safeguarded, and the possibility of mechanical or other breakdown of the air conditioning system indicated increased reliability should be sought by separating the crucial water-cooling part of the system into several sections that could function independent of one another.[23]

When the two groups met again 20 days later, they immediately discovered the separate power estimates they had come up with did not agree. This was "due mainly to the display console power" called for, including enough to operate the cooling blowers that each console must have. When they added up all the various power requirements that came to mind, it became clear that the total for one installed AN/FSQ-7 computer would be between 500 and 600 kilowatts. Also, they should plan for future expansion.

If external power were to be cut off unexpectedly, what "prime mover" should they provide to cope with the emergency? Should they have storage batteries that would immediately go on line to power the motor-generators until the prime mover was powered up? Some diesel-powered generators were reputed to be able to go "from complete rest to full voltage in 5 seconds." In any event, for the XD-1 prototype they need not decide immediately on a "prime mover," because there would be opportunity later to try out such units. It was more important to design the system "so that any single power supply can be by-passed in an emergency."

For normal operation, turning power on and off would be far more complicated than flipping a single switch or pushing a button. To avoid damaging the equipment, they would begin by cycling on, in the course of one minute (a long time for electronic equipment), the 120-volt regulated alternating current to be supplied to "all filaments in the system, except the auxiliary or servicing equipment, and . . . to the plates of the dc supplies." Then the negative direct-current circuits would be turned on. After they were monitored, the positive voltages would be switched on.

Turning off the equipment called for even more complicated procedures, the men reasoned.[24] They also concluded that duplicates of all DC power supplies should be provided, and this raised the question of what their standby condition should be: remain idle until needed, be operating in parallel with their counterparts and therefore be available if any failure of a counterpart should occur, or be run "at reduced load to supply approximately half of the system and in emergency . . . be reconnected to supply the whole system." It was not practical to make all decisions at one sitting; there were always ramifications to be explored.

Obviously, fuses would be required, and therefore distribution relays must be wired in, to switch to an alternative power source if a power fuse should blow. These relays would also be available "for maintenance, for marginal checking, and for switching standby dc power." But it wasn't immediately clear which would be more desirable, to install the distribution relays on each of the major frames of the computer, or to install them all in a centralized location. Such decisions, too, could wait while further information was acquired.[25]

Depending on the engineering area under investigation, the challenges posed by Project Grind took different forms. In the case of marginal checking of the reliability of the internal circuits of the computer, the MIT engineers' experience with Whirlwind I stood them in good stead, for even though the FSQ-7 defense computer would be a more elaborate machine than Whirlwind, its basic, modular design would not be that different from the smaller-capacity machine. It made sense, the conferees agreed, to have the joint Marginal Checking Committee "submit a sample of the proposed breakdown of the arithmetic portion of the machine for detailed comment by all interested members." In this way they could proceed from their consensus on basic aims of marginal checking to "individual differences of opinion that may exist in the exact details of how logical circuits could be grouped together." They estimated that as many as 170 marginal-checking lines would have to be fed into just the central portion of the computer to confirm in detail the consistent operation of the circuits being monitored. With so many circuits to check during those moments when they were not operating in the machine's program, it would be advisable to devise an automatic switching system so that marginal checking would not interfere with the practical operation of whatever program the machine happened to be running.[26]

Since marginal checking of machine components was accomplished by varying circuit voltages and checking the responses of components in the circuit, and since there would be so many circuits and subcircuits to check, automatic checking would preclude provision of any mechanical techniques. The engineers agreed that, by forming certain groups of the proposed 170 lines,

> rather large sections of the computer could be varied for checking conditions. These rather large sections could then be divided for easier diagnosis should the over-all check fail to indicate adequate margins. It was felt by some that this group of large sections would allow rapid marginal checking most of the time and that only on the failure of a large section would it be necessary to go through all 170

> lines. There is a difference of opinion still existing as to the time element which this arrangement would provide. WWI experience would indicate that we will probably need automatic switching to provide complete daily check on the whole machine.[27]

The engineers spent the entire third day of Project Grind and part of the fourth on the problems associated with firming up the design of auxiliary and buffer drum storage. Eleven MIT and eight IBM engineers attended the third meeting, eight from MIT and ten from IBM the fourth.[28] The men from both organizations were already agreed that information going into the computer from radars or from human operators at their consoles, as well as information coming out of the computer, must pass through temporary storage on external magnetic drums. From there the men went on to determine that five drums probably should be required for every installed Q-7 computer, and each drum should accommodate six "fields" (categories) of stored information.[29]

On the fourth day eight MIT engineers sat down with between six and ten IBM engineers to go over the design details of the consoles that would display information to the human operators who would be monitoring air defense operations. Their object, as in all the Project Grind meetings, was to bring any problems into the open in order to arrive at effective decisions that they could all agree on. In this meeting they examined preliminary specifications of the operating parameters of the display console.

The conferees agreed that most of the human operators in the Direction Center would not require "a broad, sweeping picture" of the radar tracks of a developing aerial attack. Instead, each operator would see a display (renewed every 2 seconds) of selected tracks, which would not be too numerous for one operator to keep under surveillance. These tracks would be selected by the computer after it had identified them by tracing their developing courses. When these courses indicated potentially hostile flight behavior, the computer would have to present enough of the history of each track to permit the human operators at their consoles to make a judgment about the character, course, and direction of any such track brought under surveillance. The operators could then confirm or reject the computer's selection of that track and decide whether it merited continuing surveillance.[30]

If 16 words per track were accommodated on the track display drum, the operator should be able to obtain "18 points of history for hostiles and at least some history points for all other tracks" in order to remain aware of what was going on. The designers would have to determine later, from actual and simulated tracking tests, whether 18 points would really suffice. If more should be required, then they would add more fields to the drum system. In the meantime, seven fields would allow for 896 tracks at 16 words per track.

To handle electronically this much information, the items of data required "should probably be interleaved in a special way on the drum" in order to provide the data when needed and also to allow for crosstelling from one computer or Direction Center to another. Cross-telling would enable the operators to maintain full monitoring of any particular track, as (for example) the "hostile" aircraft passed from one radar surveillance sector to the next.[31]

To make the console system properly operational, the video screens, the filters, the use of the operators' light guns, and the psychological aspects of display rates would all have to be integrated to ensure adequate monitoring and effective defensive response to an attack. To this end, Project Grind identified numerous detailed "design specs" and either nailed them down or held them in suspension until more information and experience could be obtained.[32]

On the fifth day of Project Grind, when eleven engineers from each organization attended the meeting, some of the problems examined were pressing; others admitted of more leisurely solution. Crosstelling from computer to computer or direction center to direction center would be, the conferees realized, "a low priority item on XD-1 although XD-1 will eventually cross tell to itself, to WWI, or to XD-2." The output-drum situation was more subtle, for although this buffer system would not be needed immediately for the XD-1, its detailed design would have to be established early so that it would be compatible with output links that were being developed by other designers outside the project. Among these links were telephone lines, weapon data links, and special crosstelling provisions.

The engineers agreed to allocate one memory drum to display digital information that the human operators should have at their disposal regarding selected tracks under surveillance. The computer would write the information on any one of 256 slots that the operator could consult at will. Each slot would hold "48 characters - 5 bits each," and "some of these slots [would] contain intercept-pair data." Although electronically an operator could cycle through all the slots in half a second, this would be too fast "for psychological reasons." If a cathode-ray tube were to be used for the display, a 3-inch rectangular tube might suffice, but the size could and should be decided later.[33]

As the dominant features of the mechanical design of the machine were firmed up, two sizes of easily replaced pluggable modules were established, one containing six vacuum tubes and the other nine. Two bays of standard design would be employed, one to hold the nine-tube units and one to hold the six-tube modules. Each bay would hold 20 units, and bays could be tied together, side by side, to form "subframes" of variable size. Ducts would be provided to hold the wiring, and other ducts would be incorporated to conduct the cooling air that the tubes would require whenever the machine was up and running.[34]

The sixth day of Project Grind brought to a temporary end the intensive design scrutiny that the MIT and the IBM engineers had been giving to their joint enterprise. On this day the eleven men from MIT and the twelve from IBM concentrated on specifying tube types that would be used, then turned their attention to miscellaneous circuits and the responsibilities of the Standards Committee. Since there were still design decisions to be made, a seventh meeting was scheduled in 2 weeks. The report of the findings of the sixth day reminded everyone once again that "the object of the minutes of the Project Grind meetings is to put on record some of the decisions made and some of the reasons for these decisions" and that "any problems will be brought into the open so that decisions can be made as soon as possible."[35]

Two weeks later the engineers gathered up some additional loose ends at the seventh meeting of Project Grind. Ten attended from MIT and eighteen from IBM. They realized that design provisions for crosstelling would have to be provided for the larger air defense system of which the prototype XD-1 would be a sample part. Consequently, D. C. Ross of IBM was assigned to "write up the system that was generally accepted." "This system," it was noted, "will not be finally accepted until it can be shown that it will be all right under the heaviest load conditions."[36]

Project Grind was a one-time affair in the history of the SAGE computer, even though some subsequent single meetings were given the same label. Although Project Grind produced fewer decisions than the leaders of Division 6 had hoped for, it succeeded in accelerating and intensifying IBM's efforts in Project High, and it demonstrated the need to set up a more permanent arrangement for reaching consensus and making decisions on the design of the emerging computer.

19

Adjustments and Triumphs

Project Grind had an immediate effect on the behavior of the IBM design team. Both Norm Taylor and Art Kromer noticed it. "The general activity has reached a much higher tempo," Taylor reported after a visit to Poughkeepsie early in July. In view of their use of the arithmetic type of circuitry and associated logic they had imposed, he expected the IBM design team to be getting preliminary test data within a month. Meanwhile, work on video mapping, magnetic-drum logic, and in-out shifting registers was "proceeding at a good rate."[1]

Kromer was heartened by the seventh meeting of Project Grind, held in mid July. "A review of the activity covered," he noted, "indicated that in general work was proceeding satisfactorily on all phases of the development covered by the previous Project Grind discussions."[2]

Beyond these immediately visible signs, the accomplishments of Project Grind—which provided a tightly focused managerial perspective on what the IBM and MIT engineers had done and what they still needed to do—also led to a rapid sequence of formal administrative and programming changes in the collaborative design procedures that the two organizations were following. Both groups realized it was time for them to act upon the technical information they had been exchanging and sharing—time to start building a working prototype of the XD-1 defense computer.

Indeed, while Project Grind was going on, Taylor had completed and sent to Forrester a rather detailed schedule based on the current administrative consensus in Division 6 regarding what would have to be done, and when, if the first of two FSQ-7 prototype machines was to be delivered in the 2 years stipulated by Lincoln Laboratory Technical Memorandum No. 20. Emphasizing the importance of the project as "an essential part of the national defense effort" that could accommodate very few delays, Taylor drew up the schedule in the form of a series of bar graphs with accompanying explanatory comment, and he set forth the chronological stages of work to be done by IBM and Division 6 from design through testing to installation, ending in "delivery of a prototype system by 1 July 1955."

The schedule presented the by-now-familiar breakdown of tasks already distributed among five major groups (Central Computer, Associated Equipment, Input, Output, Auxiliary Memory and Display). It included designing and constructing "Building F" to house the prototype on the grounds of Lincoln Laboratory. It stipulated establishing design and manufacturing standards. It called for incorporating radars and a ground-to-air data link (both already available and undergoing expansion in the Cape Cod System) and for providing the communications equipment and network that the air defense system would need.[3]

Accompanying Project Grind and Taylor's schedule were eight memoranda that Forrester issued in mid July to all Division 6 engineers working on the AN/FSQ-7. These memoranda explained formally how the engineers would proceed, how key lines of administrative responsibility and authority would operate, and how a new "Scheduling Office" under Kromer would be set up to operate as an internal clearing house for information.[4] Forrester identified three formal communication channels to be used, and he explained how they should be used so that the right hand would know what the left was doing, whatever the means of communication: teletype, documents (such as M-Notes), or personal visits to IBM.

Everett would approve all basic circuit designs and would "arbitrate any questions where general agreement has not been reached." Should any departures from the agreed-upon schedule become necessary, they must be approved by Forrester or, in certain designated cases, by Everett, and only after Taylor had made written comments. Tasks accomplished on time could be reported orally, but anything that appeared likely to be late should be reported promptly in writing, along with the reason for the delay, the time of expected completion, and any possible effects on other jobs. Any jobs which happened to be so scheduled that "some steps may be more than two or three weeks apart" might, at the discretion of Kromer's Scheduling Office, require interim progress reports estimating whether the job was on schedule and whether the next scheduled date appeared likely to be met.[5]

Crucial to the engineers' understanding of the design of the overall computer, of its logical operations, of its component parts, and of how they would proceed to build it in detail were the "block diagrams" and the outline statements that preceded them. One of Forrester's administrative memos explained that groups of MIT and IBM engineers working together would first spell out "the general logic for the equipment," then prepare the block outlines. Either Forrester or Everett must approve the block outlines. On the basis of each block outline, IBM engineers would then develop "the detailed block schematics. When these are completed, they will be gone over with MIT engineers before adoption as the official block schematics."

Once both Project High and Division 6 engineers had finished their review, Forrester or Everett would review and approve the schematics and, "acting for the Lincoln Laboratory, . . . arbitrate any questions where agreement cannot be reached" before closing the procedure by giving final approval. Information of these approvals of the schematic block diagrams would, in turn, be sent to "Division 6 group leaders, Group 62 staff, Project High, Division 2, AFCRC via Kromer and the Planning and Control Office, and the Lincoln Director's Office."[6]

But none of these procedures accomplished a transfer of responsibility from MIT to IBM. As Forrester laid it out, this transfer would be a separate action requiring a letter to Coombs at Project High. The letter would be signed by Forrester or Everett and would accompany a formal, explicit, detailed description of pertinent "Equipment Design Data."[7]

Forrester's organizational memoranda to his engineers presented a thoroughly rational and orderly world view. Whether the procedures would be followed in practice and whether they would accomplish the desired results remained to be seen. In any case, they could not compete for excitement in the laboratory with a coincidental, emerging, increasingly dramatic technological event in Division 6—an event that took 3 months to mature.

In May, a month before Project Grind convened, there had been news from Ken Olsen, who was in charge of the Memory Test Computer project, that the magnetic memory was "installed and operating" in the computer built primarily to test the performance of magnetic-core storage. Taylor had added to Olsen's report the remark that the MTC had performed so impressively during the first 2 weeks after its core memory was installed that it was time to carry out "a thorough marginal-checking study and establish in a very quantitative way the quality of our magnetic memory."[8]

Thus, when the selected engineers from Division 6 and Project High were immersed in their Project Grind meetings, the design of the magnetic-core memory was still firmly in the hands of the Division 6 engineers. Detailed specifications for the cores to be used in the AN/FSQ-7's high-speed memory were being readied by Dave Brown's Group 63 for delivery to IBM on July 1, 1953, and 250,000 "good cores" were being ordered from General Ceramics. A like amount were being ordered from RCA. If all went well, these should be delivered by November 15, 1953.[9]

Forrester and Everett had been ready to move when the news of the heartening preliminary performance of the Memory Test Computer arrived. Before the end of May 1953, they shifted Olsen to a new job "in charge of the WWII systems section."[10] By June 15, this much was clear to Forrester, Everett, and their Digital Computer Laboratory Group Leaders: "Now that MTC is showing signs of success, an increasing number of outsiders will be requesting detailed information on MTC."[11] By the time the seventh and last Project Grind meeting convened, on July 15, 1953, Division 6 engineers were examining specific proposals for "making MTC an integral part of the program to test AN/FSQ-7 (XD-1) units." During the first week in August, Division 6 engineers set to work transferring the MTC memory to Whirlwind I.[12]

Figure 19.1
A magnetic memory bank.

It was in the historical context of these dramatic technical developments during June and July of 1953 that the Project Grind engineers could regard it as "fairly well established that magnetic memory [for the Q-7] will contain 4,096 registers, 33 bits each (including one parity bit) in a 64 × 64 × 33 core array using a 2:1 current ratio selection scheme." They would also make allowance for adding a second duplicate core memory later on.[13]

Events in the laboratory came to a climax in August of 1953, when the engineers gave Whirlwind I its new, superbly performing magnetic-core memory. They had surgically removed it from the Memory Test Computer in the Whittemore Building (at the edge of the MIT campus) and wired it into Whirlwind (in the nearby Barta Building). As its performance in the MTC had indicated, the memory proved to be a major technological advance when operated as an integral part of the Whirlwind machine.

All the signs appeared to indicate that the MIT engineers had eliminated the industry-wide Achilles' heel that had dominated the first decade of computer design: the lack of a reliable internal memory of acceptable capacity.

In the midst of their jubilation over solving the internal memory problem that had been dogging the project ever since Whirlwind had been

Figure 19.2
Magnetic-core memory.

Figure 19.3
Magnetic-core memory in Whirlwind.

Figure 19.4
Jay Forrester with a 64×64 core memory plane. *MITRE Corporation Archives.*

conceived, and in the midst of their efforts to settle the remaining design details of the FSQ-7, most of the engineers paid little attention to a deceptively minor adjustment in the managerial organization of their laboratory: the creation of still another office, an office that "grew to be a very powerful management tool." Kromer referred to it briefly in his October collaboration summary: "To improve the coordination of engineering activities at both locations, a systems office has been set up here and at IBM." It was subtly symptomatic of the direction events were taking that Kromer had begun his report with an unexceptionable reference to "concurrence" reached by IBM and MIT engineers on certain design features to be used. This word had been quietly and carefully chosen by the head of the new office, Jack Jacobs.[14]

Although it was not apparent at the time, a delicate, fragile, and crucial diplomatic change was underway at this managerial level of operations. On the formal administrative level within Division 6, Kromer's

Scheduling Office was already in existence; however, that office did not take care of the substantive engineering issues that were being met head on by the joint groups of engineers from the two organizations, and that had to be resolved. So it made pragmatic engineering sense to set up another administrative mechanism of cooperative and negotiated decision making—one that, with a minimum of friction, would work in two directions. In one direction it would control the design of the FSQ-7 in a genuinely operative way at the working level and establish a design consensus. In another direction it would subsequently win endorsement and confirmation of these engineering decisions from higher administrative levels.

A discouragingly large number of separate organizations were necessary participants in coordinating and hardening the computer's design. These included Division 6 and Division 2, between which there was ongoing political tension behind the scenes as George Valley sought (persistently, subtly, and unsuccessfully) to bring Jay Forrester's group under his informal and formal influence and control. Another party whose confirmation of the computer's design was needed was the upper administrative hierarchy of Lincoln Laboratory. On the IBM side, confirmation was needed from Project High and from the IBM administrative hierarchy above it. Then there was the "paying customer" behind both Lincoln Lab and IBM: the federal government, acting through the military agency of its Lincoln Project Office and the associated Air Force administrative hierarchy to which the Project Office reported. Finally, there were the private contractors and subcontractors working for and with these various organizations.[15]

It was in these circumstances during the autumn of 1953 that the low-profile Systems Office came into existence in Division 6 to expedite progress toward the same goal to which Project Grind had been single-mindedly devoted: building, as rapidly as was possible, a continental air defense computer of unprecedented capacity, speed, and reliability. The function of the new office was to provide (as Jacobs and Astrahan described it 30 years later) "a forum in which consensus about the main features of the design in all aspects of the system could be obtained." This engineering consensus was then set forth in documents to be formally approved by the Air Force, which was underwriting this "crash program." The Air Force, lacking the expertise that Lincoln Laboratory had been created to exercise, had little alternative but to place its money where its faith lay—ultimately, in MIT's Lincoln Laboratory—when implementing the charge, given it by the National Security Council, to defend the nation.[16]

Following the lead of Division 6, IBM created its Engineering Design Office. Both IBM's and MIT's administrators realized that they were scouting untrodden managerial territory to which existing military specifications and customary industrial practices largely failed to apply.[17] Following this course of action, the engineers of Division 6 and Project High—"without answering the question of who was in charge," as Jacobs noted—managed to agree on the designs of the basic circuits to be used in "flip-flops, gate tubes, pulse amplifiers, register drivers, cathode followers, etc."

"Concurrence" became the enduring name for the procedure, and it enabled the engineers, in a rational and competent way, to "proceed with detailed design, block schematics, and layout of the various frames of equipment."[18] A minimum of structure and a maximum of communication were injected into the arrangement. Thus, engineering drawings began to pass through three levels of acceptance, monitored and expedited by Jacobs's Systems Office: "a) preliminary status, b) concurrence between the two engineering organizations, and c) final authorization and release for use in construction of the model."[19] Level c, in particular, was deliberately and carefully implemented by Jacobs's office as quietly, inconspicuously, and matter-of-factly as possible.

More than organizational arrangements and facilities were needed, of course, and Taylor was pleased to find by September that "an intense effort on the part of IBM to meet the schedules is very evident." Ever watchful of the overall pattern of events, Taylor reminded his engineers in Group 62: "It is very necessary that the MIT role of helping to freeze basic circuits keep up with this increased tempo, but it is also important that we minimize suggested modification in the logic of the machine if we are to avoid delaying the IBM activity." This meant, for example, that if they were to expedite placing orders for the component parts that would go into the dozens and dozens of plug-in modules that dominated the architecture of the machine, then the design decisions of the component subcommittee of the Central Standards Committee must be (and were) streamlined in order that IBM might place purchase orders for equipment more rapidly. To this end, "either Paine or Watt [from MIT] have agreed to be on call in Poughkeepsie at all times during the next four- or five-week period. They will have the authority to accept substitutions in components wherever it is necessary to make a compromise in order to meet some particular schedule."[20]

All these managerial efforts, and others too numerous to detail here, were intended to accelerate the tempo of the R&D program to which the two teams of engineers were committed. It became essential, in consequence, that they adhere to the tight schedules and procedures to which they had committed themselves. Inevitably, such a situation created the possibility that a schedule might slip and fail to meet its programmed deadline. Sensitive to this eventuality, the Division 6 administrators had convened the Hartford and Project Grind meetings in order to reach decisions that would accelerate the tempo in order to keep the design work on schedule. Overhauling and streamlining the management procedures would also help to avoid such slippage. But none of these efforts could guarantee against slippage, and the possibility remained a chronic threat. The wonder is that the slippage that did occur was not even greater in a program of such technical and managerial complexity, with so many organizations participating at so many levels. The SAGE R&D program was unusually ambitious and rather unorthodox in this respect, partly because time was extremely short and the international stakes were so high.

20

Slippage, the XD-1, and the XD-2

Half a year after Taylor's planning schedule of June 1953 was issued, IBM's engineers at Poughkeepsie carried out a careful review of the progress they were making and discovered that Project High was falling behind. Collaborating MIT engineer Steve Dodd brought the bad news to Division 6, and David Brown reported it at the next regular meeting of the MIT Group Leaders, on February 1, 1954. "The tendency to underestimate the time schedule," Brown reported after reviewing the situation with R. P. Crago and J. M. Coombs of IBM, "is general, and failures to meet the schedule are noted even after one month. Procurement is also causing some delays. Responsibility and coordination for various aspects of the program are not clearly defined. . . ." IBM's solution was to offer a "more realistic" schedule that would deliver the first prototype system to Lincoln Laboratory about 4 months later than originally scheduled. Furthermore, a "strong effort" would be required in order to avert a longer delay.[1]

The response in Division 6 to Dodd's news was prompt. Forrester, Everett, and their Group Leaders agreed at that same meeting, after some discussion, that Project High should be assured of their support. But Division 6 should take no further action until they knew what specific countermeasures and remedies Project High's administrators were contemplating.[2]

Forrester wasted no time. He presented the problem to the Lincoln Steering Committee's meeting that same day, and Lincoln Lab's Director, Albert Hill, asked Forrester and Division 2 Director George Valley to "lay out their needs for discussion at the next Steering Committee meeting."[3]

Four days later, Everett, Taylor, and Dodd met with Coombs and Crago to review the problem and consider a proposal prepared by the Project High staff to get the program back on schedule. The proposal recommended four actions, three to be taken by IBM and one by Division 6.

Project High, for its part, would reduce to a minimum its "work on drawing and print handling without adversely affecting reliability," would hire additional subcontract engineers, and would attempt to persuade corporate management to assign more engineers to the project. The "bailout" action that Project High proposed Division 6 undertake gave the MIT engineers pause. Division 6 was asked to assume nothing less than "complete responsibility" for the engineering design of parts of the system which the MIT Group Leaders thought had "received little or no effort from IBM." Project High's request included "design of the outputs frame, design of the digital data transmission system, design of the display generator frames," preparation of "the site for installation of transformers, M G sets, power supplies, air conditioning, fulse [*sic*] ceiling, et cetera for XD-1," and assignment of "a few engineers to Poughkeepsie to assist . . . in design of the memory."[4]

At their February 8 meeting, the MIT Group Leaders paid as much attention to what the Project High leaders had *not* proposed as to their requests. Although one way to avoid late delivery would be to ship the XD-1 from Poughkeepsie without making preliminary tests, IBM had not mentioned that possibility. Nevertheless, the Group Leaders recognized the possibility that MIT might have to do the testing in order to ensure that the machine would arrive before the end of November. At the same time, they realized this: ". . . we must not undertake so much work for IBM that we will be unable to meet our primary responsibilities. Emergencies can be expected here which will prevent committing all our engineering to the IBM program."[5]

As the leaders of Division 6 considered the five items that Project High was asking MIT to take on, several things became clear: They could undertake all but the design of the outputs frame. They would very likely need help from Lincoln Divisions 2 and 7. Because they would really be acting "as an engineering department of IBM," they should be sure to pass procurement costs on to IBM. They would be wiser to do the engineering design without taking on the next step of the detailed engineering involved in packaging the design (i.e., specifying the details of the hardware configuration). Meanwhile, John Proctor (the leader of Group 60) should report to Forrester the kind of workload these actions would impose on the drafting room where the engineering blueprints were generated, and Division 6 should look into "the possibility of hiring people here to do the packaging."[6]

Under the circumstances, the division of labor that Project High proposed was not unreasonable, since the XD-1 would be the prototype and not quite the production model of the air defense computers that IBM was contracting to build for the Air Force's Combat Control Centers. Rather, the XD-1 would go to Lincoln Laboratory, where it would be used in testing an expanded and more ambitious form of the Cape Cod System. IBM would keep the second prototype, the XD-2, to help it get started producing the production models of the FSQ-7 that would go to the Combat Control Centers.

In any event, Division 6, after clearing it with the Lincoln Steering Committee, accepted the idea that Lincoln Lab would have to carry a larger share of the testing load, even though this meant that more people would have to be added to the fiscal year 1955 budget. Forrester explained to the Lincoln Steering Committee how the schedule that had been set in 1952 had "called for completion of installation and testing by July 1955, at which time the XD-1 prototype was to be ready for use by the Cape Cod groups." Since the program had already slipped 4 months by February of 1954, the practical way to avoid an even later delivery date than November 1955 was for Lincoln to take back some of the work originally assigned to IBM. Welcome or unwelcome, it was a practical way to "relieve the overloaded condition of the IBM engineering in Poughkeepsie," as Forrester diplomatically phrased it. It also meant that Division 6 would have the satisfaction of having under its direct control the details of the design of the central display—the computer's output—that depicted the combat situation in each sector. (The engineers used "sector" in its military sense, meaning a subdivision of a defensive military position. Although the eventual goal of the Transition System was to provide a truly continental air defense system, that goal would have to be accomplished by subdividing the country's perimeter into a network of radars and computer command sectors joined by telephone lines.) And while the complex process of preparing the site for installation of the prototype at Lexington would be their responsibility, it would also be theirs to control.[7]

One of the classic consequences of a slipped schedule, the "ripple effect," impelled Forrester to point out that later delivery of the XD-1, instead of relieving pressure to erect the building in which it would be housed, actually increased the pressure to finish it sooner. One way to make up some of the lost time, he explained, was to "bring equipment to Lexington early and do some of the testing here which otherwise would have been done in Poughkeepsie."[8] But permission from the Air Force to construct Building F to house the XD-1 was not soon forthcoming.

IBM's corporate management responded favorably to Project High's request for more manpower and assigned an additional 130 engineers to the program, for it was not at all resigned to the schedule delay. Rather, it was committed to taking whatever action was required to meet the new November delivery deadline and avoid further slippage. Eighty of the engineers would work on the XD-1, the rest on designing the production model.[9]

All these actions taken by MIT and IBM at the end of the winter of 1953–54 failed to stop the slippage, but this was not immediately detectable. Project High's leaders had agreed to issue biweekly reports commenting on the status of the work, yet by the middle of March these had not been posted. At the time, this was an equivocal indicator of the progress of events, for IBM offered two different reasons for the delay, Group Leader David Brown reported in his capacity as recording secretary of the Division 6 Group Leaders' Meeting. "Sufficient manpower has not been available," Brown noted, and "IBM is reluctant to issue in writing the status of the program," partly because they did not want copies of such reports to reach either the Air Force or AT&T's Western Electric personnel who were working in the Air Defense Engineering Systems (ADES) office that had been organized to implement building the Transition System (later called SAGE). To break this minor logjam, Division 6 would have to work out a policy solution with John Coombs of Project High. Forrester and Everett realized that, as Brown put it, "ADES and the Air Force look upon the Lincoln Laboratory as having the design responsibility for XD-1 and XD-2."[10] Coombs agreed to issue the first of a series of monthly reports early in April, and MIT agreed not to pass them on to other agencies. At the same time, MIT would "discuss these reports with outside agencies when such discussion is pertinent. IBM will be informed of any such discussions in advance so that it can send a representative."[11]

When MIT received the first monthly reports containing time schedules from IBM, in the middle of April, Steve Dodd optimistically suggested that IBM was "trying to get back on the original time schedule and . . . doing fairly well." Taylor proposed, more pragmatically, that "the only way IBM can meet the original schedules is to go on three shifts." Furthermore, if MIT obtained further factual evidence, "it should go to people in the IBM organization high enough to take steps to put the program back on schedule."[12]

But the tenor of events was perhaps more accurately reflected in Coombs's official, passive, courteous request to learn how IBM might take part in the "installation, maintenance, and operation" of the XD-1, since primary responsibility and control belonged to MIT.[13] Two other straws in the wind apparently drew no detectable special attention to their significance early in May of 1954. The first was an informal request for assistance that would enable Project High to retain senior engineers who might be assigned to other projects if advanced development work related to the Q-7 and coordinated with MIT's work could not be found for them. From Coombs's point of view, and perhaps from MIT's, "retaining senior FSQ-7 design personnel in Project High" was "essential for the future of the FSQ-7 program."

In addition to this conservative and orthodox managerial judgment, a second straw manifested itself when David Brown reported: "The first XD-1 frame has been assembled approximately four weeks behind schedule. . . . The present status of the job indicates that XD-1 will be delivered approximately four to eight weeks after the scheduled November date." To offset this disquieting news, Brown, in his Group Leaders' Meeting report, related that the tempo of activity in construction and test work was encouraging.[14]

Early in June of 1954, at a monthly meeting the Air Materiel Command held "to review the status of the FSQ-7 prototype program," it became clear that the program had been slipping 2 weeks per month over the preceding 3 months and was some 6 weeks behind. IBM's administrator in charge of defense engineering, C. F. McElwain, responded that at the next meeting IBM would be able to tell them if it could meet the November 15, 1954 target date for delivery in order to have the computer ready for air defense tests by July of 1955, as planned.

Once again Everett, Taylor, and Dodd set about determining the true state of affairs at Project High, and once again their efforts elicited expressions of concern to no avail. A month later, IBM confirmed its inability to meet the November deadline, advanced the delivery date to January 1, 1955, and asked Division 6 for more help in "system testing of AN/FSQ-7 (XD-1) at Poughkeepsie."[15]

In July, realizing that under these circumstances he could plan no definite course of action, Forrester began talking with "several representatives of IBM management," only to conclude (as he reported regretfully to his Group Leaders) that "IBM seems to have failed to recognize the magnitude of the job and establish the organization necessary to handle it."

His patience at an end, and without waiting for IBM to take the remedial steps it was promising by July 23, Forrester went directly to an upper management level of the corporation and arranged to meet with McElwain and J. Zollinger on July 26.[16]

Forrester's criticisms, which irritated and worried IBM's top managers (they were fearful that Forrester might recommend another source), appeared to have some effect. McElwain explained that he would step in and become more active in managing the program. He would spend about 90 percent of his time on it, and he would be in Poughkeepsie most of that time "to represent both engineering and manufacturing." There would be weekly meetings at which he, Coombs, Crago, and others from IBM, including the head of the new Kingston plant being erected to build the production models of the Q-7, would discuss the program's status with Taylor, Dodd, and others from Lincoln Lab. In addition, IBM would assign a "strong representation" to Lincoln to handle XD-1 "maintenance, technical aspects, design changes, and the production control office." The representatives would be "men from IBM Engineering Design Office or men of equivalent stature."[17]

Project High was reorganized. H. D. Ross was assigned responsibility for the two prototypes, M. M. Astrahan for the production models, and N. Edwards for engineering and advanced development. All three previously had been simply Associate Managers of Engineering Design. Now each was assigned personal responsibility for a particular project area, ostensibly "because of a gradual shift in emphasis on the project, a shift from predominantly design work to predominantly packaging, releasing, fabricating, and testing." In addition, the Engineering Design Office was abolished, having served its "original purpose of pooling the experienced engineering managers."

It remained to be seen whether these changes would be more than cosmetic. Norm Taylor learned that McElwain's new role at Poughkeepsie had not been publicized and that no public announcement would be made. "McElwain's new position is regarded as an additional coordinating function, not a change in organization," he reported.[18]

As it turned out, neither McElwain's participation nor the reorganization of Project High produced the hoped-for turnabout. No gains were evident by late September, and some of the more difficult tasks still lay ahead. Meanwhile, Project High continued to develop its plans for an advanced research effort that would assign "three men to transistor circuitry, three men to magnetic-core circuits, six to ten men to magnetic-core memories, one man to liaison with MIT's advanced development program, and one man to review of older components."

When Edwards asked what MIT thought of these plans, Forrester, Everett, and the Group Leaders realized that. although they were sympathetic toward IBM Poughkeepsie's desire to exploit the creative talents of its engineers, it was more important to put first things first and recognize that "many challenging problems and urgent needs of the prototype and duplex central programs must have highest priority." Project High's leadership apparently did not agree, nor did it appear to be maintaining the kind of focused control of its efforts that Forrester and Everett in Division 6 were accustomed to maintaining.

In spite of all these problems, the XD-1 machine was taking shape at Project High, and testing of it was being carried on around the clock. Since one-third of the test engineers were from MIT, the leaders of Division 6 had a clear view of these developments. What they saw was disturbing:

> That part of the machine now operating includes approximately 4000 tubes. Approximately one half of the orders [fed into the machine] are functioning, but considerable refinement of circuit details will be required before smooth operation can be achieved. XD-1 is behind schedule and some of the more difficult parts of the job, especially problems associated with the terminal equipment, lie ahead of us. Unfortunately, some IBM representatives in Poughkeepsie feel that their contract with the Air Force does not completely cover the job to be done. Although a cooperative attitude is found at most levels in IBM, some reluctance to expediting the program has been encountered in Poughkeepsie. N. H. Taylor has discussed this with C. F. McElwain in Endicott. J. W. Forrester would like specific instances of this failure to cooperate brought to his attention. It should not be necessary for MIT to ask IBM's New York office to send assistance to Poughkeepsie.[19]

The situation continued to worsen, from MIT's point of view. A delay in awarding the contract for the air conditioning of Building F resulted from "IBM's introduction (at a late date) of a cost-redetermination clause in the contract," and IBM had yet to ask for bids that would encumber funds "being made available for lighting Building F." "It appears," David Brown reported in the minutes of a mid-October Group Leaders' meeting, "that neither XD-1 nor Building F will be ready on January 1, 1955. The date for delivery for XD-1 must receive serious consideration."[20]

Division 6 acknowledged the inevitable and, given no choice, agreed to accept delivery of the XD-1 prototype in January, realizing that even January delivery was increasingly unlikely. There was at least one bit of consoling news, as of November 12: "The central part of AN/FSQ-7(XD-1) was running, . . . with a 6 microsecond memory cycle." However, unanticipated technical problems had arisen at Poughkeepsie, and Taylor had accumulated a list of "22 specific difficulties with XD-1 and XD-2" before the end of November. Furthermore, preliminary performance of the auxiliary but essential magnetic drums was uncertain, and delivery of the drums was a month behind schedule. If they were to postpone delivery of the XD-1 until after Poughkeepsie had received and tested the drums, March 1 would become the optimistic rather than the realistic delivery date, and once again destructive "ripple effects" would begin to cripple other activities vital to the development of the nation's air defense umbrella.

"It appears that Division 6 should do more planning for the SAGE system and not get committed further to the XD-1 [preliminary testing] program," reported David Brown in the minutes of a Group Leaders' meeting. "As much of the XD-1 job as possible should be handled by IBM. We must, however, be in a position to take over in case IBM fails."[21]

The delay in delivery of the XD-1 to Building F in Lexington was doubly serious in that it portended a delay in the delivery of the production-model FSQ-7 machines, whose design depended heavily on data which the XD-1 would provide when harnessed to the radars of the Experimental Subsector in air defense flight tests. Delaying these tests and the production program threatened the safety of the nation at a time when intelligence sources suggested that the possibility of foreign attack would be greater than previously estimated.

By November of 1954 the situation had become serious enough for Forrester once more to take to higher administrative levels at both Lincoln and IBM the need for more forceful supervision at Project High—the same point that Kromer had first argued, with less overwhelming evidence, just before the Project Grind meetings in the summer of 1953. During lengthy discussions with IBM corporate managers on November 11 and November 16, Forrester "frankly pointed out . . . that the time schedule for FSQ-7 production and availability was in jeopardy due to the likelihood that engineering information would not be ready in time." From Lincoln Laboratory's perspective, the primary obstacle was weak leadership at Project High. John Coombs had failed to provide strong guidance, and "a very large gap existed between the IBM management and the High Street attitudes."[22] In these discussions and in a follow-up letter that set forth his recommendations in greater detail, Forrester suggested several actions that he believed would remedy the managerial shortcomings. Project High needed a stronger and more forceful leader, he pointed out—one able "to inspire individuals to place the ultimate end result of successful field operation above all lesser considerations." He urged the creation of a "joint IBM-MIT Steering Committee" that would meet weekly not only for policy discussions but also to take a "sampling of matters at all levels of detail." Also, a "strong planning and scheduling operation" should be organized and supervised by a strong executive who would possess "experience in the transition between research, development, production design, manufacturing, field installation and trouble-shooting." This executive should report to the manager chiefly responsible for "maintaining the pace of the FSQ-7 program." That manager would see that planning was done and implemented, but he would not himself carry out any of the work, for "the conflict of interest between the doing and the monitoring is one that cannot be shared in a single individual."[23]

Forrester reminded the IBM administrators that, in view of Lincoln Laboratory's dependence on federal R&D monies, especially once the Experimental Subsector was engaged in intercept tests and production of the FSQ-7 was underway, the lab simply could not engage in production activities at the same time. By established legal custom, production was the responsibility of the prime contractor and its subcontractors. Lincoln Lab's "responsibility for technical direction and coordination of the entire SAGE System program" meant that it had to use its staff for "those aspects of technical guidance which [could not] be effectively delegated to other organizations."

As a matter of fact, Lincoln Lab's past acceptance of tasks at Project High's request was a precedent to be avoided in the future. The "proper procedure," Forrester argued, was the one Lincoln Lab had established with Western Electric and the Bell Telephone Laboratories, under which "even such things as the writing of test specifications" (a Lincoln responsibility) would be done at Bell Labs under Lincoln's "guidance" and for Lincoln's "approval." A similar mode of operation should prevail in the relationship with IBM "for those activities which directly and indirectly arise from the FSQ-7 production contract."[24]

IBM responded promptly, taking a number of actions that implemented Forrester's recommendations and also explaining these in a meeting in Cambridge in the MIT office of Edward Cochrane, MIT's Vice-President for Industrial and Governmental Relations. Diplomatically restructuring the chain of command affecting Project High, IBM inserted an additional level of authority into the project, appointing Zollinger as Director of Defense Contracts and McElwain as Director of Defense Engineering and Manufacturing. Although McElwain's was nominally a staff position, it gave him genuine "authority over the entire FSQ-7 program at IBM," thereby effectively linking IBM's prototype and production efforts.

By setting up a joint IBM-MIT Coordination Committee at the executive level and giving it the "primary purpose . . . to discuss and solve problems, rather than exchange information," IBM brought Division 6 instrumentally into the chain of command at the action level, regardless of what might appear on the organization charts. On November 9 this committee began to meet weekly, and it soon became clear that no area of activity in the joint endeavor would escape its scrutiny.

At the meeting in Edward Cochrane's office, the IBM representatives brought out a delivery schedule for the XD-1 covering the period January–July 1955. Acceptance tests were planned for the following October, and an official demonstration was scheduled for February of 1956.[25]

The arguments were not all on MIT's side, as Tom Watson Jr. recalled years later in his memoirs. Forrester did not appear to appreciate sufficiently the complexities of the manufacturing process that was IBM's life's blood, so of course he would be impatient—an impatience Watson feared might have jeopardized his corporation's securing the contract "to manufacture and service the dozens of computers" that would be required for the operational SAGE System.[26]

In its collaborative work with Lincoln Laboratory's Division 6 to produce the prototype of the air defense computer, IBM was in a position both technologically advantageous and commercially difficult. In the exploitation of a commercially promising revolutionary technology, competitors start from a common base line. To meet the competition, IBM had to maintain between its military and commercial operations a balance that would accrue to the advantage of both, and to transfer its newly gained engineering experience to its commercial operations as expeditiously as possible. Did this effort mean that Project High was being shortchanged in the process? It was hard to say. Competent, well-trained computer engineers and programmers were in extremely short supply nationwide, and yet transfer of theoretical knowledge and know-how from SAGE to commercial machines might well make the difference in a tight, competitive, commercial race that might well lie ahead. Nor were the issues simplified by the circumstance that in 1954, as IBM was getting itself ready to manufacture the AN/FSQ-7, it also began work for American Airlines on an airline seating reservation program that was to become the famous SABRE (Semi-Automatic Business-Related Environment) system.[27]

IBM was not alone in restructuring its administrative procedures in response to the crisis. In the closing months of 1954, Division 6 set about rearranging some of its activities to match the fact that designing and building the two prototype computers was increasingly giving way to concentrating on the production-model Q-7 versions and on how their design might be modified by the performance of the XD-1 in handling the Electronic Subsector flight-intercept tests.

Forrester transferred the liaison responsibilities of Group 62 (Norm Taylor's air defense computer group) to a newly formed Production Coordination Office, which became "Group 66" under Art Kromer. The new office would continue to coordinate FSQ-7 activity inside Lincoln Laboratory and would provide "liaison with outside agencies (armed services, industrial concerns, etc.)."[28]

Group 62 would concentrate on designing further equipment for the XD-1 and the Experimental SAGE Subsector, and it would donate some men to a reconstituted and expanded Systems Office headed by Jack Jacobs, with Ben Morriss as second in command. This new "AN/FSQ-7 System Coordination Systems Office" then took on a number of tasks that Group 62 and Steve Dodd's Group 64, "Production FSQ-7 and WWI," had handled. "It is through this group," Forrester explained to the Lincoln Steering Committee, "that we expect to insure the transmittal of the necessary design changes into the production program based on changing system requirements and on results from XD-1 installation and testing."[29]

However useful, necessary, and heartening these administrative adjustments might appear, they did not change the engineering realities, which had become focused on delivering a tested prototype to Lincoln Laboratory. At the final Group Leaders' meeting of the year, held two days after Christmas,

> the question whether or not XD-1 is now in good enough condition to be shipped to Lexington was discussed. The Daily System Test Summaries written by IBM indicate considerable trouble. No marginal testing has been done on the system for the last six weeks. Considerable difficulty has been reported with tubes. Trouble location is requiring more than 50 percent of the time.[30]

The reaction in Division 6 was to propose a rigorous, detailed, exhaustive test procedure that would be carried out at Poughkeepsie before delivery was attempted. The tests, carried out on December 28 and 29, produced results "probably better than had been expected. Margins are very good. The major trouble is due to poor solder connections in the base pins of a particular tube type." The central computer section of the XD-1 was ready to ship to Lexington.[31]

On January 5, 1955, it left Poughkeepsie. By January 12, power was being applied to the reassembled machine in Lexington. On January 19, "all eleven of the available programs for use with only one [magnetic core] memory bank were successfully operated." The engineers were encouraged to begin to turn their attention more seriously to the details that would be involved in integrating the machine with the radar net of the Experimental SAGE Subsector. Several months later, the Experimental SAGE Subsector would begin the next series of intercept flight tests.[32]

Even before the XD-1 was moved to Lincoln Laboratory, Forrester had entered into discussions with IBM over installing the XD-1 at Lincoln and dividing between IBM and Lincoln the work that would follow. While delivery of the machine was in process, he finished revising a letter he had begun a month earlier. Copies of it went to a large number of officials in IBM, Lincoln Laboratory, the Air Force, and the Air Defense Engineering Services (ADES) Project Office (which was staffed by both Air Force and Western Electric personnel). Forrester understood that a change in the relationship between Lincoln and IBM must follow the building of the two prototype machines, and that this change would have to be set in a practical policy perspective.[33]

Forrester described the XD-1 as the "Combat Information Central of the SAGE System Experimental Subsector . . . comparable in scope, amount of equipment, problems, and time schedule with the first production subsectors at Maguire and Stewart." Accordingly, the XD-1 system should be handled so as to "coincide as nearly as possible with the way in which the installation of production systems" would be handled. This meant that IBM would have "full responsibility for manpower and management to install, debug, test, and modify the XD-1 within the framework imposed by the fact that others [would] simultaneously be installing equipment, testing computer programs, and training operators." Lincoln, Forrester informed his readers, would function as coordinator, responsible for scheduling week by week the activities that would involve two or more groups and responsible also for working out the "technical incompatibilities" which might emerge in the process. Lincoln would specify and direct those "system physical changes" that seemed "necessary to ensure operational air defense success." Western Electric, in order to train large numbers of supervisors and acquaint them with "various aspects of the system installation," would join the Bell Telephone Laboratories in providing a large group in Lexington to help install the Experimental SAGE Subsector. These two AT&T affiliates would assign engineers to prepare system installation and test procedures, and they would supply the men to "manage the installation of phone equipment, assist in writing operational computer programs and . . . observe and assist in other activities."[34]

IBM should anticipate maintaining a crew at Lexington, Forrester warned, even "beyond . . . actual physical installation," until the system met the "reliability objectives for air defense operation" and until the "necessary logical and equipment modifications have been made to meet the requirements for an effective air defense system." The reliability standards that he proposed included "30 consecutive days of operation . . . with a maximum of one hour daily of prescheduled preventive maintenance and a maximum of 30 minutes per 24-hour day of unplanned down time."

The periods allowed for installing first the prototype at Lexington and later the production model at Maguire Air Force Base were "essentially the same," Forrester noted. But the schedule of the former would be tighter. For one thing, the engineers would be adding parts of the prototype system as they became available. For another, their advance planning for the XD-1 could not be as complete as that which they would be able to apply to the production-model Q-7, and it would be unable to draw on the prior installation experience that would be available to the latter.[35]

Obviously IBM would gain valuable experience, because installing the prototype as though it were the production model would provide essential information in several directions. Forrester identified five distinct advantages that would accrue: Future managers of the Maguire, Stewart, Syracuse, and other installations would acquire needed training. Their first experience in installing would occur "while only one is in progress rather than several simultaneously, as will happen in the production program." Here would be their first opportunity "to coordinate tightly engineering, manufacturing, field service, and relaying of field difficulties to supporting engineering and factory groups." Another advantage would be "a first-hand opportunity to observe FSQ-7 deficiencies which require redesign to meet air defense requirements." Finally, there would be "a chance to anticipate the equipment and management problems which will otherwise appear for the first time at Maguire and Stewart to jeopardize the entire production program."[36]

Forrester recognized that his ideas went beyond IBM's anticipations and might require additional contractual arrangements. If so, these should be completed immediately in order to prevent any further delay in the progress of the production prototype. The program presented a "major challenge," he admitted, one that would "tax the IBM organization very heavily" and doubtless require a 24-hour, seven-day work week. But if it were "handled smoothly and on time," he said in closing, they could reasonably anticipate that the "triple coincidence of Maguire, Stewart, and Syracuse" would be "carried off quite successfully."[37]

Although work on the XD-1 at Lexington was ahead of schedule in some respects by the middle of February 1955, the slippage had not been stopped completely. The magnetic drums to be used with the in-out equipment were behind schedule. That caused the IBM and MIT engineers to begin to explore together the possibility of shipping the drums to Lincoln early in March, even before they were fully pre-tested. Once again a "ripple effect" emerged: the schedule for delivery of the drums for the other prototype, the XD-2 located at Project High, was even further behind, and IBM's suggestion that work on the XD-2 might be slowed down in response to that delay would, the MIT engineers realized, only jeopardize progress on the XD-1. The reason was simple: the XD-2 was expected to test out the data-input equipment that would later be used by the XD-1 for the intercept trials.

Since IBM was thinking not only of slowing down the building of the XD-2 but also of asking for a month's delay in providing the first drums for the XD-2, there were unhappy implications for the development schedule that had been set up for the production-model FSQ-7 machines. On top of that, the entire issue was complicated by the fact that "MIT has technical responsibility for the [Q-7] program and at the same time is attempting to enforce the time schedules." Before the end of February 1955, Forrester began to look into the possibility of having the Air Defense Engineering Systems office (operated by the Air Force and Western Electric) take over the effort "to check progress, enforce time schedules, etc. ADES might suggest [with more clout than Division 6 possessed] that a second source is needed for magnetic drums."[38]

The display consoles that were part of the output equipment of the computer posed similar slippage problems. When IBM in February also proposed sending the XD-2 display console system to another manufacturer (Hazeltine) for testing, MIT realized immediately that it was time to review with IBM the development plans for the XD-2 once more, because "such a move would seriously interfere with the training program contemplated for XD-2."[39]

Although Forrester's appeal to IBM's higher corporate managers had produced definite remedies, the underlying philosophies behind the two organizations' collaborative efforts on the engineering level had not changed radically. By the end of February 1955 another conflict had appeared, this one over "what kinds of things can be changed without allowing a delay in the time schedule." Forrester, Everett, and their Group Leaders were convinced this problem and its solution must be given explicit statement:

> Two steps for carrying out a change must be recognized. The first step is to establish the necessity and the technical detail of the change. The second step, which includes production schedule changes, is the one which causes difficulty. When these difficulties arise, MIT will state in concurring on a change that the production schedule is not yet determined, but that the change is required by some stated date.[40]

It had already become clear to the Division 6 Group Leaders early in January of 1955 that IBM had become reluctant to "discuss technical planning and time scheduling with MIT representatives."[41] Before the middle of the month it was evident that design agreement over various details of the production FSQ-7 machine was also slipping behind schedule. "Concurrence," which the Division 6 engineers had anticipated would be reached by early January, had been postponed until the middle of March. Here was another problem for the Joint IBM-MIT Coordinating Committee to take up at its January 12 meeting.[42]

Perfecting an engineering design is the story of encountering and overcoming a seemingly endless series of problems, and when one takes the measure of the pattern of problems that emerges it is easy to lose track of the progress being made. In the case of the Q-7 and its two prototypes, progress was occurring in spite of the slippage. By June of 1955, most of the drums system had been checked out, and other parts of the system had been installed and tested or were being installed. The central computer portion of the XD-1 system had been "operating steadily" for several months, "assisting in the testing of new parts, performing useful work for the Group 61 programmers and in obtaining performance data on its own various circuits." Performance problems that surfaced as more complicated programs were fed into the machine led to significant improvement of the XD-1's core memories.[43]

In July of 1955, Norm Taylor reported in a biweekly Lincoln-ADES meeting that performance of the XD-1 was "good in most areas." Some components were still behind schedule, but Taylor was optimistic that all but the display-console problems would be taken care of by October. The display-console schedule was lagging "because of late console deliveries and required changes in console layout specifications."[44] The display consoles were scheduled for delivery during July and August, and the electronic performance of the central computer and its associated equipment was to be checked out by October. The next 6 months would then be "used to check out the computer program for Air Defense operations."[45]

The Division 6 engineers were always looking ahead, often farther down the road than the IBM engineers. And they were aware that another enterprise of far more profound consequence than these to-be-expected hardware-development issues loomed ahead—an undertaking at least as large as designing and building the Q-7 and its twin prototypes: devising, testing, debugging, and perfecting the software that would enable the continental air defense system to carry out its mission.[46] This was to be carried out in two phases: first experimentally connecting a handful of radars to Whirlwind I in the Cape Cod program, then linking a more elaborate array to the XD-1 in the Experimental SAGE Subsector program.

21

Technical Memorandum 20: The Pattern for the Future System

It had been one thing to use one computer and one radar, as the Bedford tests had done in April of 1951, to direct an aircraft to intercept a moving target under carefully controlled experimental conditions. It was another to use that computer and several radars to control the movements of a number of airplanes under normal air traffic conditions in a broader expanse of the sky. This was to be the next step in gauging the practicality of a computer-controlled air defense system. Before such a step could be completed, institutional sanctions and facilities had to be in place, and a pattern for the expanded system, integrating experience and conception, had to be prepared. To that end, about 4 months after the first successful interception (April 20, 1951) cleared the way to proceed with the Cape Cod System, the final report of the Project Charles study group had been submitted, and Project Lincoln had been established to carry out the study group's recommendations.

Once Project Lincoln was formed, the continuing Bedford experiments became part of its responsibilities. George Valley and his group, operating as "Division 2, Aircraft Control and Warning," continued to conduct the tests in harness with Jay Forrester's Digital Computer Laboratory's air defense team, soon to be incorporated into Project Lincoln as "Division 6, Digital Computer." They proceeded with the design and installation of the model system called for next in the Project Charles report: the Cape Cod System. Continuing and building upon the crucial Bedford tests of 1951, which had followed the earlier air traffic control studies, they realized that using the digital computer to direct airplanes in flight was "as much a systems problem as a computational problem; in other words, the use of a computer requires new concepts of the whole system if maximum benefits are to be obtained."[1]

The Cape Cod System would serve not as a prototype of the projected air defense system but rather as a "proving ground" capable of being developed into a "complete model air defense system suitable for tactical evaluation" and flexible enough to "permit installation and testing of new components as they develop."[2]

As the physical system gradually expanded during the years that elapsed between the Bedford system of 1951 and installation of the first operational system in 1958, the designers found themselves subdividing their experimental design efforts conceptually into several experimental test sequences, each more elaborate than the previous as it incorporated more functions. By early 1953 they were calling the first phase of the experimental Cape Cod tests the "1953 Cape Cod System." During 1954, when alterations and improvements were made in order to prepare the equipment for a more ambitious series of tests, the "1954 Cape Cod System" emerged. At the time, for planning purposes, the engineers came to think of these two as experimental test systems that would culminate in the "Transition System" (as they called it for a while). In historical sequence the two Cape Cod test programs were followed by the Experimental SAGE Subsector, which emerged as the experimental prototype of a sector of the Air Force's continental air defense system. Subsequently, the Air Force, in an administrative action taken by the Air Defense Command, renamed ESS the Experimental SAGE *Sector.*[3]

As Air Force administrators became conversant with the tests and their goals, they coined the label "Semi-Automatic Ground Environment System." As the new air defense computer began to take shape, they designated it the "AN/FSQ-7." "SAGE" and "Q-7" became common terms; "SAGE" replaced "Transition System" as the name to use, and "FSQ-7" or the easier-to-say "Q-7" replaced "Whirlwind II" (the name that Forrester and his MIT engineers had coined, mindful that the new machine's tentative design was descended from the machine that Project Whirlwind had developed in the Barta Building).[4]

It made good engineering sense for the physical system to grow gradually. As it became larger and more complex during 1952 and 1953, the attention of those in charge at the engineering level shifted from the performance of hardware components to the performance of subsystems and then to that of larger systems. Performance parameters, in turn, demanded ever more sophisticated software programs to operate the hardware systems.

Early in 1952, George Valley (as head of Division 2) and Jay Forrester (as head of Division 6) were apprised by their Group Leaders of the status of aircraft-guidance equipment. A North American B-25 medium bomber was being outfitted with an eight-digit very-high-frequency data link by the

Air Force Cambridge Research Center. The link would present to the pilot compass-heading data sent out from Whirlwind I. At the same time, a Douglas B-26 light bomber belonging to MIT's Instrumentation Laboratory would be provided with a compass-heading data link connected directly to the airplane's electronic automatic pilot (also being provided by the Instrumentation Laboratory). A third airplane, a Lockheed F-94 "Starfire" jet interceptor, was being readied. It would be equipped with an APG-33 radar unit, an Instrumentation Laboratory electronic autopilot then under development, and a Cambridge Research Laboratories eight-digit, compass-heading data link. The first two aircraft should be able to participate in preliminary Cape Cod tests in February. The third would not be ready before summer.[5] Such were the plans at the beginning of 1952.

In February of 1953, Group Leader Bob Wieser was receiving such technical reports, not generating them, and they reflected more elaborate engineering planning constructed according to an increasingly sophisticated system approach. One such report described the "overall organization of the Cape Cod Information and Direction Center" (IDC) that would be employed in the "1953 Cape Cod System" during the autumn of 1953 if all went well. Two levels of operation would be under test, according to the plan that David Israel of Division 6 described. One would call for the IDC to focus on assembling and processing data from a group of radars surveying the skies of a subsector. Tracking would be initiated and monitored by the Whirlwind machine, aircraft would be identified as worth keeping track of or ignoring, and weapon-direction orders would be issued to initiate interception.[6] The second level of system operation would involve a model Sector Command Post located at the IDC site and making use of the same computing facilities. The SCP would identify and evaluate threatening "enemy" flights, decide which defense weapons to commit, and order them into action. In the 1953 version the operations of the two levels would not yet be completely integrated in the way described in the comprehensive engineering plan, TM-20, issued in January of 1953 and describing the proposed "Transition System." Instead, the tests would require only such major parts of a Sector Command Post as would be appropriate for monitoring experimentally a smaller chunk of the sky than a normal Air Defense Sector.

Under these limitations they could expect to demonstrate the performance of the Sector Command Post and the Information and Direction Center, each in fully operational condition. They would be able to handle the hypothetical situation in which the IDC would be fully operational, while an SCP *at a different location* would be presumed to be "knocked out" of operation. This situation would require the IDC personnel to take on the duties of the knocked-out SCP. The entire scenario would be directed from the Barta Building under both "All Clear" and "Battle in Progress" conditions.[7]

During the intervening year, 1952, numerous tests had been carried out with the Ground Observer Corps of aircraft spotters to see how well this European, World War II approach would work in harness with the Cape Cod System. The results were dismal. The GOC could not compete with or provide significant supplementary information to the radars-and-computer approach. The state of the art had passed the GOC by.[8]

While the GOC was passing into technological desuetude and oblivion as far as the Lincoln engineers were concerned, more attractive and sophisticated engineering challenges beckoned, such as that being posed by "slowed-down video." SDV, when perfected, would put radar data in a form that could be sent over a telephone line to the computer, but before entering the computer the SDV signal would have to be translated into the binary numbers that the computer required. These numbers would be stored temporarily, the engineers decided, in a "Buffer Drum" memory obtainable from Engineering Research Associates. From there the numbers could pass into the computer at times of the computer operations program's choosing.

Translating SDV into binary numbers would require suitable conversion equipment, which was being developed by Division 2 and Division 6 engineers working together. The plan was that, while Division 2 would be designing and building "the SDV discriminator or receiver," Division 6 would be developing the counters and the control hardware to go in the computer, as well as the programs to finish the conversion.[9]

Such engineering activities, whether on the component level or the more elaborate systems level, were being undertaken not simply at random and as the need arose but pursuant to the overall plan indicated in the Project Charles report. Thus, late in 1952, the same men who were at the time engaged in designing and constructing the Cape Cod System were hard at work drawing up their next major plan, titled A Proposal for Air Defense System Evolution. The provisional name they gave it was "the Transition Phase."

The men were George Valley, H. W. Boehmer, and J. V. Harrington of Division 2 and Jay Forrester, Bob Everett, and Bob Wieser of Division 6. On January 2, 1953 they formally issued their plan under the title A Proposal for Air Defense System Evolution: The Transition Phase

(Technical Memorandum 20). The proposal was accepted by the Air Force in April of 1953 as the system "most nearly meeting the desired improvements in the ground environment of the Air Defense Command." In the summer of 1954 it received its permanent name: Semi-Automatic Ground Environment (SAGE) system.[10]

The circumstances in 1952 that generated Technical Memorandum 20 are worth reviewing, because they reflect how technical development—regarded as important by high-level policy makers in the American military establishment during the 1950s—was being carried on. These particular circumstances were at the same time political, technical, and complicated, for the TM-20 proposal that Lincoln Laboratory submitted was a response to several needs that administrators perceived as increasingly urgent during the latter half of 1952.

In part, TM-20 was a response to a challenge implicit in another Air Force-sponsored air defense program underway at the Willow Run Research Center of the University of Michigan. In part, TM-20 demonstrated the progress Lincoln Laboratory was making in developing the ADSEC-MIT proposal, now 2½ years old. In part, it was a compilation of the testing experience MIT engineers had gained to date, and it embodied in a detailed "blueprint" the first centralized military command-and-control system that would incorporate the electronic digital computer. It was also a response to sharp and uneasy Air Force concern, for before the end of August 1952 Valley had been reporting to the Lincoln Steering Committee that the Air Defense Command (ADC) was "very interested in improvements to the air defense system beyond those provided by Quick Fix but operating in 1956 . . . before the estimated completion dates for the Lincoln digital system."

Valley's Division 2 had thereupon responded by setting to work on a proposal which would give dates and cost estimates "for an Interim Fix made up of gap-fillers with SDV data transmission and conversion to rectangular coordinates for display." By December of 1952, ADC's sister organization, the Air Research and Development Command, was asking for "a description of the system proposed by Lincoln for the period between Quick Fix and the future or Charles System." All parties saw that such a system, if operational by 1956, would fill the gaps between the "quick fix" modifications of the present, the inadequate existing system, and the future centralized system which ADSEC and Project Charles had envisaged.[11]

It was in this rather elaborate way, exhibiting the consequences of the "Midas touch" of institutional administrators asserting their efforts to control events, that the informally discussed "interim fix" became the "Transition Air Defense System." That autumn, it took the joint efforts of Valley and Forrester, as heads of Division 2 and Division 6, to produce the detailed master plan for "a system containing digital computer data-processing but utilizing existing large radars and gap-fillers to be installed." From their point of view, such an advance would be a quite normal step in the evolution of the future system, one which would employ improved radars tied into the same computing centers. The first draft of the plan which they prepared was discussed with the Air Research and Development Command in December. Then it was revised and issued as TM-20 immediately after the new year began.[12]

The knowledge and the experience that both of these Lincoln Divisions were gaining from the Bedford tests, augmented by their joint planning and construction of the Cape Cod System, contributed directly to the formulation of the TM-20 proposal for the evolutionary system. It became known at first as the Lincoln Transition System. The prototype of this system that they planned to build (which would later become known as ESS) would incorporate the prototype of the production-model command-and-control computer that Division 6 was designing specifically for air defense.

Thus, in a sequence that developed quite naturally from the knowledge the engineers were acquiring from experimental tests, the 1953 Cape Cod System became an interim experimental model, demonstrating, as Jay Forrester explained near the end of 1953, "in one integrated whole the basic principle of the Transition System."[13]

The Lincoln designers called their new Transition System "the first major upgrading" of the nation's air defense system, an upgrading achieved by installing the digital computing element part of the proposed future system.[14] Everett had noted in the December minutes of the Lincoln Steering Committee that it would be a "normal" step forward, and the January second draft of TM-20 reiterated this point, adding that it was a "necessary initial step" toward the Future System recommended by Project Charles.

The designers presented their proposal as a provisional design, one that they knew was subject to alteration and modification whenever required by the needs of the Air Defense Command or by advances in the state of the art gained from experience with experimental systems. Obviously, guidelines established by national fiscal policy could modify the enterprise at any time.

At the start the system would offer "maximum technically feasible capacity for aircraft tracking, control, and defense" by its use of existing radars supplemented by gap-filler sets, all connected to "digital computing equipment for processing data, presenting displays, and calculating weapon orders." Most attractively, the proposed Transition System would provide, stated TM-20, "a major system improvement in the shortest possible time," and at the same time assure the Air Defense Command "most of the functions" it held essential to an adequate air defense in the immediate future.[15]

An important feature of this transition system would be its flexibility, since the greater part of its operations would be controlled by programming instructions inserted into the computer. By changing instructions, the designers explained, "the organizational structure of the system can be changed without any change in the electronic structure." In this way advanced weapons and radars could be accommodated, the capacity of the system could be increased, its control activity could be redistributed among the various tasks to be accomplished, and new attack and defense procedures could be introduced at will.[16]

Not only was the proposed Transition System essential, in their view, to the Future System called for in the Charles Report, and not only was it sufficiently flexible to allow controlled expansion and improvement; in addition, there were other important and attractive characteristics. The system would, they were convinced, "meet most of the ADC needs, including time schedules." It would allow any degree of centralization or decentralization that might be desired. It would strike a balance between men and machines while reducing operator errors and ensuring "a high level of operator alertness." And it would "provide up-to-date . . . track and weapons information on display scopes at Defense Force and Air Defense Command headquarters."[17]

The program that TM-20 proposed was anchored to a firm experiential and philosophical base, the designers felt. "A background of systems planning . . . proven by system construction and successful operation" was mandatory, they insisted, if a major change in the nation's air defenses were to be accomplished. To demonstrate their credentials, they reached back to the origins of Project Whirlwind and to Forrester's and Everett's application of the digital computer to antisubmarine warfare, expressed in 1947 in Servomechanisms Laboratory Report L-2.[18] They reached back also to the data-link work of the AFCRC, begun by Jack Harrington and his associates in the Radar Systems Laboratory in the years immediately after the end of World War II; that work, when combined with Whirlwind I, resulted in the "Bedford tests hookup," which in turn was the basis for the Cape Cod System being developed in 1953. The proposed Transition System was thus undergirded by the experience gained from these earlier efforts, and it was further supported by IBM's background in computer development and production.[19]

The details, the Lincoln designers acknowledged, had not been "fully coordinated with Air Defense Command requirements." Before this could be done, they explained, a number of the more important technical problems had to be resolved and "convincing experimental evidence . . . presented to show how the system would perform." The Transition System's flexibility would permit adapting it to both ADC planning and revised requirements in the future. "The optimum time to crystallize the system into the form desired by the Air Force," they declared, would be during the coming year, 1953.[20]

In the spring of 1953, when these persuasive arguments were coupled with the technical achievements of the Bedford tests (which had continued after the April 1951 breakthrough) and with the emerging Cape Cod experimental system and its inherent promise, Air Force leaders became convinced of the wisdom of adopting the "Transition System" approach that Lincoln advocated. Nor did the cumulative effect of that approach stop with the proposed Transition System, for in design and operating essentials the Transition System would determine and would indeed become, when modified to accord with Air Defense Command needs and experimental data, the actual continental air defense system.

That the design of the Transition System (soon to be called SAGE) would become embodied in the Air Force's continental air defense system became evident 2 years later in the details of the Air Defense Command's Operational Plan of 1955, which described SAGE as "that portion of the air defense system that provides the means for the semiautomatic processing of data and weapon control." These functions were already explicit in the Transition System. For example, in both schemes—the contemplated Transition System and the later, actual SAGE air defense system—the continent was divided geographically into interconnected air defense units, each of which possessed a sensor network tied to data-analysis-and-command centers.

The system, whether called "interim," "Transition," or "SAGE," ultimately became a communication and combat information-and-analysis military command unit, unique in size and scope, in which data were forwarded from sensors to digital computers for processing, not only to provide analysis of the air situation but also to dispatch commands for counteraction and interdiction.[21]

Several decades later, when Wieser could see the whole mosaic in place, he observed that the intercept tests themselves had progressed naturally through three phases, from the original Bedford tests through the Cape Cod System, and that each of the phases made a distinct contribution to the development of SAGE: the Bedford tests "provided elementary proof-of-principle evidence to support continued development," the 1953 Cape Cod System "demonstrated feasibility of SAGE," and the 1954 Cape Cod System provided essential "additional design data for SAGE" that the engineers would then build into the system that would defend the continent.[22]

If the Bedford tests marked the beginning of the beginning, supporting (as Wieser noted) "continued development," the issuance of TM-20 marked the end of that beginning and the commencement of a program to actualize the system envisaged by ADSEC and Project Charles. Not a plausible description of the possible but a rational description of the probable, reflecting experiential confidence and conviction, TM-20 proved to be a timely technical statement with immediate political and administrative consequences, and it influenced the determination of policy at both the Air Force level and the presidential level.

22

The Air Force Decides

That technical accomplishments can influence if not dictate the determination of policy was clearly demonstrated when, after the issuance of TM-20, Air Force Headquarters decided to concentrate solely on the program that Lincoln Laboratory advocated. As a technical engineering program and as a policy-making administrative action, TM-20's detailed description of a centralized transitional continental air defense system offered persuasive technical considerations that influenced the Air Force in reaching its decision.

The decision was not reached without controversy within the Air Force or concern within Lincoln Laboratory. As Lincoln Lab's Bob Wieser recalled in October of 1987 when speaking before a Whirlwind/SAGE symposium at Boston's Computer Museum, "the decision to build the SAGE System did not fall out of building and demonstrating the Cape Cod System. Competing systems existed and there was a lot of missionary work to do to get our ideas accepted."[1]

The real challenge came from a program underway at the University of Michigan's Willow Run Research Center. The university had been engaged in a series of air defense R&D programs since 1946. The first of the series was Project Wizard, a defense against ballistic missiles, later expanded to include sonic and supersonic bombers. In June of 1950, under a subcontract with the Boeing Corporation, research was started on a "ground based guidance system" for the BOMARC, a ground-to-air missile then under development for the Air Force by Boeing.

In 1951, the University of Michigan entered into a prime contract with the Air Force for Project Miro, a program "intended to improve the U.S. Air Force ground reporting system through research and development, especially in conjunction with the BOMARC missile being developed at the same time." Three components were included in the proposed ground reporting system: a data-gathering system composed of ground radars and a ground observer corps, a data-processing system that would convert radar data to digital pulses, and a weapon-assignment center that would use the processed data to intercept attacking forces. From the Miro Project came ADIS (Air Defense Integrated System for Surveillance and Weapon Control), a program to design and develop an effective air defense system, but on a less ambitious scale than the Lincoln system.[2]

Functionally and geographically, ADIS was similar to the Lincoln Transition System, but it proposed "decentralized data processing and weapon control at the Air Defense Direction Center (ADDC) with centralized threat evaluation and weapon assignment at the Air Defense Control Center (ADCC)." The Lincoln system, on the other hand, would combine all these operations in the Air Defense Control Center.

The Michigan system, although developed independently, was similar to the British Comprehensive Display System in its philosophy and in the details of its technique, but it was not so far advanced.[3]

Like the British system, ADIS was projected to use only a single radar or closely associated group of radars. Consequently, it could not provide a control network covering a wide geographical area. The engineers at Willow Run were, however, planning modifications that would permit the transfer of data "from place to place electronically." General Benjamin W. Chidlaw, Commander of the Air Defense Command, favored the Willow Run system for the "here and now," and his Command was planning a test of ADIS by the end of 1952 in the hope that, if it was successful, their approach could secure the support of Air Force Headquarters.[4]

The two programs appeared to be competitive alternatives moving in the same direction. On the one hand, Lincoln was bringing IBM into its program and preparing its TM-20 proposal for the Transition System, while the Air Research and Development Command (ARDC) was assigning to its Cambridge Research Center responsibility for "engineering functions . . . as soon as development-production implementation" of the proposed MIT interim system occurred.

On the other hand, the Air Materiel Command (AMC) was proposing to procure for ADIS "limited production quantities of all components of the system through a single prime contractor." Responsibility for the "engineering functions" was assigned to ARDC's Rome (New York) Air Development Center.[5]

The fact that Rome Air Development Center was attached to ARDC rather than to AMC simply called the attention of knowledgeable observers to the reluctant and competitive transfer of administrative authority and political power that was then taking place between AMC and its younger sister Command, ARDC, with regard to supervising Air Force R&D. So the competition was not restricted merely to two different technical approaches to air defense or to two educational and research institutions. Included, also, were local, state, intra-service, and even possibly industrial rivalries, all of which would be likely to have troublesome political and economic ramifications.

The leaders of the Lincoln program had been aware for several months of the challenge posed by the Michigan program and its supporters within the Air Force. A meeting in New York with IBM in mid October of 1952 had opened with Jay Forrester's observation that there had been considerable discussion concerning early production of ADIS as a stopgap air defense system. Forrester expressed doubt that the Air Force would take such an unwise step "now or in the near future," since the system was still "in the paper planning stage." He acknowledged, however, that if the Michigan program had the potential to be effective it should be encouraged.[6] Many years later, Forrester recalled that at the time he believed it did not have such a potential, he thought the estimates of time and money needed to bring it into being were much too low.[7]

Concerned lest IBM become involved in both the Michigan program and the Lincoln program simultaneously and spread its technical resources too thin, Forrester also expressed his concern about reports that were circulating regarding the possible use of the IBM 701 computer in the Michigan system. James W. Birkenstock, executive assistant to Tom Watson Jr., sought to reassure Forrester that, although IBM could not "for policy reasons . . . refuse a 701 machine to Michigan," it would make clear to both the University of Michigan and the Air Force that it could "only provide limited manpower for adapting the computer to ADIS." He believed that a firm statement to this effect would "adequately protect IBM's previous decision to put their major effort on the Lincoln Air Defense Project." Forrester was skeptical, but he recognized that a request by the University of Michigan for a 701 could not be refused.[8]

Perceived by Lincoln Laboratory as a threat, the University of Michigan's program undoubtedly stimulated and influenced the preparation of TM-20. The MIT proposal was carefully framed to describe in detail the status of the Lincoln program and to encourage military policy makers to make an accurate comparison with the Rome-Michigan program. The proposal should, Lincoln Lab's leaders believed, "correct some mistaken ideas of the capabilities and delivery schedule for the Lincoln System" and counter those within the Air Force who, unfavorably comparing the Lincoln program to the Michigan program, were urging that the Rome-Michigan system proceed to limited production. Fortunately for MIT, no decision would be handed down by Washington until Lincoln Lab's proposal was received and considered.

In the meantime, to strengthen Lincoln Lab's case, the Group Leaders of Division 6 were instructed in November of 1952 to place greater emphasis on the Cape Cod intercept program in order to accomplish a demonstration in September of 1953. This demonstration was seen as an event of major importance, since it would provide information for important basic decisions. The Group Leaders were cautioned at the same time to keep in mind that Lincoln Laboratory and the University of Michigan had "a common objective—an adequate air defense system."[9]

Seven days after TM-20 was issued in January of 1953, MIT took precautionary action to improve its competitive position and resolve the threat. In a letter to Air Force Secretary Thomas Finletter, MIT president James Killian, responding to the rumors that ADIS was being given serious consideration and might even have been accepted by the Air Force, requested a technical evaluation of the Lincoln program. Killian was disturbed, he said, by the continuation of competitive and possibly redundant development programs and, as he recalled later, by his conviction that the University of Michigan was making a serious effort to compete. The evaluation should be initiated by the Secretary of Defense, Killian wrote, since the program was tri-service under the terms of MIT's agreement with the Department of Defense. An overview of Lincoln and of SAGE should be undertaken, and "particular attention" should be paid to "the relationship of its program to Air Defense Systems based on centralized digital computation." The review, Killian suggested, might be conducted by the committee recently appointed under the leadership of Mervin J. Kelly of the Bell Telephone Laboratories to investigate the state of the nation's air defenses. Repeating the offer he had made a year earlier, Killian expressed MIT's willingness to withdraw on the grounds that the air defense project was not really compatible with MIT's educational mission.[10]

Killian's letter precipitated an immediate reply from Secretary Finletter, a reply which was hand-carried to Cambridge by an Air Force representative.[11] There was apparently some truth to the rumors that the Pentagon feared MIT might withdraw if the University of Michigan program were continued.[12] Finletter was agreeable to periodic reviews (the first to take place during the late summer of 1953), but not by the Kelly committee, for that body's charge had been to deal with "broad policies" rather than details. He was willing, however, to have the review conducted by a later group chaired by the president of Bell Labs. Kelly also was agreeable, expressing his willingness to lead "a technical review group convened by the Air Force with the participation of the Army and Navy through the Senior Advisory Committee to Project Lincoln." Sensitive to the issues involved, Kelly further suggested that the Air Research and Development Command "convene a meeting of senior officials from Michigan and MIT and present to them preferably in writing, Air Force policies and decisions affecting their present development programs."[13]

In response to Killian's reiterated offer to withdraw, Finletter repeated his argument of the year before: MIT had to continue its commitment, "unequivocally" assured that the Air Force held no doubts concerning the necessity of MIT's participation. Recognizing the true reasons for Killian's concern, Finletter pointed out that there were other approaches to air defense which had to be explored and other programs which might produce usable equipment before Lincoln did. Such efforts had to be supported. But he emphasized they would not be allowed to "detract from the Lincoln program."[14]

Provost Jay Stratton of MIT accepted the hand-carried reply on behalf of Killian, his reaction indicating the accuracy of Finletter's interpretation of Killian's true intent. Stratton found Finletter's words "generally encouraging" but not definite or specific enough concerning the Air Force's continued support for air defense programs, "financially or otherwise." He told Lieutenant Colonel Peter J. Schenk of the Office of Special Assistant to the Chief of Staff that the reasons for delaying the review were understandable, but that the executive officers of MIT and Lincoln Laboratory feared that before any review could be made the Air Force might make decisions favoring large-scale production of competing systems—decisions that would be irreversible even if future tests proved the Lincoln system to be superior.[15]

Two weeks later a more detailed explanation from the Commander of the Air Research and Development Command arrived on the desks of both James Killian and Harlan Hatcher, president of the University of Michigan. Lieutenant General E. E. Partridge wrote that the urgency of the need for a "greatly improved Air Defense Environment," to be available by 1955 or as soon after as possible, prevented the Air Force from committing itself to "a large scale production program on either system to the complete exclusion of the other." Both, therefore, would continue to receive support until enough "factual information" had become available to enable the Air Force to decide on the best program. That point would be reached within the following several months, he hoped, but definitely, because of the budget cycle, by the ninth month. The system that after proper and adequate testing first met the requirements would be "recommended for quantity production for installation and use by the Air Defense Command." Conceivably, Partridge warned, "decisions may have to be made which will result in the installation of both systems in a compatible arrangement within the ADC." It was essential, he emphasized, that the Lincoln system as well as the Michigan system be able to work with the unmanned BOMARC and with the F-102 and such other manned interceptors as would still be operational when the new air defense system became operational. This would be a primary Lincoln responsibility, he implied, accomplishment of which would require total cooperation and coordination among all the organizations involved in the air defense effort.[16]

Partridge had put his finger on a separate technical problem that the engineers were still all too busy to give proper attention to: that of integrating new and old weapons systems into the air defense grid. It was a problem that many continued to ignore, but it would not go away.

Early the following May, well before the "factual evidence" ARDC's commander had called for was in, the "dual approach" proclaimed in January was rescinded by the Air Force's announcement that the Lincoln Transition System had been selected to implement a "unilateral approach to the solution of the R&D problems in the field of Air Defense Electronic Environments."[17] The groundwork for this decision had been laid in the middle of March when Partridge and his Vice Commander, Major General Donald L. Putt, had visited the University of Michigan to discuss with President Hatcher and his associates the future of the Michigan air defense program. The two groups discussed the draft of Hatcher's reply to the Partridge letter of May 6 in which Hatcher questioned continuation of the ADIS program, since it and the Lincoln Transition System were "so closely allied and had components . . . so similar." University of Michigan administrators, forewarned that the Lincoln System was preferred, were prepared, unless treated and funded equally, to withdraw gracefully from the program.[18] They indicated, however, a willingness to continue to "work on any Air Force Projects," including "development of any items in work" that might contribute to "solution of the Problem by Lincoln." The contents approved, it was agreed the letter should go forward to ARDC headquarters in Baltimore.[19]

The attitude of the University of Michigan's administration obviously made easier the decision, reached on April 10 in discussions between Air Force Headquarters in Washington and ARDC Headquarters in

Baltimore, to concentrate solely on the Lincoln system. The decision was informally brought to the attention of the Lincoln Steering Committee some nine days before the formal announcement of May 6 went out.[20] The reasons for a unilateral approach were compelling. Not only were the two systems becoming more and more similar, but the cost of supporting two laboratories of the magnitude of Lincoln Laboratory was more than the Air Force felt it could afford. Revised schedules had indicated that sequential employment of the two systems was impractical, for both were scheduled to become operational at approximately the same time. Furthermore, the Air Force was encountering difficulty in interesting a suitable contractor to undertake production of the ADIS system. Lastly, the argument was advanced that a single approach was essential to "prevent . . . confusion of design parameters for weapons systems."[21]

There was no major challenge to the selection of the Lincoln Transition System, even from those who favored a less centralized system. The real proof lay, of course, in building one of each, but that clearly was well beyond available financial resources even if presidential approval were granted. The decision was obviously not based solely on technical grounds, and it could not have been unexpected. One could reasonably infer that the January decision to support both programs was reached primarily to allow time in which to discuss the matter with Michigan authorities and ease the university's withdrawal. The competition had pitted an untried potential giant (the University of Michigan) against a warborn giant with the vast experience of the Radiation Laboratory, the Servomechanisms Laboratory, and MIT's other wartime contributions; the odds favored the latter. The Air Force also presumably recognized that, in seeking funds for Lincoln Laboratory, it would not experience the difficulties it would encounter in seeking funds for the University of Michigan.[22]

MIT may have possessed greater influence and the Northeast greater political "clout," but Lincoln Laboratory also had the decisive technical advantage. Once Air Force Headquarters accepted the concept of a centralized air defense system, no other choice was equally supportable. As MIT's TM-20 technical analysis put it, some 2 years' "field experience in digital-computer analysis of radar data and aircraft control" had disclosed many of the problems to be encountered, but also many possible solutions. Basic operations in air defense had "already been individually performed by experimental pieces of the Transition System," and plans for the Cape Cod System, which would be "an integrated model," appeared to be well advanced indeed.[23] The knowledge accumulated during those

2 years, together with the contributions of the Air Defense System Engineering Committee, Project Charles, and Project Whirlwind, combined to undergird the Lincoln proposal for the transitional phase. Furthermore, a step toward production had already been taken when IBM had been brought into the program to work alongside Division 6 in "the preliminary planning and logical design of control-center equipment to perform as outlined in Technical Memorandum No. 20."[24]

The focus of planning efforts had already begun to move from an air defense computer to an air defense system, and with the Air Research and Development Command's decision in January of 1953 to give "early support of production phases" to the proposed system the air defense effort had begun the transition from its research-and-development phase to a development-and-production phase. Consequently, in the autumn of 1953, when the National Security Council (with presidential approval) endorsed the development and installation of a technologically advanced air defense system capable of meeting future threats, Lincoln Laboratory, the Air Force Cambridge Research Center, and IBM were already well embarked on the development and production of such a prototype centralized system.

23

National Policy Decisions

The actions taken by the Air Force in adopting the system outlined in TM-20 clearly demonstrated that technical accomplishments could influence if not dictate the determination of policy. Proving equally effective at the national level, Lincoln Lab's accomplishments as evidenced in TM-20 were a persuasive consideration in the conclusion the National Security Council reached in the autumn of 1953: that a continental air defense system was feasible and warranted support from the highest level of government. Endorsement at the presidential level gave a command impetus to the program that it had not possessed before, and for some in the Air Force this marked the real beginning of SAGE as a military program. It also offered a solution to the controversial issue of massive retaliation versus continental defense, which had carried over from Harry S Truman's administration to Dwight D. Eisenhower's.

A brief review of this issue indicates how fundamental and how slow was the shift in national policy to which ADSEC and Lincoln Lab's program contributed. Until alternatives could be weighed and complicated options could be sifted, a stable, measured policy consensus identifying the best course of military action was not attainable. With the end of World War II, the Truman administration had acted promptly to return the United States to a peacetime level of activity. The armed forces were cut back to a level commensurate with the need to maintain occupation forces in defeated enemy territories and to defend the homeland under conditions of peace—all in a climate of relative fiscal austerity.

Planning an appropriate peacetime military establishment was complicated externally by the onset of the Cold War and internally by a resurgence of the struggle within the War Department to secure an independent identity, role, and mission for the Army Air Forces. In 1947 Congress authorized the reorganization of the military establishment under the National Security Act, and the Air Force emerged as a separate and independent member of the new Department of Defense.

The reorganization of the national military establishment simply added a third party to the chronic rivalry between the Army and the Navy. Within the newly established Department of Defense, controversy over areas of responsibility and authority soon came to the fore, and defending the nation against air attack became an especially sensitive area of policy concern for all three services, particularly after the Soviet Union's first atomic test.

In March of 1946, before the Department of Defense was established, a major structural reorganization of the Army Air Forces occurred. Three major operational Commands were created: the Strategic Air Command, the Air Defense Command, and the Tactical Air Command. This was but one of a series of reorganizations during which SAC, the nation's primary offensive force, remained the constant, while others phased in and out under different designations and organizational affiliations. Whether AAF or USAF, the Strategic Air Command continued to be the dominant and preeminent arm. Other Commands labored under lower priorities, victims of fiscal limitations and of a prevailing confidence that offensive strikes constituted the best defense. The policy was clearly stated in a military report of February 1946:

> The best defense appears to be to convince the entire population of enemy countries that this country is prepared to retaliate immediately on any aggressor and will answer any unprovoked attack by wholesale devastation produced by atomic bombs, biological agents, and lethal gases of great intensity.[1]

This doctrine of "massive retaliation" was enunciated at a time when the United States possessed both an intercontinental strike capability and a monopoly on the atomic bomb.

The emphasis on SAC did not totally eliminate attention to the needs of the more defense-oriented arms of the Air Force. Concern over the lack of adequate and technologically advanced aircraft warning and control systems persisted, and proposals to secure this end were always present. Major Peter Schenk of the Air Communications Office, after conversations with John Marchetti of the Cambridge Field Station (forerunner of the Air Force Cambridge Research Center), proposed in 1947 that an early-warning radar fence be constructed.[2] General Carl Spaatz, Air Force Chief of Staff, implementing an earlier pronouncement by Secretary of Defense James Forrestal, proposed SUPREMACY, an aircraft control and warning system for which Congress would be asked to appropriate some $388 million. The plan virtually died aborning, however, for neither the Bureau of the Budget nor Congress would approve such an expenditure in 1948, an election year. Spaatz's proposal faded from view, replaced by a simpler and less expensive warning and control system offered by Major General Gordon Saville, commander of the Air Defense Division, a new office created on July 1, 1948, within the Directorate of Plans and Operations at Air Force Headquarters.[3]

The Air Force's efforts to construct an effective air warning and control system were attempted within the limits imposed by the fiscal restraints of the Truman administration. There was no feeling of immediate crisis, but the efforts were pursued in a climate of some urgency due to the threat of Soviet expansion. During the crisis in March of 1948 that led to the Soviet blockade of Berlin and the Allied airlift that followed, the American Military Governor in Berlin, General Lucius Clay, warned Washington that "we have or are reaching a crisis in our relations."[4] The American people, though concerned, continued to feel secure behind their atomic shield.

The threat of armed conflict perceived in the hostilities of 1948 took on more serious and ominous overtones the following year, when the Soviets detonated an atomic device 4 or 5 years sooner than had been anticipated. This action converted urgency into crisis, stimulated greater popular fears (as evidenced by the subsequent civil defense activities) and evoked a sharp reaction from American military and political leaders. General Omar Bradley, Chairman of the Joint Chiefs of Staff, addressing the American Forestry Association in 1949, emphasized the necessity for completing a "radar interception net around the western hemisphere."[5] Congress appropriated funds for a radar program.

The following year, the fears that Soviet activities had aroused were intensified by the outbreak of the Korean War. The invasion of South Korea in June of 1950 by Communist North Korean forces, especially when viewed alongside the Communists' seizure of power in China in 1948 and the USSR's moves in Europe, convinced many that an aggressive Communist conspiracy directed from Moscow indeed confronted the Western democracies. War had become a clear and present danger.

Understandably, concern over the state of the nation's air defense increased markedly. Representative Carl Vinson of Georgia, Chairman of the House Armed Services Committee, immediately took action to monitor the air defense programs underway. In a *Newsweek* interview, retired

Air Force Chief of Staff Carl Spaatz argued in favor of a radar network, interceptor aircraft, and ground-to-air missiles.[6]

A Sword of Damocles seemed suspended over the American homeland, shaping the emotional environment within which the Air Force had to formulate a response. The initial action of Air Force Chief of Staff Hoyt Vandenberg and the Air Force's Scientific Advisory Board, on the technical level, to appoint the Air Defense System Engineering Committee (ADSEC), was followed by the mounting of other technical investigations. Both the Air Force and the Department of Defense began to seek immediate corrective measures and long-range solutions.[7]

The reaction was not limited to government-sponsored technical investigations. Even as such investigations were in process, the debate resurged and spilled over from the governmental into the public arena, where influential journalists and members of the scientific community joined in the argument over which would provide maximum security against a surprise air attack: a defensive "Maginot Line" or massive retaliation.

Writing in 1961, Harvard professor of government Samuel P. Huntington suggested that the issue of an appropriate defense policy had become caught up in the controversy over the development of the hydrogen bomb and the basic questions of military strategy which that issue reflected. Lincoln Laboratory, he noted, because it was engaged in the development of an air defense system, had been "viewed with suspicion and even outright opposition by those elements in the Air Force associated with the Strategic Air Command and primarily concerned with the development of the forces for strategic deterrence."[8] According to Huntington, the opinion of those in the Air Force who were opposed to emphasizing air defense or even admitting it to equality with strategic retaliatory policy was expressed by the Air Force Vice Chief of Staff in mid October of 1953 when, as the controversy was nearing resolution, he warned "that the 'safest and surest defense was a strong retaliatory force and to neglect this force' because of preoccupation with last-resort defenses would be suicidal."[9]

Proponents of massive retaliation were convinced that the only real defense of the nation against attack was the ability to strike back immediately and decisively with atomic weapons, bringing the enemy to his knees by the total destruction of his resources. The weapon to accomplish this was SAC, which, under the command of General Curtis LeMay, was emerging as "the primary organizational embodiment of strategic deterrence."[10]

Any program that threatened the achievement of what was seen as SAC's full potential was regarded by many SAC supporters as detrimental to national security. Even those not opposed to SAC and its defense role but willing to consider other applications of atomic weapons and other means of defending the nation were regarded as not only dangerous to the nation's security but possibly even disloyal. One such dissident, J. Robert Oppenheimer, recognized as a scientific authority because of his role in the development of the atomic bomb, came to be regarded as the epitome of the opposition. His reservations on the development of the hydrogen bomb, compounded by his support for the development of an effective air defense system, made him the bête noire of "big bomber" advocates.

The intense dislike for Oppenheimer and what he seemed to represent was clearly disclosed during the course of his loyalty hearings before an Atomic Energy Commission tribunal. In addition to challenging the development of the hydrogen bomb, he had been involved in two other studies which SAC supporters saw as inimical to the Air Force's position as the nation's dominant nuclear force, thereby threatening its control over the nation's nuclear stockpile. One of those studies—Project Vista, conducted at the California Institute of Technology in 1951—had considered the possibilities of tactical application of nuclear weapons; the other, conducted at but not sponsored by Lincoln Laboratory in 1952, had considered the appropriateness of ongoing air defense R&D. Both studies had been seen as "complementary aspects of the same policy . . . designed to place tight curbs on strategic air power, if not to eliminate it altogether."[11] For those who supported the development of the hydrogen weapon, Huntington concluded in his analysis of the debate, "this was a case of once-successful innovators opposing subsequent innovations which might lay competing claims on resources."[12]

The Soviet Union's atomic test compelled a reappraisal of the United States' foreign and military policy. Begun by the Truman administration and carried over into the succeeding Republican administration, it emerged in the autumn of 1953 as President Eisenhower's "New Look."[13] It was set forth in the National Security Council's paper NSC 162/2.[14]

The principal difficulty that faced both Truman and Eisenhower was their commitment to curtail the national budget, a commitment that imposed severe limits on expenditures for the national military establishment. The difficulty, however, was political not economic, for involvement in the Korean War had shown, as one analysis summed it up, that "the problem was not the capacity of the economy to carry larger expenditures, but rather the ability and desire of the administration to undertake the task of persuading the taxpayers and their representatives that the gain from larger expenditures would be worth the costs."[15]

Five months of intense debate marked the interval between the Soviet atomic test and the first, measured response by President Truman. The American decision in January of 1950 (the month that George Valley coincidentally discovered the existence of the Whirlwind computer) had been to begin an even more determined effort to explore the feasibility of developing an H-bomb.[16] At the same time, Truman had directed his Secretaries of State and Defense to "undertake a reexamination of our objectives in peace and war and of the effect of these objectives on our strategic plans, in the light of the probable fission bomb capability and possible thermonuclear bomb capability of the Soviet Union."[17]

The report of the joint State-Defense study group reached the president on April 7, 1950. Five days later it was forwarded to the National Security Council for further "consideration [and] a clearer indication of the programs . . . envisaged in the Report, including estimates of the probable cost of such programs."[18] Six months of deliberations within the NSC followed. On September 30, NSC 68 was given presidential approval as "a statement of policy to be followed over the next four or five years."[19] It "emphasized a general objective, leaving to Executive leadership the flexibility and discretion it would need to achieve that objective."[20]

The joint study group, the National Security Council, and the president were in agreement that "the probable fission bomb capability of the Soviet Union" had "greatly intensified the Soviet threat to the security of the United States." Furthermore, within 4 or 5 years the USSR could possess the ability to deliver "a surprise atomic attack of such weight that the United States must have substantially increased general air, ground, and sea strength, atomic capabilities, and air and civilian defenses to deter war and to provide reasonable assurance in the event of war, that the nation could survive the initial blow and go on to the eventual attainment of its objectives." In turn, this contingency required the intensification of American efforts in the fields of diplomatic and military intelligence and research and development.[21]

Since it was a general policy statement, NSC 68 did not discuss any particular defensive countermeasures, but it contained the general observation that "effective opposition" to a Soviet atomic ability required "among other measures greatly increased air warning systems, air defenses, and . . . a civilian defense program which had been thoroughly integrated with the military defense system."[22] Clearly, this statement implied support for ongoing efforts to improve and modernize the national air defense system. It is likely that the preparers of the paper were familiar with the various investigations and military R&D programs in process.

On September 4, 1952, the Secretaries of State and Defense and the Director for Mutual Security were directed by the president to reexamine "the amounts and allocations of resources" to national security.[23] In the memorandum that accompanied their response, they emphasized that more resources should be applied to continental and civil defense programs.[24] Submitted on January 16, 1953, only four days before Truman left office, and issued as NSC 141 the day before he departed, the report was positive and direct. It made three salient points: (1) The ability of the United States to defend North America against atomic air attack was "extremely limited." (2) By 1955 the Soviet Union would possess "the capability to make an air attack upon the United States which would represent a blow of critical proportions." (3) "Defensive measures" in contemporary programs would not "provide an effective defense against mass atomic attack."[25] Deterrence, secured through "an effective land, sea, and air offensive capability," the authors concluded, offered the "best defense" and should not be lessened by "defensive considerations." Nevertheless, other measures that would "materially increase" the nation's "capability to meet the atomic threat in 1955 and thereafter" should be undertaken. Such measures included "additional small radar installations for low altitude coverage, semi-automatic computers for the air defense filter centers, radar picket ships."[26] It is not unreasonable to conclude that the findings of ADSEC and Project Charles and the technical accomplishments of Lincoln Laboratory's air defense program were beginning to influence national defense policy recommendations.

The report also contained a recommendation for the "installation of a distant early-warning fence," at an estimated cost of about a billion dollars.[27] Such a radar net had been recommended by an informal study group that had met at Lincoln Laboratory during the summer of 1952. Elements within the Air Force that favored offensive over defensive capabilities had sought to suppress the study group's findings, but their arguments did not dominate NSC 141.[28] The proposals for additional defensive measures included in NSC 141 were, however, accompanied by the cautionary reminder that "programs such as these would materially increase the effectiveness of the continental air defense system, but would not guarantee an absolute defense against atomic attack."[29]

Tactical political considerations doomed immediate acceptance of the recommendations, for "expenditures of very substantial additional resources," adding approximately "eight and one-half billion dollars through 1955 over and above the cost of existing programs," would be required, the report acknowledged, to improve continental defenses.[30] The increases demanded would have been very painful politically, and the outgoing Democratic administration appears to have been quite content to pass the problem along to the incoming Republican administration. The Eisenhower administration, however, was equally committed to budget austerity by conviction and campaign pledges, so upon taking office the new president had no political choice but to formally reject the recommendations offered in NSC 141.[31]

Despite this prompt rejection of the recommendations carried over from the Truman administration, the first year of the Eisenhower presidency proved to be the pivotal year in the formulation of a continental air defense policy. Earlier efforts toward a solution had been hesitant and indecisive. The Soviet atomic test of 1949 had provoked significant technical activity and had stimulated discussion, but the search for an effective policy had been limited by political considerations to investigation and discussion.

Hesitancy and indecision ended with the Soviet Union's demonstration of hydrogen bomb capability on August 12, 1953, which "broke the back of resistance to air defense within the Administration."[32] Six days before the Soviet thermonuclear test, the Joint Chiefs of Staff had "identified continental defense and massive retaliation as the two principal military problems facing the country." Twenty days after the test, the new chairman, Admiral Arthur W. Radford, told his first news conference that the Soviet H bomb compelled the United States to "review and strengthen its air defenses."[33] Radford's statement marked the beginning of the end of the long debate over the issue of an appropriate defense against the threat of an air attack on the continental United States. The end of the debate came on October 30, 1953, when President Eisenhower approved National Security Paper NSC 162/1, a major policy statement establishing the basis for what was later to be dubbed the "New Look."[34]

When the Eisenhower administration addressed the problem of air defense, it had before it a plethora of advisory opinions and documentary materials. Divided and perhaps confused, the new administration's policy makers appointed more investigatory bodies. A committee of businessmen was formed to "study the problem from a business viewpoint." These "Seven Wise Men" advocated a "go-slow policy." A second group was chaired by Lieutenant General Harold Bull, a trusted wartime associate of the president. Both committees emphasized the "need for action."[35]

All the members of the Bull Committee were acquainted with the air defense program conducted by Lincoln Laboratory, for project administrators Al Hill, Malcolm Hubbard, and George Valley had discussed the program with Bull and his committee members in June of 1953. Arthur Fleming, head of the Office of Defense Mobilization, had been "informally told about the work of Lincoln" by James Killian, and a visit by Fleming to the laboratory was scheduled for October.[36]

The members of the Bull Committee agreed with the earlier conclusion of a study group chaired by Mervin J. Kelly of the Bell Telephone Laboratories that "a defense system approaching invulnerability" was "probably unattainable." They nevertheless concluded that "a reasonably effective defense system can and must be attained." Finding such a system to be under development at Lincoln Laboratory, they described it as necessary to the overall continental defense, although of less immediate urgency than the seaward extension of early-warning radar nets and the preparation of plans to "insure the continuity of essential functions of government."

Although the Bull Committee did not accord Lincoln Lab's system topmost priority, it did express a very high regard for that system's potential:

> The Lincoln Transition System for the integration of radar and other information into a semi-automatic traffic handling, display and weapons control center at each Air Division will increase overall system capacity tremendously, and permit optimum utilization of our interceptors and other weapons. It also holds some promise of filling one of several essential requirements in a future defense system against intercontinental ballistic missiles. Further, as a peacetime by-product, it will be an accurate and positive means for the control of all air traffic.[37]

While these committees were engaged in their investigations, the president and his advisors were reviewing the recommendations and the supporting arguments submitted by investigatory bodies appointed during the final days of the Truman administration. The study group headed by Mervin Kelly recognized the need for development of an effective air defense system but advised that "crash programs" be avoided. This caveat, a later student of the period concluded, was a reference to Lincoln Lab's program. Kelly's committee had visited Lincoln in March of 1953 and had witnessed a "full-scale operation with test aircraft."[38]

Another study was conducted by a special State Department Committee on Disarmament. Chaired by J. Robert Oppenheimer and including among its members Vannevar Bush and Allen Dulles, this group advised that "development of continental air defense . . . be speeded up."[39]

Defense of the continent against air attack was but one facet of the larger problem of U.S. military and foreign policy and strategy. American civil and military leaders had to reconsider and reshape national strategy in order to meet, both at home and abroad, the increased dangers arising from the Cold War, the Korean War, and other hostilities of the atomic age. Project Charles in 1951 had considered continental air defense, Project Vista in 1952 the application of atomic weapons to tactical use, and Project East River, also in 1952, civil defense measures. The deliberations and findings of such ad hoc study groups and the routine activities of the Department of Defense and other departments and agencies concerned with national security entered into the deliberations of the National Security Council as it formulated position papers for presidential action.[40]

In addition, the defense of Europe had again become a matter of concern. Sir John Slessor, Royal Air Force Chief of Staff, while visiting Washington in July of 1952, had argued for a radical change in the North Atlantic Treaty Organization's strategy, on the ground that "American atomic advantage was a war deterrent and a war-winning factor dominating all others."[41] Slessor was speaking on behalf of the British Chiefs of Staff, who, in reviewing "military factors bearing upon the world strategic situation," had concluded that "the defense of Western Europe with its traditional emphasis on large aggregations of ground forces" was in "urgent need of revision." Britain and the other members of NATO had concluded that they simply were unable to meet the costs of establishing and sustaining the massive ground forces necessary to contain the Soviet threat. Hence, the broad problem confronting the Eisenhower administration included not only defense of the American continent but also a practical defense strategy for Western Europe.[42] It is little wonder that the American president, caught between budget austerity and increased military expenditures, ruefully informed members of Congress that his "dilemma" was "giving him sleepless nights."[43]

The final stage of the discussions that were to end with the formulation of Eisenhower's "New Look" in continental defense had begun in a White House sun room in May of 1953, several months after NSC 141 had been rejected. Between May and October, the recommendations of "Operation Solarium" had gone to the National Security Council for review and analysis, and the NSC had acknowledged in September that defense of the continent was "clearly inadequate."[44]

By the end of October 1953, 2½ months after the Soviet Union had demonstrated thermonuclear capability, these discussions and recommendations had enabled the National Security Council to issue a revised statement on "Basic National Security Policy." Reported out on October 30 and receiving presidential approval, the NSC paper emphasized both the USSR's increased ability to deal a "crippling blow" to the nation's "industrial base" and its "ability to prosecute a war." However, an "effective defense . . . could reduce the likelihood and intensity of a hostile attack," although it could not "eliminate the chance of a crippling blow."[45] Therefore, the United States had to maintain "a strong security posture with emphasis on adequate offensive retaliatory strength and defensive strength." Such a posture required "a massive atomic capability" and "an integrated and effective continental defense system."[46]

Among the items proposed to realize the defensive capacity were additional anti-aircraft forces, conventional and guided missiles for the Army and the National Guard, superior fighter-interceptors for the Air Force, and "increased expenditures for radar warning and control, as well as communication networks."[47]

Three policy papers issued by the National Security Council with presidential approval had reflected the course of the debate on the defense issue: NSC 68 had implied the need, NSC 141 had stated the need, and NSC 162 had avowed the need. In the course of the deliberations that had led to the preparation and issuance of the last of these policy position papers, the president and the members of the NSC came under intense political pressure. Technical programs were proving that a more effective air defense system was possible, but would the taxpayers be willing to spend the vast sums required? Inter-service and intra-service rivalries had to be resolved, as did differences within the various groups advising the president.

The issue became even more difficult to resolve when it was argued in the press by leading military journalists and critics. Even James Killian and Albert Hill became involved. In an effort to move the issue off dead center, they had prepared for publication in the November issue of the *Atlantic Monthly* a response to the critics of the air defense development program that was underway. The intensity of such debate within the scientific community had been evident earlier in the Atomic Energy Commission hearings on the loyalty and reliability of J. Robert

Oppenheimer. No longer, however, was the emphasis on massive retaliation or a "Maginot Line." Now it was on "the right *balance* of defensive and offensive investment."[48]

The issuance of NSC 162 did not completely still the controversy over the issue of continental air defense, but the die had been cast. The air defense program conducted at Lincoln Laboratory under Air Force contract was given formal authorization at the highest level of government, and the fears expressed at the lab and at MIT concerning government support for the program were laid to rest. The government's commitment to continental air defense was made even more evident by implementation of the recommendation of the Lincoln Lab summer study group for installation of a "distant early warning line" across the continent's northern reaches.[49]

The NSC policy statement did not quiet further discussion of the nation's defense posture. In the years that followed, Sputnik and other Soviet technological accomplishments created threats that exceeded the capacity of the originally conceived air defense system of which SAGE was a component, not because the concept underlying the system was at fault, but because effective defense weapons were lacking amid the rapid progress of the technical state of the art. It appeared that defense forces could detect, could identify and track, but could not interdict. Or could they? Was an effective air defense system possible? While debate continued, the years continued to pass, and the changing continental defense system was not put to the test in actual battle conditions. Instead, readiness was all.

24

Implementing TM-20: "Who's in Charge?"

By 1953 both the foreign policy climate and the technical engineering climate had changed several times over since 1947, when Jay Forrester and Bob Everett had first presented in some detail the idea of using an electronic digital computer at the center of a military command-and-control system. On that occasion they had been talking to the Navy about anti-submarine warfare, and in view of the circumstances of that time it is not surprising that their proposal fell on deaf ears.[1]

By 1953 these same men responsible for the technical development of the air defense system as described in TM-20 saw themselves—then 3 years into the enterprise since George Valley had first laid eyes on Whirlwind—pressuring the R&D process to accelerate it as rapidly as possible in order to bring into being the first computerized command-and-control system, one designed specifically for aerial defense of the North American continent.

Without waiting for the Cape Cod System to succeed or fail, and without waiting for national policy issues to be settled, Forrester, Everett, Valley, and three other Lincoln Lab electrical engineers had proposed in TM-20 the "Transition Phase" as the next step to take in the overall R&D program that Project Charles had recommended and that they were carrying forward.

At the start, and in the eyes of the six men who authored TM-20, not to mention the engineers of Division 2 and Division 6 who worked under them in 1953, TM-20 was important principally because it described in some technical detail the proposed new instrument of military command and control of any future aerial battle sector. This instrument would be the computer they were designing under the name "Whirlwind II," along with its attached radar network. The computer would possess "a computing speed of 30,000 to 50,000 single-address operations per second and an internal storage capacity of 2000 to 8000 numbers or instructions of about 24 binary digits each."[2] That was the computer they visualized in 1953.

Their R&D philosophy at that time was the same philosophy they had held in 1950 and they would hold in 1957. It postulated overlapping sequences of design, build, and test efforts, dominated by a policy of planning ahead in detail. Ideally, these overlapping sequences would make optimum use of time and would give sufficient weight to both experience and foresight to permit IBM to begin manufacturing the computer even before Lincoln Lab and IBM engineers had finished their R&D testing. There was no time to economically "fly before you buy," because the United States was virtually defenseless against any massed air attack.

To implement this R&D policy, the next step beyond the Cape Cod System R&D tests would be tests with the Experimental SAGE Subsector (ESS), the model air defense system that would link the R&D phase of the project to the production phase by harnessing the prototype of a computer faster than Whirlwind I to a larger radar network. These would all ultimately come under the control of specially trained Air Force personnel operating under the guidance of Lincoln Lab engineers.

As anticipated in both TM-20 (1953) and the Operational Plan (the Air Defense Command's SAGE "Red Book," which followed in 1955), ESS would be brought into existence step by step and tested in such a way as to provide the further information the men would need to start building, in overlapping sequences, the modular, full-scale sectors of the continent-wide system that the Air Defense Command (ADC) would operate against any real attack. Nor would they have to wait until tests with the ESS were finished before setting to work on the continent-wide system.[3]

A design goal of the Transition System that had been set forth in the 1953 proposal was that it "meet most of the ADC needs, including the time schedules. A single computing center should handle three hundred to five hundred aircraft tracks (track-while-scan, smoothing, prediction) and at the same time calculate midcourse guidance and return-to-base instructions for a hundred weapons simultaneously."[4]

The authors of TM-20 had recommended several alternative sector arrangements for the Air Force to choose from. One of these called for three computers to be used per sector (figure 24.2). These could be flexibly located at one command-and-control center or distributed among as many as three centers per sector, as cost and military wisdom might determine. Two computers should handle all normal situations. If less than two computers were operating for any reason, degraded performance should be expected in that sector.[5]

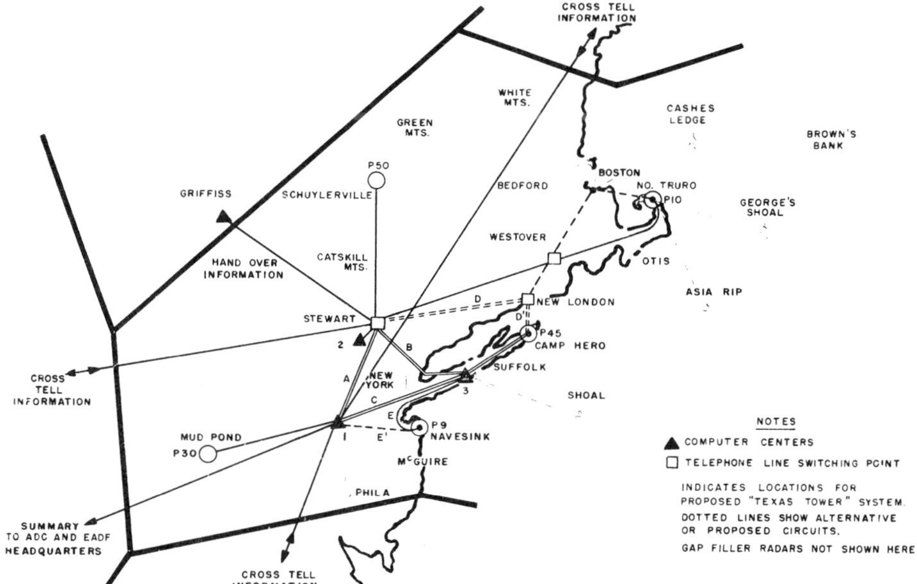

Figure 24.1
"Air Defense Sector Network"

So the thinking ran early in 1953 as the men envisioned the repetitive modular structure of the air defense system. The basic units of the system would be the local command-and-control centers for the sectors. The command-and-control centers should be modular, so that several of them could be interconnected to constitute a Sector Command Center, where the assignment of weapons and other duties could be carried out with computer assistance. Going beyond the Sector Command Centers to the next wider level of geographic control in their thinking, the authors of TM-20 had suggested an "Air Defense Force command post" that would use computer-processed radar data to coordinate command activities affecting more than one sector. Finally, they had proposed a computer-equipped "Air Defense Command Center" at the highest Command Headquarters level. Providing up-to-date summary information from all sectors, it would present the "big picture" that top commanders would need.[6] As they saw it in 1953, a practical way to proceed toward this ideal was for Lincoln Laboratory to collaborate with the Air Force in setting up in New England a small, regional, modular, model "Air Defense Sector Network" that would bridge the gap between the

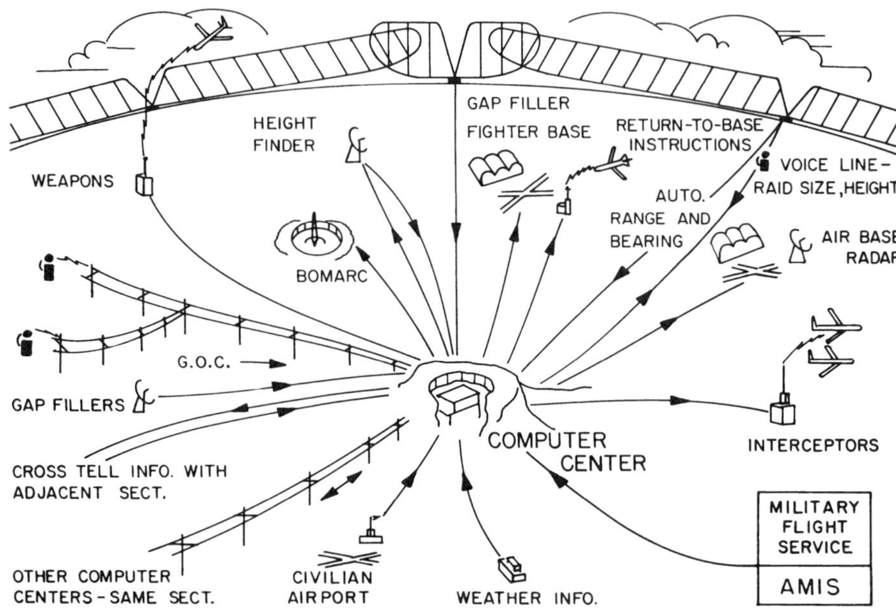

Figure 24.2
"Air Defense System, One Computer System."

even smaller experimental Cape Cod System (then being brought into being) and the larger "Transition System" proposed in TM-20.[7]

The Lincoln Lab planners spent much of 1953 estimating costs and manning requirements, determining the feasibility of the siting and dispersal of technical facilities for the initial regional sector, and examining the design implications of fresh technical information as it became available from a number of sources. These sources included the intercept tests of the Cape Cod System that involved Valley's Division 2, Forrester's Division 6, and IBM's Project High.[8] During 1953 they held discussions with the Air Defense Command and with the Air Defense Engineering Service (ADES) project set up by the Bell Telephone Laboratories to assist ADC with the logistics of creating the continental communication network needed by the proposed air defense system.[9] Out of these conversations emerged a less ambitious, less costly, but more practical scheme than TM-20 had proposed. Fewer Direction Centers would be built, but each would have two computers, one running on standby to take over almost instantly the tasks of the other should it falter or fail for any reason. This "Duplex Central" arrangement was a practical compromise. On

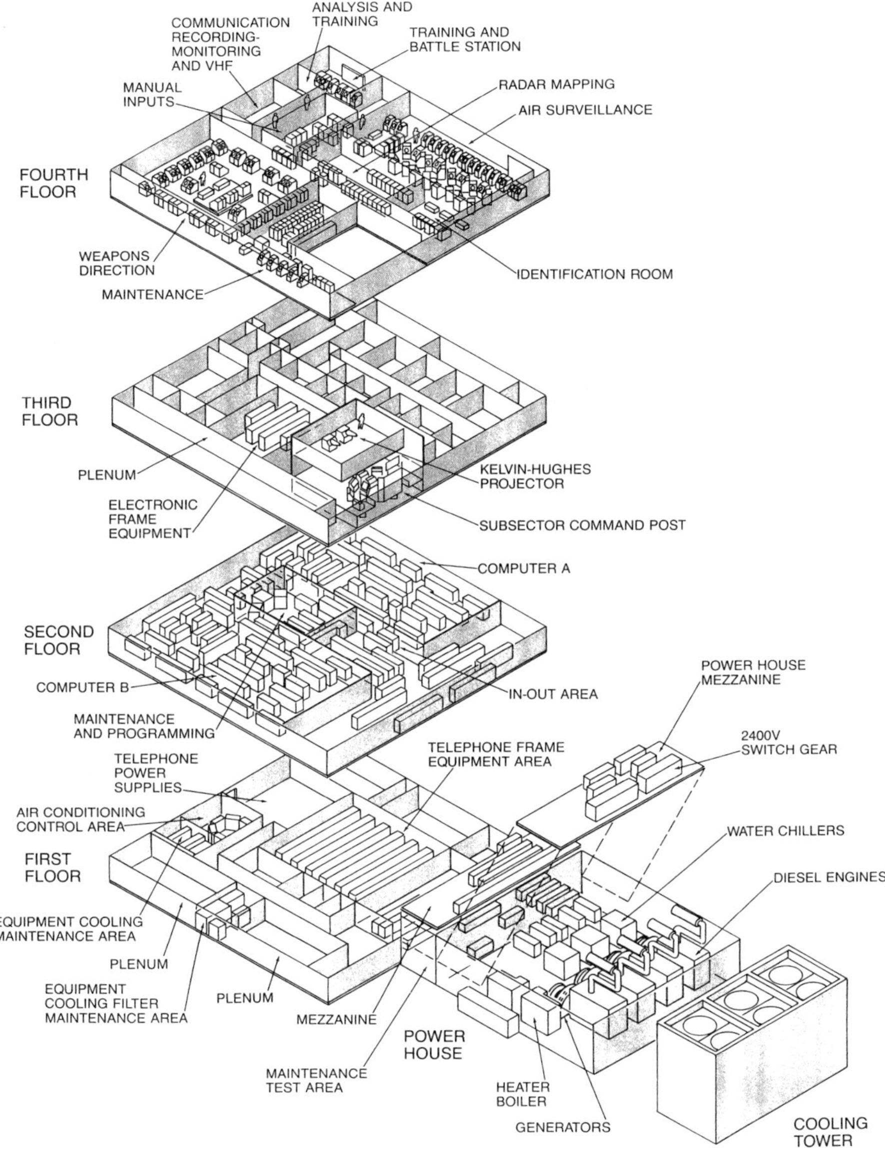

Figure 24.3
Plan for SAGE Direction Center.

the one hand, the more Direction Centers there were, the more targets an attack would have to destroy in order to cripple the system; on the other hand, the duplex arrangement would increase the basic equipment needed per sector by less than 10 percent while doubling reliability and allowing for transfer of the defense burden to other, still undamaged control centers or sectors if the need should arise.[10] As plans for the regional, experimental "Electronic SAGE Sector" network took shape, however, the decision was made to use only one computer, the XD-1 prototype, in a "simplex" arrangement sufficient for ESS trials.

By the end of August 1953 (the same month that Whirlwind began to operate with reliable, magnetic-core storage in the Cape Cod tests), the Air Defense Command, in a parallel managerial effort, was examining in technical detail the actions it must take to begin to move the Air Force toward installing and trying out the more ambitious Transition System that TM-20 had described. Preliminary plans that Division 6 and Division 2 had put together during July for installing the more technically complex system in the 26th Air Division Sector were discussed at an August conference with ADC. "ADC will study the proposal covered by L-113," reported Art Kromer to Division 6 engineers after the conference, "and secure the necessary additional information to permit them to present the complete philosophy and overall picture of the Transition System for Air Defense on a nationwide basis to higher Air Force and Department of Defense agencies in order to secure approval for adoption of this system as an operational plan."[11] But the future course of events would not be so straightforward.

In September of 1953, Art Kromer (in charge of liaison with IBM) and Norm Taylor (head of Group 62, Computer Development) estimated that it would take 3½ years from the time the Air Force gave formal contractual authorization until the first AN/FSQ-7 computer could be installed, tested out, and operating, ready to carry out mock or real air intercepts for the Air Force. With this schedule in mind, the Division 6 engineers concentrated during the winter of 1953 and the spring of 1954 on firming up, with IBM, the design details of the XD-1 prototype of the "Q-7" air defense computer. To be installed in a Model Direction Center (Building F, built for the purpose and located next door to Lincoln Laboratory), the XD-1 would operate as the "brain" of the experimental regional-network model of the "Transition System."[12]

Before TM-20 was a year old, it came under vigorous attack from R&D policy makers at Bell Telephone Laboratories (commonly referred to as Bell Labs or BTL) and at BTL's manufacturing affiliate, the Western

Figure 24.4
Lincoln Laboratory, circa 1953.

Electric Company, which had been brought into the newly formed Air Defense Engineering Service (ADES).

The Air Defense Engineering Service was not simply an Air Force idea. Both the Air Force and MIT had begun informal conversations during the spring of 1953 with Mervin J. Kelly, president of Bell Labs, to obtain the help of the national telephone network in providing telephone-line communication links among the radars, computers, and sectors of the proposed continental air defense system.[13]

Mervin Kelly was by no means uninformed about the technical affairs or the political and administrative situation in which MIT and Lincoln Lab were involved. During the winter of 1952–53, when both the University of Michigan and MIT still held air defense contracts with the Air Force, the Office of the Secretary of Defense had appointed Kelly to head a committee that would include three industrial members and three academic members and would examine "measures being taken for continental defense." The Kelly Committee had visited Lincoln Lab in March of 1953.[14] Before the end of April, Air Force General Donald Putt had "talked to Kelly who is willing that Western Electric take on certain responsibilities for the Transition System," and Kelly and his staff met with James Killian and Jay Stratton to discuss preliminary arrangements.[15]

Not long after October 30, 1953, when the White House authorized development of a continental air defense system, an Air Force study group formally recommended that a contract be negotiated with Bell Labs and Western Electric "to provide for technical assistance, and other services, for which the Air Force [lacked an] existing capability." The Air Force accordingly issued a letter contract in January of 1954, and in May it agreed formally to establish the ADES project. In September, an ADES Project Office was officially opened in New York to coordinate the work of Bell Labs, Western Electric, and the various Air Force Commands.[16]

The policy momentum of this chain of events gave BTL's technical managers an early opportunity to take a close look at TM-20, and it put muscle behind a sharply critical report they issued in December of 1953. They objected to various technical details of TM-20 and to its overall philosophy of managing R&D. Their analysis made it clear they did not share the R&D philosophy to which Forrester, Everett, and Valley had been committed since before Lincoln Laboratory came into existence.

The Bell Labs report redefined the major technical military functions of the Transition System to be "Observation, Command, Guidance," emphasized the need to design a system more effective than one based on still-incomplete Cape Cod test results, expressed a preference for decentralized data processing instead of doing so much of it at the central computers in a sector, and raised a number of questions about "cost, personnel requirements, subdivision of military functions, purpose and areas of application, degree of departure from past practice, and proper size of basic building block."[17]

History never repeats itself identically, but for those running Lincoln Laboratory and MIT there must have been a sense of déjà vu. The basic policy issue that had been settled by the withdrawal of the University of Michigan from technical competition in the spring of 1953 appeared to be rising again in December in slightly different form, having been given substance by the fresh criticisms from Bell Labs and Western Electric.

In January of 1954 the Lincoln Steering Committee discussed the report, agreed as a precautionary measure to take "a serious look" at Cape Cod testing before Bell Labs representatives visited Lincoln Lab early in February, and concluded that Lincoln should also diplomatically consider developing an alternative "decentralized or smaller Transition System," but at a lower priority than that accorded the Transition System.[18]

Meanwhile, in addition to this political response and on the technical level, Lincoln Laboratory engineers mounted a prompt rebuttal. Addressing in particular BTL's basic philosophy of the "Scope and Flexibility of Air Defense Ground Environment," the Lincoln Lab engineers described the central differences they saw between their approach and BTL's approach to setting up an operational continental system:

> We prefer to think of the Transition System, not as a certain specific information flow channel (the present large radar, SDV link, orthodox digital computer, display and interceptor of the BTL report), but as a matrix into which all detectors, many types of data transmission, missiles, picket ships, and weapons can all be fitted. Effective use of the gap filler and "Muldar" type of short-range all-altitude radar seems, for example, to be well fitted to the Transition System but ill fitted to alternate proposals in the BTL report.
>
> The Transition System must have flexibility to cope with substantial changes in threat, *observing equipment,* forms of data transmission, and weapons, and to serve and include all of these, not just one combination. . . .
>
> As pointed out in TM-20, the Transition System is the transition between the present system (with present radar sets) and the Future System (with improved Observation as well as Data Processing). . . .
>
> Since not all possible forms of this improvement can now be foreseen, we must question the philosophy of Pages 13 and 14 of the BTL report, where it is implied that flexibility for future development will be needed in the Command and Guidance phases but not in the Observation and Data Processing phases.[19]

Discussions with "the group from Western Electric and Bell Telephone Laboratories" on February 4 and 5, 1954, left the leaders of Division 6 with the impression that BTL and its engineering affiliate were primarily "interested in evaluating the transition system for its effectiveness in air defense, rather than in techniques and components."[20] In any event, the men in charge of the program at Lincoln Lab had no intention of discarding their "Transition Phase" approach because of the BTL objections.

This policy disagreement between BTL/WE and Lincoln Lab over the former's role in carrying forward air defense R&D and deciding how and when to proceed through the latter part of the R&D phase to the production phase was not immediately settled. Instead, it proved to be only a preliminary skirmish of a larger contest being waged simultaneously on contractual, administrative, and technical fronts. This larger contest arose partly because Western Electric's managers were dismayed by the compressed timetable to which Division 2 and Division 6 had committed themselves. It also arose because various echelons in the Air Force had their own ideas about who should maintain program control. Furthermore, the widening scope of the air defense program afforded opportunities for other influential organizations in the public and private sectors to attempt to seize "a piece of the action." The prize at stake was the Transition System and policy control of SAGE research and development practices.

The maneuvering that ensued was also a consequence of longer-range trends in the conduct of national military affairs. The American government-and-industry "team" approach that had emerged during World War II was altering the conduct of peacetime military R&D, causing it to depart from its traditional form and giving wide latitude to innovative managerial and policy actions of the sort that MIT, BTL, and the Air Force found themselves caught up in.[21]

The developing contest among Lincoln Laboratory, Western Electric, and various Air Force Commands reflected genuine concerns and policy differences regarding the growing magnitude, the increasing complexity, and the rising dollar costs of air defense R&D, especially from 1953 on, after it received National Security Council sanction and while the system that Lincoln Lab was developing was approaching a development-and-production phase. At the same time, as more private and governmental organizations became involved, possessing greater political and policy-making influence at higher and higher managerial levels in industry and government, the technical merits of the solution that Project Charles had endorsed when recommending that Lincoln Laboratory be created were increasingly obscured by higher-level contests for a proper balance and distribution of contractual authority, administrative responsibility, and policy-making control.

Since there were differing opinions as to the proper balance, various diplomatic coordinated actions were taken as different parties jockeyed to improve their position. For example, in January of 1954, while BTL and Lincoln Lab were locking horns, Air Defense Command Headquarters forwarded to the Air Force Chief of Staff an "ADC Transition System Program" put together by "the cooperative endeavors of Air Defense Command, Lincoln Laboratory and Project ADES." The Commander of ADC was careful to give the program his diplomatically qualified endorsement. The cover letter put him on record as recognizing "that the approach of simultaneous development-production represents a certain degree of calculated risk." Acknowledging that on

December 31, 1953 General Putt had written a letter emphasizing "the importance of maintaining full momentum on the Transition System," the nominal author of the letter, Major General Jarred V. Crabb, Chief of Staff of ADC Headquarters, took care to point out that the time schedule imposed was "somewhat optimistic by normal action standards," but that it was "an achievable one" which would provide "the greatest improvement . . . in the shortest possible time, if all components function as forecast, and the correlative program for radar improvement is vigorously pushed."[22]

Crabb's cover letter was a reminder that the building of the Transition System (which became SAGE) was not to be a simple affair, either technically or managerially, and that getting on with the technical job as Lincoln Lab wanted it done depended more than once on extraneous managerial maneuvers and infighting. Thus, when an Air Force study group late in 1953 had recommended the Air Force negotiate a contract with Western Electric to establish a ground communications network for the air defense system, they also added a routine proviso that the Air Force keep "complete control of all aspects" of the program and perform "directly all phases of the work for which [it had an] existing capability."[23]

Curiously, when agreement on a letter contract was reached in January of 1954, a month after BTL issued its sharp criticism of the Transition System, there was no precise statement of the nature and scope of Western Electric's duties. Indeed, these were not to be legally defined for another year. In the meantime, the question of who was to be in charge was being settled, and this situation both affected and reflected the technical R&D controversy in which Lincoln Lab and BTL/WE were by this time embroiled over the "Transition System" concept.[24]

There was still another policy-making dimension to this whole affair. Since the Air Defense Command had been given administrative responsibility for the letter contract (as Western Electric had preferred), it was also responsible for preparing with the company the official Statement of Work, which was subject to review, of course, by Air Force Headquarters in Washington. Both ADC and Western Electric administrators agreed that the scope of the company's responsibilities should be sufficiently broad to include the authority and duties of a systems engineer operating under the administrative supervision of the Command.[25] These provisions, too, sounded quite routine. But to cement this arrangement, it was suggested during preliminary discussions between Western Electric and the Air Force that the company be made the prime contractor for the proposed continental air defense network. As Albert Hill and the Lincoln Steering Committee realized, IBM would then become a subcontractor, and Lincoln's role would be reduced, while the roles of Bell Labs and Western Electric in ADES would be enlarged. In January of 1954, Hill drafted a letter to the Air Force pointing out that if this course of action were to be taken then perhaps MIT and the Air Force should re-examine "Lincoln's responsibilities and authority."[26] Hill planned to circulate the letter informally.

At this juncture, the Air Materiel Command of the Air Force entered the action and challenged ADC's assumption of contractual supervisory authority, pointing out that the exercise of such authority properly belonged to AMC under existing Air Force mission assignments. At the same time, AMC recognized the complex intermix of responsibilities among development, procurement, and operational use of a continental system. AMC proposed a compromise: instead of having a single Command in charge, set up a joint project office.[27]

The compromise was accepted in principle by Air Force Headquarters and by the ADES Project late in the spring of 1954, but it proved to be an elusive objective for the remainder of the year, partly because some of the Air Force members of the new ADES Joint Project were *not* reluctant to grant Western Electric the duties and authority of systems engineer. In fact, in the proposed Statement of Work they were willing to give the company responsibility for "management-services" and to empower it with positive authority to establish the air defense system.[28] MIT's and Lincoln Lab's administrators became increasingly concerned over the course events were taking.

At the level of engineering analysis, Western Electric and Bell Telephone Laboratories (WE's corporate parent) continued during the spring and summer of 1954 to voice adverse technical criticism of the sort that had been expressed in BTL's Report #3 (December 1953) to the Air Research and Development Command. In an opening skirmish in April (which was really between Western Electric and Lincoln Lab), the Air Force asked what level of funding should be given to BTL's December recommendation #4: that Lincoln Lab investigate an air defense network less centralized than TM-20's Transition System. The Lincoln Steering Committee had diplomatically agreed in January to look into the BTL recommendation, then had given it a low priority.

In the LSC's April discussion of the issue, Forrester, after observing that "there appears to be a growing feeling outside that Lincoln is doing much more work on Item 4" than in fact it was doing, suggested characteristically that "as long as Lincoln is committed to Item 4, . . . something effective should be done even if this means giving up some less important work." But his suggestion was not taken up, perhaps because it could be interpreted as an oblique criticism of the conciliatory, diplomatic brush-off tactic with which Lincoln Lab earlier had sought to neutralize the BTL report's criticism, or perhaps because Lincoln did not wish to appear that accommodating to ADES and the Air Force. Earlier in that same April meeting the LSC had been discussing "Lincoln-BTL-WE relationships in general," presumably with some unease, in the light of the events that were transpiring. Whatever may have been their degree of unease in April, it may have prepared them to be less than utterly surprised by events that broke in June.[29]

On the first Saturday in June, a high-level Air Force meeting, chaired by ARDC's Deputy Commander for Technical Operations, Brigadier General Floyd B. Wood, heard representatives from Western Electric's ADES group, with the sympathetic support of General Wood and Colonel Richard M. Osgood (head of the AMC Joint Project Office in New York), propose a new schedule for the Transition System. Apparently only Albert Hill was there to speak for MIT, and if he spoke his views were ignored. At the next meeting of the Lincoln Steering Committee, Hill reported the following:

> Western Electric proposed that two Direction Centers and one Combat Center be produced and installed for field testing. Production would then be stopped or reduced to a low level pending completion of these tests. After field tests had proved the adequacy of the Transition System and had proposed desirable changes, production would be restarted at an increased rate in an attempt to meet the present schedules for installation of the entire national system.
>
> Colonel Osgood and the JPO were asked to investigate rescheduling FSQ-7 production.[30]

Several days later in New York, Colonel Osgood called a two-day meeting, reported Art Kromer, one of the two representatives attending from Lincoln Laboratory.[31] Attending the first day, in addition to Kromer and R. E. Rader from Lincoln, were six representatives from Western Electric, one from Bell Labs, and seven from the Air Force. The Lincoln Lab representatives were quite outnumbered and were low enough in administrative rank at Lincoln to be attending supposedly in order to be told what Lincoln was to do.

A fait accompli was in the making, as Osgood's opening remarks made clear. He explained, said Kromer, how the Saturday meeting at the Pentagon (at which Hill had been the only representative from Lincoln Lab) had "resulted in the conclusion by all present that the schedule for the AN/FSQ-7 Duplex Centrals should be altered so as to have a relatively low, early rate of production followed by a field evaluation trial for the System and, subsequently, to have production at a higher rate than previously contemplated so as to provide the total quantity of equipments for the country at approximately the same termination date as the present schedule would provide."[32]

If the old schedule were to be followed, Osgood continued, by the time the Air Force had the first two or three Direction Centers running, it would already be committed to build 25 such centers across the continent at a cost of about $550 million. In addition, "the ADES people indicated that they did not believe the XD-1 System would provide operational experience under conditions which were close enough to field-operating conditions to provide any significant data regarding the performance of the overall system," said Kromer in his report to Everett, Forrester, Valley, and Taylor.

Along with the proposed change in schedule went a new name: Semi-Automatic Direction Center System. The new proposal would not employ the XD-1 prototype destined for Lincoln Lab's Building F; instead it would employ the first two FSQ-7 computers placed in Air Force Direction Centers, "the Data Processing system, including equipment at the radars, the transmission lines between the radars and the centrals, and also the Data Processing equipment and transmission lines for Outputs from the Direction Central to the points which connect to ground/air links for weapons."[33]

That Western Electric and Colonel Osgood's Joint Project Office were bent on taking over the closing development phase and the opening production phase of the program became clear as Osgood outlined the purpose of the present meeting: "to establish a new overall schedule for design, manufacture and installation of equipment and associated buildings for Direction Centers based on sound principles and experience so as to provide a workable Defense capability for the Air Defense Command."[34]

Western Electric representatives explained that their company's proposal ought not cause any delay in the delivery schedule, "but would, in fact, provide better equipment as soon, and perhaps sooner, than the present program" for Lincoln Lab's Transition System. The Bell System's experience had shown them that "equipment of this type, if it is placed into production and installed without adequate system field experience, results in a very extensive field modification to eliminate all of the 'bugs' and problems which arise as the equipment starts to be used by the eventual customer—in this case the Air Force." Kromer found it significant that they had no scheduling charts available at the conference, although later one of the Western Electric men "indicated that, in his opinion, this modification period would be something in the order of a couple of years for the early systems and might taper down to approximately one year for the systems installed at the tail end of the program."[35]

A measure of the Joint Project Office's soft-shoe tactics was the fact that IBM did not join the conference until the morning of the second day, when J. E. Zollinger, J. Fraser, and R. Whalen arrived. Osgood had left the meeting the previous noon and had been replaced by A. G. Wimer from ARDC, who was ready with three questions "for IBM to study and prepare reply and proposals for to the JPO": First, how soon could IBM provide the first two Duplex FSQ-7's and, in place of the third Duplex originally planned, provide a "Combat Center set of equipment"? Could they improve on the delivery times presently scheduled? Second, "assuming a complete stoppage of production following the first two Direction Centers and one Combat Center, . . . what is the best schedule rate that IBM could achieve thereafter?" Third, how much could IBM slow down and still "maintain a nucleus of production capability" in their own plant and those of vendors and subcontractors while the first three units were being tested and improvements were being made, pending resumption of full-scale production in October of 1957? If IBM should have any alternatives which might "fit the intent of the conference," the Joint Project Office would be glad to consider them as late as June 25 (2 weeks thence).[36]

IBM's spokesman, Zollinger, replied that he was "very surprised," reported Kromer. Over the past 2 years he had come to feel that the Air Force was interested in having an air defense system as soon as possible and that it considered the Lincoln Transition System the best way to go. Suddenly it appeared that "the fundamental concept of the Transition System is now felt to be wrong," and the policy of improving the system "on a piecemeal basis by adding equipment as rapidly as it can be made available is now felt to be an unsatisfactory or undesirable approach." Would not a better way to proceed be to have those who were questioning the Transition System "place their points of concern out on the table for frank and open consideration by all parties involved in the program"? Would not this provide a faster solution, especially since it now appeared that the "value of a single AN/FSQ-7 Duplex Central as a primary building block for Air Defense—and the benefits that could be gained by installing each of these units and bringing them into the System at an early date—is now minimized or discounted completely"?[37]

S. P. Schwartz of Western Electric responded that ADES had few misgivings about the XD-1 itself and regarded it as "primarily a laboratory tool" best left under the control of engineering groups to try out changes and new ideas. But "one of the big questions, to be determined by a field evaluation test, is the capability of man and equipment to function" together in field trials. The same hardware could not be used for both of these purposes in the little time allowed. To Kromer, aware as he was of Lincoln Lab's philosophy of R&D, ADES seemed "unwilling to extrapolate from Cape Cod System and XD-1 performance." Instead, "they seem to be looking towards something approximating 100 per cent of the eventual capability of the equipment, rather than looking toward increasing the present Air Defense capability in a succession of steps."[38]

The meeting ended with the announcement that Western Electric and Air Force personnel in ADES, along with Rader from Lincoln Lab, would get together the next day to prepare a report of the conference.[39]

IBM was not the only party to object. The new commander of the Air Research and Development Command, Lieutenant General Thomas S. Power, ruefully commented on the impassioned phone calls he had received from Lincoln Lab, the Air Defense Command, and the Burroughs Corporation. Burroughs was under contract to manufacture "the critically important radar-data coding device" to convert radar information to a form readily transmittable over telephone lines to the central computer. John Harrington had initiated its design before his transfer from the Cambridge Research Center to Valley's Division 2 of Lincoln Laboratory.

The leaders of Lincoln Laboratory speculated as to whether the Air Force was losing confidence in Lincoln Lab's competence, and on the deleterious impact that delay would have on staff morale. The Air Defense Command argued that, since the decision to go ahead with the Lincoln system had already been made, the Transition System program should be pushed forward without further delay.[40]

On June 16, MIT counterattacked at a meeting in the Pentagon attended only by representatives from MIT, Lincoln Lab, IBM, Bell Labs, and Western Electric. No Air Force personnel were present at this bare-knuckles affair. Whether they were not invited or whether they discreetly elected to stay away is not clear. Representing MIT's administration was

Vice-President for Industrial and Governmental Relations Edward Cochrane. Albert Hill was present, as were George Valley, Malcolm Hubbard, and C. Robert Wieser. The only architects of the experimental and operational Transition System present were Valley and Wieser; none of the others who had been centrally involved with the system from the start—Forrester or Everett or Taylor—was there to address its fate or the wisdom of bypassing it in setting up field trials. J. E. Zollinger and C. F. McElwain were attending from IBM. There were two representatives from Bell Labs and four from Western Electric, including F. R. Lack and S. P. Schwartz.

It was the day before a scheduled meeting with Secretary of the Air Force Lewis, and Lack wanted to know "what all the shooting was for." He had assumed that the forthcoming meeting was simply an opportunity for Western Electric and its client, the Air Force, to review the schedule that lay ahead. Cochrane responded that was his understanding, too, but the sense of urgency behind the time scale for mounting the air defense system was at issue here. Who could foretell how soon such a system might be needed? If they could be sure of the answer, *then* they might think of a more leisurely time scale such as that which ADES was proposing.

But Western Electric was proposing to *slow down* delivery of the central computers after the first two were delivered, Hill interjected, although he understood that Colonel Osgood of the Joint Project Office was under orders to speed up the program, not delay it. And IBM had been asked how it could speed up delivery of the first few units, Zollinger added.

Valley asked whether there was a misunderstanding here about Western Electric's intentions. Was it their intention to speed up the program? Zollinger confessed his confusion over the fact that, as he understood it, IBM had been asked to build the first three computers "and then stop construction for two years." Lack replied that what Western Electric desired was an orderly program of the sort that was applied to manufacturing automobiles. "Cochrane pointed out that this was not a good analogy," to which Lack responded that the proper way to proceed was to allow an interval for "debugging." Cochrane shot back: "What debugging?" And when Lack pointed out that they would be dealing with untried equipment, Hill pointed out that to call it untried equipment was to fail to understand the importance of the prototype sector employing the XD-1. Lack countered that equipment in a laboratory basement would not provide the answers they needed. "How do you know it is in a lab basement?" Hill asked. "Didn't you know that GIs will be running it? Hadn't you better find out what the XD-1 installation is?"

And so the discussion went, gradually exposing the fact that the Western Electric managers were not intimately acquainted with how Lincoln Laboratory was carrying on its research and development. As it began to appear that Western Electric was not as well informed about the R&D details as they had assumed and that MIT was not about to give up without a fight, the men from Western Electric began to back off. It was the beginning of the end.[41] For these reasons and others not apparent in the available records, Air Force Secretary Lewis decided, on the next day, at a meeting attended by Cochrane, Zollinger, and Lack, to "go ahead on the undelayed production schedule," and Western Electric "agreed to abide by this decision."[42] Two weeks later, on July 2, 1954, Cochrane presided at a conference at MIT attended by representatives from MIT, Lincoln, Bell Labs, Western Electric, IBM, and the Air Force Joint Project Office. There they re-examined and identified the formal responsibilities of all the parties and determined how they would proceed. MIT (and thereby Lincoln Laboratory) was clearly in charge. F. R. Lack of Western Electric "indicated that Western Electric is ready to proceed with the program but wants a clear definition of the functions and responsibilities of ADES set forth, since it would reflect directly upon the size and kind of an organization they would set up."[43] The skirmish was over, and the Transition System appeared to be back on track.

But four days later, Cochrane (who usually did not attend Steering Committee meetings) told the Lincoln Steering Committee that TM-20 had become "obsolete and incomplete." A new and detailed description of the Transition System must be framed in close cooperation with the Air Defense Command. It was agreed that Valley and Forrester would lead Lincoln Lab's operation of preparing it.[44] Eight months later, the "SAGE Red Book"—Operational Plan, Semi-Automatic Ground Environment System for Air Defense (Formerly Designated The Transition System), Short Title: SAGE—was issued by the Air Defense Command.[45]

In the meantime, under the new arrangement, Western Electric was assigned responsibility for managing "the timely implementation of the major portion of the SAGE system" under the supervision of the Air Defense Engineering Service Project Office.[46] A co-managerial relationship was established in which Western Electric and Air Force members of the Project Office were "jointly responsible for bringing the program into being."[47] Western Electric organized its own ADES support group to carry out its contractual responsibilities and drew its members from its own numbers, from Bell Labs, from AT&T, and from the Bell System Operating

Telephone Companies. More than 1300 people were ultimately brought into this support group, which put together the communications network for the SAGE operational sectors.[48]

Western Electric and its allies in the Joint Project Office may have lost the skirmish, but they had not yielded the war. The company would repeat its attempt to delay the delivery schedule when, in 1956, the preparation and installation of the FSQ-7 computer's operating program developed by Division 6 required the delivery schedule to be stretched out, postponing some deliveries.[49] ARDC, too, persisted in its efforts to establish a more centralized management structure and to make Lincoln Lab a more cooperative team player. Indeed, from the point of view of ARDC's Command Headquarters, the air defense situation was getting out of hand, and Lincoln had to be brought under control by the creation of a centralized management structure if an integrated and unified continental air defense system was to be attained. To this end, ARDC sought to increase, at Lincoln Lab's expense, the influence and authority of the Western Electric Company, operating in conjunction with the military side of the ADES Project Office.[50] But such a goal was easier to describe than accomplish.

From the day it had been given administrative responsibility for the Lincoln Lab contract, ARDC Headquarters had been perturbed by its lack of authority. Perturbation became dissatisfaction if not irritation as the SAGE System moved into the development-and-production phase. A December 1954 ARDC staff study argued that Lincoln Lab's talents were being wasted on engineering activities. Greater controls were mandatory. The only way to secure these, the authors of the study concluded, was to change the basic documents that defined Lincoln Lab's mission and responsibilities; in this way ARDC should be able to bring the laboratory under greater control and focus its attention on "the evolution of radical concepts for the future."[51]

ARDC after much negotiation did secure new charters, but for its Command Headquarters the unsatisfactory relationship continued.[52] In the face of Lincoln Lab's and MIT's determination to retain maximum freedom of operation, ARDC simply wasn't able to secure the "more definite control and direction by the Government" that it sought.[53] Unfortunately for ARDC, in view of the apparent sympathy for Lincoln Lab's position displayed by Air Force Headquarters, the outcome was virtually predetermined.

25

The 1953 Cape Cod System

When the design of the Transition System was submitted to the Air Force in Technical Memorandum 20, at the beginning of January 1953, its antecedent in the developmental chain, the Cape Cod experimental system, was already under construction. Some parts of Cape Cod were already operational as extensions of the Bedford tests (begun about 2 years earlier), and by that January several other significant steps had been taken.

Small radars had been installed in the Massachusetts towns of Rockport and Scituate and had been connected to the Whirlwind computer through slowed-down video (SDV) data-transmission links. (These links were used to carry on track-while-scan studies on aircraft in the airspace where radar coverage overlapped.) A data link between the computer and a Ground Control Station at Truro (on Cape Cod) had been designed and was to be installed that summer. SDV links and various component plug-in assemblies, such as flip-flops and gates, were being manufactured in small quantities. An auxiliary memory drum intended for installation in the spring had been delivered by Engineering Research Associates, and an order had been placed with that company for a "magnetic drum for buffer storage" to be delivered early in 1953. Finally, the men had prepared and tested computer programs to inspect and data-screen the entire system.[1]

By the middle of March 1953, as a consequence of the engineering planning carried out by Division 6 and Division 2, the Cape Cod System was beginning to emerge as the experimental proving ground that Project Charles had had in mind. The design goals covered a wide range, from testing new components to developing the system concepts needed to operate a high-track-capacity system. The engineers would pragmatically validate the required muldar (multiple radar) concept by directing live aircraft in response to live radar data, and in this way they would develop the specifications for the command-and-control computers that a truly operational military air defense system would require.[2]

The Cape Cod demonstration experiments employed some equipment that had been in use since the Bedford tests of 1951 and some newer items. The older equipment included the Bedford MEW radar, now nearly 10 years old and scheduled for summer replacement by an Air Force CPS-6B search radar located at North Truro. Smaller radars at Scituate and at Rockport were already in use for tracking tests, and an MPS-4 nodding-beam radar was being used as a height finder. The latter was directed according to voice instructions given over the telephone to operators on the site.

Since the digital radar relay link (built before Project Charles came into existence) could transmit only a small part of the picture the radar generated, it would be replaced during the summer by the newer SDV link being experimentally developed for use with the North Truro radar. The arrangement was flexible in several ways. The data that reached Whirlwind came in over telephone lines and could be recorded on magnetic tape. The engineers could then use the recorded data for further analysis, planning, and the training of ground-crew personnel. The data also permitted analysis by the engineers to see what improvements could be made and what flaws could be removed. Furthermore, the emerging Cape Cod arrangement of radars could be used either as "a network of small radars" or as "a large radar (GCI Station P-10 at Truro, Mass.), with some of the small radars used as gap fillers."[3]

Whirlwind I itself was given two broad functions to carry out in the Cape Cod System. The integration of these functions made the intercept approach semi-automatic. Whirlwind's programs acted on the received radar data to produce vectoring instructions that could be used for midcourse guidance of manned interceptors, and they provided displays that the human operators could select to direct the intercept operation. These activities of the machine, when broken down into further detail, produced a carefully orchestrated sequence of operations:

> In processing data, the computer automatically performs the track-while-scan function, which consists of 1) taking in radar data in polar coordinates, 2) converting it to rectangular coordinates referred to a common origin, 3) correlating or associating each piece of data with existing tracks to find out which pieces of data belong to which aircraft, 4) using the data to bring each track up to date with a new smoothed velocity and position, and 5) predicting

> track positions in the future for the next correlation or for dead reckoning if data is missed. Once smoothed tracks have been calculated, the computer then solves the equations of collision-course interception and generates and displays the proper vectoring instructions to guide an interceptor to a target.[4]

The intercept procedure was semi-automatic. At the same time that human operators decided which aircraft were targets and which were interceptors, and could initiate surveillance of any track, initiation of tracking was automatically being carried on by the machine as a consequence of the sorting of the scanning data being acquired. The operators had at their disposal photocell "light guns," which they could place over any blip on a scope and, by pressing a button, tell the computer to begin tracking that blip (figure 25.1). Manually setting a selector switch would tell the machine whether to treat the blip as an interceptor or as a target.

As Wieser explained it to those attending scheduled demonstrations of the system, in such ways "the human beings make decisions and

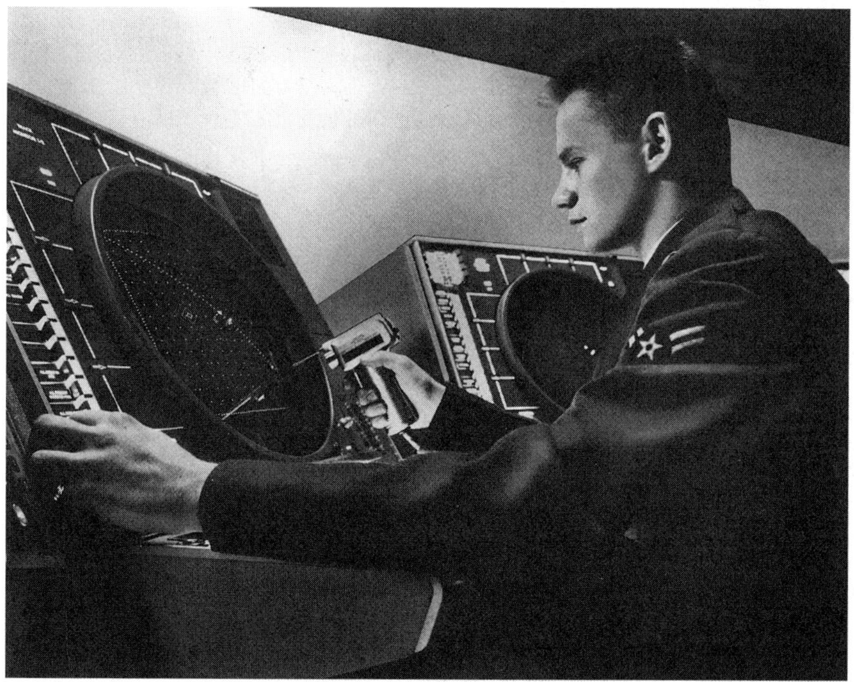

Figure 25.1
An airman with a light gun. *U.S. Air Force.*

improvise, while the computer handles the routine tasks under their supervision." To assist the operators, varieties of information, including alphanumeric symbols and geographic positions of observed aircraft, could be presented on the display scopes (cathode-ray tubes) at the operators' will.[5]

The two demonstrations Wieser described for visitors that day in March of 1953 vividly illustrate the state of the art they had reached by that date, when they were still relying on the Whirlwind computer:

> The first is an interception, in which a target aircraft is sent on a simulated raid. An interceptor is then assigned and both the target and the interceptor are tracked automatically. The computer generates collision-course vectoring instructions to guide the interceptor to the target. These instructions are displayed numerically on a cathode-ray tube and relayed to the interceptor by radio telephone. . . .
>
> The second experiment illustrates automatic tracking of up to sixteen aircraft. The initiation of tracking is manual (by means of the light gun) in part of the operating area and is automatic in part of the area. In the automatic initiation zone the computer stores all radar returns which do not fit existing radar tracks and waits for further data to form a new track. At present, the arbitrary criterion for a new track is two pieces of data received in five consecutive radar scans. Once a track is established, a vector (arrow) is displayed on the scope to show the direction and approximate speed of each track. An additional feature of the program allows the operator to call for information (by means of his light gun) on any track. The computer then displays numerically the track position, speed, and heading.
>
> These experiments are only a few of the many that have been carried out. . . .[6]

The plans the collaborating groups of the two divisions made during the year to set up the Cape Cod System Tests were elaborate and detailed. Their object was to concentrate on setting up and expanding the system by installing and testing the equipment and running realistic air defense experiments. Ten more small radars were to be added to the air surveillance net over the course of the spring and summer, bringing the total to 12. Eleven 16-inch computer-output display oscilloscopes were to be installed, bringing their total to 14. A teletype link between the direction center in Cambridge and the Boston Air Route Traffic Control Center was to be established, as were telephone lines providing a direct connection to local fighter bases to transmit "scramble" orders.

Before the summer's end, a 10-kilowatt UHF digital ground-to-air link was to be in place and operating.[7]

During the autumn, assembly of the system was to be completed by installing separate display controls that would permit generation of a display from the computer's magnetic drums. There would also be automatic "tote boards" for presenting fighter status, tracks carried, hostile aircraft, and similar data. When completed, the 1953 experimental system would, its designers confidently predicted, "permit operation with larger numbers of aircraft over a larger area," enabling the system's operators to "carry on more of the functions of a complete air defense system."[8]

Not only were the 1953 Cape Cod tests scheduled to become more elaborate as they progressed; at the same time, components and subsystems would continue to be tested while system tests were proceeding and while the entire system moved into its final stages of construction. TM-20 called for three system tests to be carried out during the spring. The first would check the functioning of the "computer program for tracking from radar network data," an action to be made possible by installing the magnetic auxiliary and buffer drums. The second would check the "automatic correlation of radar data with flight plans received direct from ARTCC and AMIS . . . a first step toward automatic identification." The third would check out "computer programs for automatic assignment of interceptors . . . to targets."

These spring tests scheduled for 1953 were to be followed by three more during the summer that would establish track-while-scan surveillance of up to 100 aircraft simultaneously, carry out live tests in order to check identification procedures in a real-time mode, and run multiple interception tests assessing automatic assignment of interceptors. In the autumn there would be still more complex tests, including multiple automatic interceptions that would call upon the computer's capacity to guide up to 25 interceptions simultaneously while checking the performance of the system against simulated raids. The autumn tests, the designers believed, would "demonstrate the essential elements of a complete air defense system."[9]

By this testing program, as it was laid out in TM-20 early in January of 1953, the Lincoln Lab designers intended to phase the Cape Cod System into the Transition System. But the fruition of such plans depended, as always, on the work done, and the work to be done required detailed operating goals to be set and met. "The Lincoln Laboratory has decided," said Wieser in a mid-January 1953 Biweekly Report circulated routinely among all staff members in Division 6, "to demonstrate to the Air Force the performance of Lincoln's proposed air defense system, and it is most important that this demonstration be held by September, 1953. Needless to say, this is a tight schedule which requires an all-out effort from all. . . ." To this end, Forrester brought engineers from Groups 61 and 64 together in several meetings to shape specific plans setting July 1 as the "freeze" date for construction. This would allow time for both program testing and operating practice. The buffer drum probably would not be available by that date, so a temporary alternative would have to be found that would let them meet the testing deadline. In any event, they should attempt to incorporate as many Transition System features as they could build within that time.[10]

By April of 1953 the engineers were ready to install in an appropriate rack on Whirlwind I the MITE (multiple input terminal equipment) they had designed to convert incoming SDV signals to pulses that the computer circuits could use. The MITE transformed into binary-coded pulses the telephone-line signals coming from the North Truro CPS-6B radar set, as the radar fed range and azimuth data into the appropriate counters connected to the in-out register of the computer. This arrangement had the advantage of reducing the sampling load at the in-out register. After preliminary testing, the engineers' plans called for them to hook MITEs into flip-flop buffer circuits being provided as temporary (but reliable, they felt) buffer storage until the buffer drum from Engineering Research Associates could be installed. Once the buffer drum had been installed and checked out, as scheduled for the autumn of 1953, MITEs could be tied in to the drum to convert data coming not only from North Truro but also from at least two smaller sets. Such was their thinking in the spring of 1953.[11]

Meanwhile, 24-hours-a-day testing of the recently arrived auxiliary drum from Engineering Research Associates boded well for the scheme of using special auxiliary and buffer storage drum systems to assist the computer in handling and interpreting data. The testing demonstrated that the usual "break-in" modifications of the drum-system designs were in order, and in the middle of February 1953 E. S. Rich of Wieser's group spent a couple of days with the engineers at Engineering Research Associates in St. Paul, Minnesota, discussing the "temperature sensitivity in the new ferrite-core heads" and other problems. Accomplishing the necessary redesigning and fabrication would put improved equipment in the Lincoln Lab engineers' hands in April. Then they could begin to tie in the drum systems with the computer. The drum had overheated on one occasion, when some heads had run onto the surface of the drum, and new read-gate crystals from Engineering Research Associates had to be installed. Also, the strength of the output signals from some of the drum tracks had faded appreciably over time, and this "glitch" too had to be corrected. In the meantime, marginal-checking tests on the drum circuits indicated that the drum system was about ready to be incorporated into Whirlwind's routine marginal-checking procedures. From the point of view of those involved on the working level, the operation was making reasonable engineering progress.[12]

Just as challenging as these problems of organizing the system to accommodate SDV transmission equipment or the auxiliary and buffer drum systems was the task of anticipating the limitations on the performance of the Cape Cod System. Some of these were limitations that a completed SAGE system would encounter. R. L. Walquist had pointed out in the autumn of 1952 that there could be times when the capacity of the system might be exceeded: too many aircraft to track, faulty incoming data to the computer, a sudden change in the data rate, or similar abnormal situations. To expect a human operator to catch these would be to expect him to be constantly on the alert, an unrealistic expectation. If the computer were programmed to catch them instead, it would all be in the line of business, so to speak, and it would occasion no special provision beyond asking the computer to identify various quantities (whether too large, too small, or obviously garbled and incoherent) and instructing the machine to "flag" a human operator when such a quantity occurred. In this way automatic continuous monitoring of the system would be achieved, whether data were unusual or lacking.[13]

As viewed in the spring of 1953, Cape Cod was rapidly becoming an experimental prototype model, writ small, of the projected continent-wide system. "A few items are ahead of schedule and a couple of items are slightly behind," reported J. A. "Gus" O'Brien in March, referring to work being carried forward on the various phases of the Cape Cod terminal equipment program. By the end of April, Steve Dodd was reporting that the Cape Cod installation was making satisfactory progress and was about on schedule. Referring to the modifications being made to Whirlwind I to ready it for the full operation of the 1953 Cape Cod System already planned for the autumn, Dodd anticipated that "a major portion of the equipment and installation associated with the control center will be completed in three weeks, and system testing can start at that time" in order to begin bringing the various parts of the system into coherent operation with one another.[14]

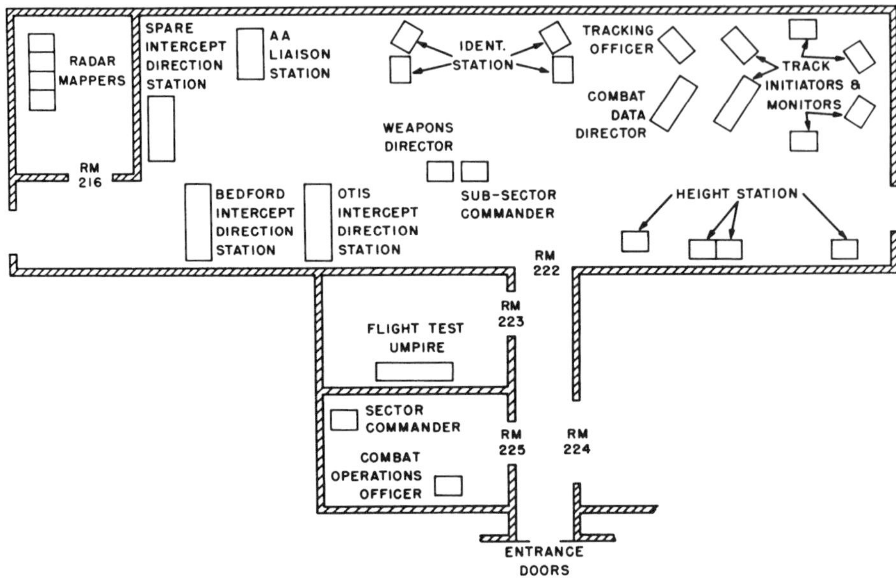

Figure 25.2
Floor plan of Cape Cod Direction Center.

The design of the overall Cape Cod System was kept flexible in order to test new hardware components and new software programs. Thus, its radar inputs could be monitored singly or in concert as questions arose. Live tracking of live aircraft with live radars and live interceptors would, the engineers in charge knew, "verify the soundness of the whole concept." High tracking capacity was a must, as were reliability and speed. The data that the Cape Cod experiments would turn up would set key specifications for the design of the SAGE computer.[15]

One of the arrangements being incorporated into the Cape Cod System would simultaneously indicate its feasibility for SAGE and improve and shorten test procedures. This was a "man-to-man" telephone communication network that was being installed during the summer of 1953 by the New England Telephone and Telegraph Company. It would enable operators in the Combat Center, in the Direction Center, and at selected remote locations to talk to one another. (At the start, both the Combat Center and the Direction Center were located in the Barta Building with Whirlwind I. See figure 25.3.) The network being installed was designed to accommodate expansion and modifications without major equipment alterations. It would incorporate one or sometimes two

Figure 25.3
The Direction Center of the Cape Cod System, in the Barta Building.

standard six-line 100-series Telephone Company key boxes equipped with a toggle-type key and a signal light for each of the six circuits. There was no provision for conference calls; only one two-way voice circuit could be operated at a time from any of the communication boxes.[16]

The 1953 Cape Cod System was complicated indeed, as is evident from its track-while-scan operations. The system's extensive software included two separate subprograms necessary to its track-while-scan (TWS) functions: a smoothing and prediction subprogram and an extrapolation subprogram. They would identify and discriminate between new "tentative" tracks and "established" tracks by examining successive radar inputs during the course of four or five successive radar sweeps. The subprograms would examine successive radar inputs during the course of a minute, making use of a scheme of breaking down the input from a radar into a "frame" 12 seconds long and comprising four 3-second "subframes." Tentative tracks were those which were new or very recently initiated as a consequence of track velocities calculated and built up by the smoothing equations ("NLS-2C type") employed in the smoothing and prediction subprogram invoked "during the TWS part of the first subframe." The buildup of velocities, identified by extrapolation and prediction calculations carried out by the computer, was examined "five frames (one minute) after automatic initiation" of a track, in order to make a speed comparison. Any track whose speed was above an identified value (say, 100 knots) was worth re-examining in the next radar sweep; it became, at least temporarily, an "established track." If the speed was too slow as a consequence of the findings of the two subprograms, the track was ignored; the computer registers that had been used to make the computations were then freed for use for the next automatic initiation.[17]

As these Track-While-Scan performance challenges were being met, Group 61 was running tests on the computer while it was harnessed to the growing number of radar reporting stations with which it was acquiring communication. Also, the group was designing and setting up the experimental Air Defense Center for the Cape Cod experimental net, and it was preparing and testing other computer programs to be used in the first series of tests.[18]

While Division 6 was concentrating on the computer portion of the projected system, Division 2 was equally active, focusing the attention of its engineers on the radar portion. Even before becoming a division of Lincoln Lab, Valley's group had begun to search for the radar sites and the equipment that would make up the primary sensor network of the Cape Cod System. Some fourteen sites were selected, located so as to cover, around Boston, an area extending about 120 miles in a north-south direction and about 80 miles in an east-west direction. In order to provide overlapping coverage, each site was within 20 or 25 miles of at least one other site. Each radar had a range of 30 miles.[19]

The radars installed at the sites were a mixed lot, obtained by Lincoln Lab from the Army, the Air Force, and the National Guard. It had been concluded very early that developing radars specifically for the air defense mission would take too long, so a decision was made to use those radars that might be available and could be modified to approximate the specifications considered appropriate to the air defense task.[20] By the summer of 1952, radars were in place at Rockport and Scituate. They were equipped with temporary connections to Group 61 in Cambridge in order to expedite the study of any smoothing and tracking problems that might arise when data were being received from several radars.[21] By the summer of 1953, the installation phase of the radars had given way to a period of operation and modification in which a large part of the work in progress was devoted to analysis and evaluation of performance.[22]

While Boehmer's Cape Cod group was putting together the radar net for the experimental system, Harrington's Data Transmission group was devising the SDV equipment as a replacement for the digital radar relay that had been used in the earlier Bedford tests. As their work on the system progressed, they realized that the digital radar relay had definite limitations, for it transmitted only selected significant elements in the radar picture. While suitable for use in the Bedford tests, it was, in the view of Harrington and his engineers, "hardly a suitable scheme for use in the radar network" that would make up the Cape Cod System. Re-examination of the problem by Harrington and his associates caused them to devise the SDV approach; this gave the computer not a selected picture but "the entire quantized picture" they desired.[23]

Slowed-down video was the second of three devices developed by Harrington's group for the transmission of radar data over telephone lines. It was an accomplishment that, seen from Everett's later, informed perspective, was "just as important to the concept of SAGE as the digital computer." Harrington's study of the problem had begun at the Air Force Cambridge Research Center, where a group under his supervision had been at work developing techniques for digital signal processing and for digital transmission over telephone lines. Everett, writing in the *Annals of the History of Computing* years later, noted that Valley had promptly "seized upon these two pioneering efforts"—Harrington's solution to the telephone line transmission problem and the Whirlwind computer enterprise that Forrester and Everett were leading. Valley had recognized that "they made a centralized computer-based air defense system possible."[24]

When Division 2 was formed, Harrington and his associates joined the laboratory as Group 24, at the same time retaining their official relationship with the Air Force. Consequently, Group 24 in Division 2 was a mixture of Air Force and Lincoln Lab employees. It was after the group joined Lincoln Lab, Harrington recalled, that it developed SDV for the Cape Cod System and then went on to develop for SAGE a still better device: the "so-called fine-grain data (FGD) system (later produced as the FST-2)."[25]

When laying out their schedule at the start of 1953 for what they thought of as the "1953 Cape Cod System," the engineers had set May 15 as a target date for completing installation of the necessary equipment. July 1 should then see the initiation of "shakedown testing" in anticipation of beginning serious system tests in September, when the system and its equipment should be in place. As Wieser had pointed out, the schedule was a tight one. It had called for five stations to be operating by May 15. Instead, by the middle of May the designers were willing to settle for two completely wired consoles and the framework for other operators' stations. "All the computer panels have been installed and wired into the distribution box," the Group Leaders in Division 6 were told at their May 18 meeting. "Power has been brought to most stations. The first two scopes and the first one or two stations will take several weeks to be checked out. Although considerable trouble is anticipated, the July 1 date still may be met."[26]

The entire situation was by no means secure, however. Although Whirlwind I's electrostatic storage doubled in size late in the spring of 1953 to "two banks of storage without spare storage tubes," increasing the total number of registers of internal, high-speed storage to 2048 and theoretically providing "much greater flexibility for computer programs," the tubes themselves continued to be disappointingly unreliable, short-lived, and expensive. And as the Cape Cod computer programs grew more elaborate, the storage tubes grew less reliable. "On July 25, parity alarms occurred at the rate of four per hour."[27]

There were other problems. The engineers expected a buffer drum to arrive from Engineering Research Associates by the end of May or early in June, but they realized that programming for it would be difficult to provide in time to be of use in the planned system tests. Of course, they would begin installing and testing the drum as soon as it arrived, but a realistic view of putting it to work as part of the system had to accept "late in 1953" at best. Some of the engineers suspected that the September system-operating target date would produce at best a "September Demonstration," but Forrester and Everett refused to lower their sights. The Cape Cod System was to be "an operating system," Group Leaders were reminded. Nevertheless, the schedule remained too tight to meet. July 1, the target date for "end of system construction," came and went with only a laconic notation in David Brown's "Group Leaders' Meeting" minutes to mark its passing: "Installation of the Cape Cod System was to have been completed July 1."[28]

Though the effect of Cape Cod's tight schedule was to place the work chronically behind schedule, this was a form of R&D "brinksmanship" that the Digital Computer Laboratory had been practicing in other areas at least since it had joined Valley's ADSEC project in 1950. Schedules for designing Whirlwind II (later called the AN/FSQ-7) were always uncomfortably tight, as were those for producing satisfactory electrostatic storage tubes or magnetic cores and the Memory Test Computer. The "Rough Resume of Magnetic Core Memory History" that Bill Papian wrote for Everett in November of 1953 contains this remark: "In the fall of 1950 J.W.F. began pushing for applied and basic research work . . . aimed at developing . . . ferrites," and in 1952 and 1953 there was "hectic design and development work . . . on cores, core handling and testing gadgetry (Group 63), and memory circuitry and layout (Ogden and others)."[29]

A consequence of this sort of pressure was the sequential installation and testing in the Memory Test Computer (MTC), from April into June of 1953, of 17 planes of 32 × 32 magnetic-core grids. By the third week of May, Papian was willing to report that "the MTC Magnetic-Core Memory has been operating, after a fashion, for a couple of weeks now." As the summer wore on, the operation of the core grids proved so impressive that at the end of July Forrester decided to transfer the MTC magnetic memory to Whirlwind I, hooking it in alternative parallel with one of the two banks of electrostatic storage tubes.

When Wieser was asked 30 years later when he had become sure, as an engineer, that the Cape Cod System and the SAGE approach to air defense could be made to work, he replied: "When magnetic core storage was installed in Whirlwind."[30]

26

Further R&D Planning and the 1954 Cape Cod System

As the higher-level administrative and policy-making events recounted earlier were gathering speed and force during 1954 and 1955, the men at Lincoln Laboratory continued to encounter engineering and scientific problems and to develop solutions. The overall effect of this activity was to shape, sharpen, and subject to pragmatic testing the overall technical program that integrated radars and computers.

Thus, Edward Cochrane's perception that TM-20 had become obsolete was shared within Lincoln Laboratory, and during the remainder of 1954 Jay Forrester led the effort in Division 6 to overhaul and update TM-20. As a preliminary to collaborating with Colonel Oscar T. "Tom" Halley and his special staff group (which reported to Major General Frederick H. Smith Jr., Vice Commander of the Air Defense Command), Forrester assigned rewriting tasks to various engineers in the division. Forrester's and Halley's job was to redefine the Transition System to meet the military operating requirements of ADC more closely. "ADC is writing the body of the new report and the Lincoln Laboratory is writing the appendices," Forrester informed his Group Leaders.[1]

Revision of TM-20 included not only technical updating of the planned air defense system but also a change in name. (The military preferred its own choice.) Lincoln Lab Director Albert Hill referred to this preference at the Lincoln Steering Committee's meeting of July 19 when he warned those preparing a presentation to the Joint Services Advisory Committee, which was scheduled to meet at the laboratory near the end of the month, to "avoid the use of 'Lincoln Transition System' as it means different things to different people. Better expressions are 'Semi-automatic Air Defense System' or 'Semi-automatic Ground Environment.'"[2]

That the Transition System was about to lose its name and become known thenceforth as the SAGE (Semi-Automatic Ground Environment)

System was a sign of the final acceptance of the Lincoln Lab approach to air defense as embodied in TM-20, even in those Air Force circles where a significant number, including Colonel Halley, had favored the Michigan approach. Four and a half years had passed since George Valley had begun urging Theodore von Karman to set up the ad hoc air defense inquiry for the Air Force that became ADSEC.

While work on the revision of TM-20 continued, Division 6 did not await completion of the revised report to firm up a variety of technical details of the projected New England regional model of the air defense system that they began to call the "Experimental SAGE Subsector":

> The XD-1 system will survey the area within a 250-nautical-mile radius of South Truro (Fig. 61-3 [26.1 here]). Coverage will be provided at 10,000 feet by the search radars in the XD-1 system. During initial operation of the system, there will be five long-range radar sites tied in by pre-production FGD [Fine-Grain Data] and 10 gap-filler sites using SDV [Slowed-Down Video] equipment. Ultimately, three additional long-range radar sites with production FGD equipment will be tied in to the XD-1 system.
>
> In addition to these search radars, 10 height-finder radars will be employed. Two of these will be the manual sets located at Rockport and Nantucket; the remaining eight will be semiautomatic. . . .[3]

Forrester reported to the Lincoln Steering Committee in the middle of August that completion of the "SAGE System Report . . . has been delayed one month to allow thorough editing of the entire report by ADC as well as Lincoln."[4] Actually, it was not until March of 1955 that the Air Defense Command issued its 300-page Operational Plan, which became familiarly known among those who used it as the SAGE "Red Book."[5] Thereafter it replaced TM-20 as the major technical planning document of reference.

Listed in Chapter X of the Red Book, as part of the "Experimental Program" to which Lincoln Laboratory and ADC were formally committed, was the "Experimental SAGE Subsector." It was the third and last of the major experimental intercept-testing programs from which the military SAGE continental defense system emerged before the end of the 1950s. First there had been the 1953 Cape Cod System, the feasibility of which had been indicated by the earlier Bedford tests; then the 1954 Cape Cod System, still extant in 1955 when ADC issued the SAGE "Red Book" and still using the Whirlwind I computer on the MIT campus; now there was the "Experimental SAGE Subsector," which Lincoln Lab and the Air Force expected to bring into being by stages during 1955 and

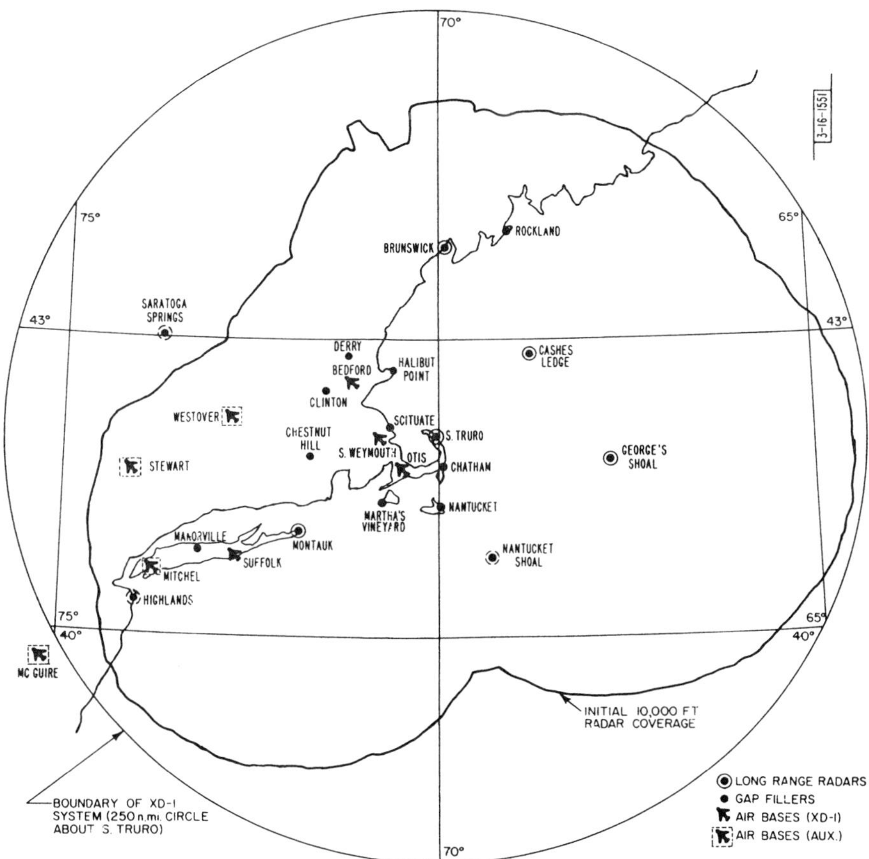

Figure 26.1
"Initial heavy radar coverage for XD-1 experimental SAGE subsector."

1956. ESS would have its own computer, the first XD-1 prototype model of the AN/FSQ-7 machine intended for use in Direction Centers and Combat Centers across the nation in the ultimate SAGE system to be operated by the Air Force. The FSQ-7 would give each Direction Center, said the Red Book, "the capability of processing information on 400 tracks, of which 200 may be weapons for which guidance instructions are generated."[6]

From a technical engineering point of view, the fact that these engineering scientists, computer programmers, and mathematicians were building a novel aerial defense system was overshadowed by the fact that they were committed to building an ever more complex and elaborate

Figure 26.2
"Map of 1954 Cape Cod System."

system. It would become a men-and-machines-interacting arrangement of mind-boggling intricacy, one in which neither man nor machine could falter without bringing down the entire array.

The Division 6 engineers under the leadership of Jay Forrester and Bob Everett were agreed that the way to maintain the system's operating integrity was to construct it of modular units, the reliability of each having been demonstrated beyond question by exhaustive testing. Elegant design and theoretical consistency were not enough. Reliable operation in practice was everything. It would not be easy to obtain.

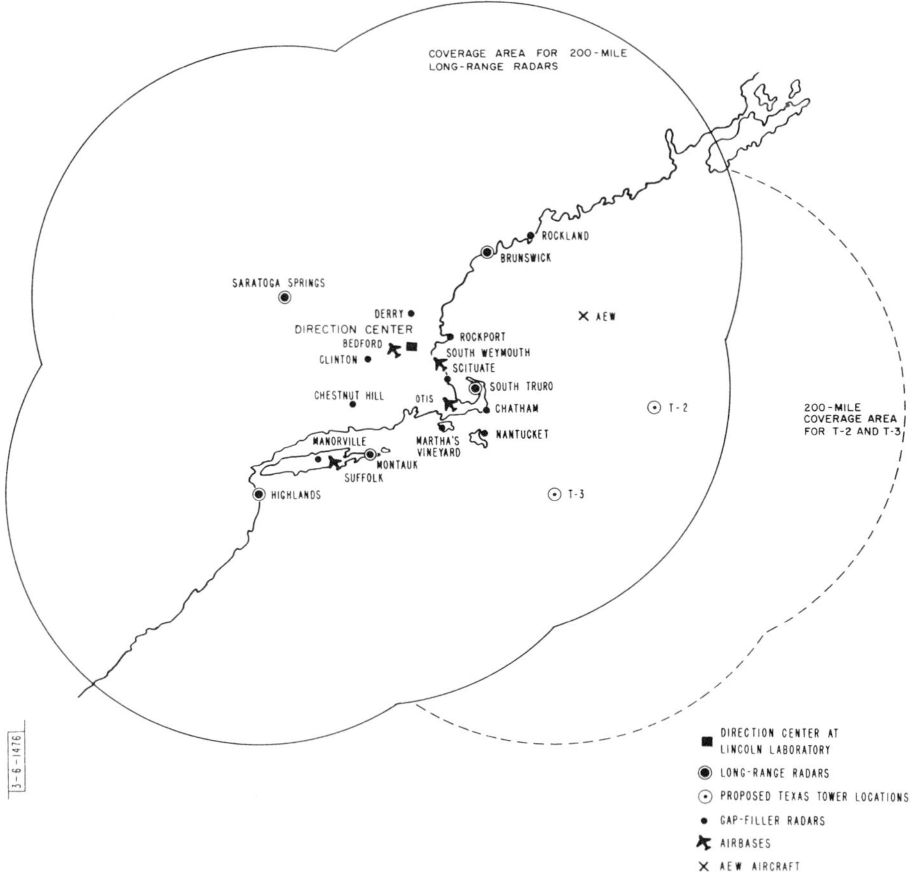

Figure 26.3
"Experimental SAGE Subsector."

One way to obtain reliable operation was to incorporate the lessons learned from the 1953 and 1954 tests of the Cape Cod radars that were tied in to the original Whirlwind computer. These intercept test systems, from the early Bedford experiments through the 1953 Cape Cod System, had demonstrated the correctness of the MIT approach to air defense. They provided substantial confirmation of the provisional trust in the Lincoln Lab project that the White House and the Air Force had earlier displayed. At the same time, they revealed to Forrester, Everett, Wieser, Valley, and Harrington, as these men and their engineers proceeded to develop the line of attack on the technical problems that TM-20 had laid out, how essential it was at this stage to learn more about the performance of the system's components, to refine the software they were developing, and to make more explicit the standard, modular operating procedures that would guarantee a working air defense system.

Such were the tasks that experience with the 1953 Cape Cod tests imposed on the next phase of activity, which the engineers came to call "the 1954 Cape Cod System" and which they came to think of as "a pre-proto-type model of the air defense system under development in Lincoln Laboratory."[7] Serious planning for the 1954 Cape Cod System got underway early in the spring of 1954.[8]

Indeed, if the Cape Cod model in expanded form were to work as expected, and if the next phase planned (the prototype "Experimental SAGE Subsector") were to be built upon the knowledge gained from working with the 1954 Cape Cod System, then the system's builders would be proceeding in accordance with the specifications of the "Transition System" described in TM-20 and subsequently modified by the Air Defense Command's "Red Book" of March 1955. The characteristics of the emerging system would determine the form of the continent-wide defense that was now being sought with increasing urgency in response to the strategic lessons of the Korean War and the uneasy truce of July 1953.

As E. O. Martin of the Bell Telephone Laboratories put it in a résumé of earlier technical reports he compiled in the autumn of 1955, the 1954 System tests were "expected to provide quantitative test data which can be extrapolated to predict Experimental SAGE Subsector and Operational SAGE Subsector performance under both saturation and non-saturation conditions."[9]

The 1954 Cape Cod System would be essentially the same as the 1953 System, but enlarged and improved.[10] An expanded radar network would include two long-range radars and six or seven gap-filler radars. The long-range radar at South Truro would be equipped with Fine-Grain Data (FGD) transmission equipment and a Mark X "Identification, Friend or Foe" (IFF) beacon, and special attention would be given to achieving the accurate measurements that could be obtained using fine-grain search-radar data.

Other operational changes included a more sophisticated computer program to accomplish automatic tracking, computer-generated displays that would more closely resemble the display format intended for the FSQ-7 air defense computer that Division 6 and IBM were developing, and improved techniques and facilities for evaluating the system. From its scanning center near the northern tip of Cape Cod, the 1954 System would have an operational radius of 200 miles and would handle 48 tracks at once, one-third of which could be those of interceptors.[11]

The opportunity to incorporate technical advances in the testing system was not the only reason for making sufficient changes in the Cape Cod System to warrant giving it a new name. During January and February of 1954, two-week programs of "familiarization and participation with the Cape Cod System" were arranged for three small groups of selected officers from the Air Defense Command and for a small number of personnel from the Air Defense Engineering Service's Project Office and the Air Force's and the Army's liaison offices. These programs solicited expert recommendations from military officers familiar with the tactical problems of intercepting enemy aircraft. Group 61 went over these recommendations in detail early in March, looking to see how they could implement the recent decision to revise the 1953 Cape Cod System. "This revision will be on a large scale," Dave Israel noted in a report to Bob Wieser, "and the resulting system, expected to be operative this year, will be known as the 1954 Cape Cod System (1954 CCS)."[12]

How much effort was applied to analyzing past achievements and problems and anticipating future technical developments was demonstrated by two more familiarization programs for Group 61 staff members, one scheduled for late March and one for late April. These were patterned after the earlier ADC presentations. The first of these eight-day programs considered the 1953 and the 1954 Cape Cod Systems and their implications for projected use of the AN/FSQ-7 computer that would replace Whirlwind I. A 21-page M-Note prescribing 33 relevant technical reports for "homework" reading was prepared.

To orient new group members, various members of the staff discussed the organization of the 1953 System and the associated allocation of personnel and duties. They described the major categories of Cape Cod operations. These included Track-While-Scan, radar input and mapping, the intercom and display-console arrangement and functions, height finding, flight size, identification techniques, intercept techniques using aircraft and subsonic missiles, the use of simulated flight tests in training Direction Center personnel, and the organization of the Combat Center. Since the next step up in sophistication would be a radar network hooked to the prototype version of the projected air defense computer rather than to the Whirlwind machine, a day was allotted to discussing the organization and operation being planned for this next-generation ESS system.[13] Although this scheme appears to have been followed in the

March familiarization program, experience with it suggested improvements that were then made to the April program, which was designed to present a coherent picture to non-Lincoln personnel. More emphasis was given to the theory of digital computers, and five sessions on the "XD-1" prototype computer were added. Once again, more attention was being given to getting ready for the future.[14]

Following the detailed planning begun in the spring, the Direction Center was rearranged during July and August of 1954. Additional mapping equipment was installed to filter the returns from the increased number of radar sites, and the radar-handling capacity of the terminal equipment was increased by adding more multiple-input terminal equipment and associated controls and by making available more buffer storage on the drums. A number of other changes were made, Steve Dodd noted in the quarterly report, including "rewiring of switches, relocation of indicator lights and switch panels, and . . . relocation of display panels, to render the system more effective and reliable."[15]

Though the major changes in the Direction Center were relatively few, the detailed changes were more numerous, and the way to handle them was to rewrite the computer program.[16] Thus, two new simulation programs were proposed in April—"one to simulate track data, the other to simulate radar data," Israel reported. At the same time, improvements were made that provided for complete simulation tests "without manual intervention at the simulation station," and changes "consistent with the plans for simulation in the XD-1 System" were made.[17] In proposing equipment changes, the Division 6 engineers were also guided by standard procedures that enabled them to avoid the confusion that tends to result when many individuals are making many changes in a complex system. These procedures, said Israel, stipulated "first, it must be satisfactorily determined that the change will not cause any undesired changes in programs, displays etc., and secondly, the changes must be processed so that all records can be up-dated."[18]

The main purpose of invoking these formal procedures for installing the 1954 Cape Cod System was to implement in an orderly fashion the technical decisions that had been made to alter and improve the system's operation. Those decisions rested on investigations into such factors as the accuracy of the radar data being received by Whirlwind. A new tracking technique referred to as "Raydist"[19] offered an attractive opportunity to measure how accurately the radar performed, even before the decision was made to suspend the Direction Center's operations long enough to set up the 1954 Cape Cod System.

The Raydist tracking system is an example of the not-unreasonable scheduling price that Lincoln Lab, led (not always willingly) by Division 6's R&D policy, was paying as a consequence of the designers' commitment to improving the quality of their emerging air defense system at every opportunity. In December of 1953, preliminary tests of this navigation and tracking system were undertaken in the hope of improving the tracking accuracy of the Cape Cod radar. Israel explained how the system worked:

> The Raydist tracking system is a cycle-counting device. In the system being used, a transmitter mounted in the aircraft to be tracked sends a continuous wave (at about 175 megacycles) to two fixed points (master and slave). A phase comparison is made at the master station of the signal from the aircraft received by the master and slave. The amount of phase difference between the received signals defines a hyperbola along which the aircraft is located. This hyperbola is a path of constant difference in distance from the aircraft to the two stations. As the aircraft moves, a phase shift occurs which corresponds to a similar change of the difference in distance from the aircraft to the receiving points. For each phase shift and corresponding change of the difference in distance, another hyperbola is determined.
>
> The addition of a second slave receiver to the system allows another phase comparison to be made and determines another family of hyperbolas. The intersection of the two hyperbolas defined by the two slaves with the master thus determines the two-dimensional position of the aircraft.[20]

For the experimental flight tests of the Raydist system conducted in December of 1953, the mobile transmitter was placed in a B-25 medium bomber, the master station was located at Scituate (just south of Boston), and slave receivers were placed at Halibut Point (north of Boston) and at Chatham (on Cape Cod). "Since only two-dimensional tracking was required and altitude could be selected for any test," the engineer Benjamin Stahl explained in a report to Group 61, "only three receiving points were necessary." After the tests were run and the data recorded at the Barta Building on a multi-channel Ampex magnetic tape recorder, programs were written during the first half of 1954; these called for the computer to compare the customary radar data with the phase-shift data obtained from the Raydist equipment by reducing their information to the "common rectangular coordinate system" of information used in the Cape Cod flight tests.

Analyses of the results by members of Group 22 revealed during the spring and summer of 1954 that the Raydist calibrations of the accuracy of the radar data were "capable of meeting the demands for accuracy of measurements as well as providing data required for radar calibration," reported Stahl. The South Truro FPS-3 radar, for example, reported both a range "too large by at least one mile" and a percentage range error that "varied directly as the distance of the target from the radar site." Further tests were warranted. Meanwhile, plans were made to incorporate Raydist calibrations of radar data in the 1954 Cape Cod System.[21]

In addition to the Raydist studies, a variety of other activities were contributing to the development of the 1954 Cape Cod System. Of more central concern were the enlarged Direction Center operations proposed for the next Cape Cod tests. The Direction Center operations of the 1954 System would require both improved hardware and new software.

In designing and writing the new programs, the Group 61 staff let itself be guided by the programming specifications prepared for each of the individual programs, such as the tracking, height-finder, and interceptor programs. Once the new programs were written, the men embarked on an extensive and comprehensive checkout of programs and equipment. They developed standard procedures for checking out a program, and they wrote a number of utility programs, thereby enabling the computer to assume many of the equipment-checkout tasks. By the end of August, the men regarded the 1954 System as virtually completed, and they anticipated that preliminary operational testing could begin early in September.[22]

On July 28, 1954, in order to prepare for the use of military aircraft to test the 1954 Cape Cod System, four of the Group 61 engineers (led by W. S. Attridge Jr. and C. A. Zraket) met with two Air Defense Command officers from the squadron assigned to work with Lincoln Lab. A training program for Air Force personnel who would be working in the Direction Center during the 1954 flight tests was to get underway in August. First, six commissioned officers and six noncommissioned officers from Section C of the 6520th AC&W Squadron would attend eight 3-hour sessions, spread over 2 weeks, in a course devoted to "a general orientation in the background and objectives of Lincoln Laboratory, the use of a Digital Computer in a Direction Center, and a detailed description of the 1954 Cape Cod System." They would also give some attention to "the XD-1 and FSQ-7 Systems," which one day would replace the Cape Cod System. Then, in a three-to-four-week program for all the members of the squadron, specific training assignments would be undertaken: "track-while-scan, height-finding, identification, AAA, weapons direction, intercept direction, etc."[23]

While this practical preparation to the 1954 Cape Cod System tests was underway, Wieser's Group 61 was taking care of other essential matters. Since these are too numerous to describe in full technical detail, and since they ranged from the minute to the broad in scope, three examples of the improvements sought in Direction Center operations are perhaps best characterized impressionistically. Taken together, these procedures typify significant portions of the R&D work that was being done to ready the Cape Cod System for more ambitious and complicated flight testing. We will begin with an example of the attention that had to be given to minute detail.

If the Direction Center was going to keep track of airborne interceptors, the Whirlwind computer would have to be programmed for that purpose and appropriate data storage would have to be identified for use by the Weapons Direction Section in the Direction Center. As many as 16 fighters would be accommodated at one time, and for this purpose 16 blocks of 9 consecutive registers in Whirlwind's memory were set aside. These 144 registers could be tagged so that the Weapons Director could distinguish a Bedford-based interceptor from a Weymouth-based one. Furthermore, the type of data held in Fighter Data Storage would provide both vector instructions and alarm conditions of the interception program while permitting operators in the Direction Center to switch information from "Intercept Direction" to "Weapon Direction" status as desired.

Since each of the 144 registers contained 15 digits, each register in each of the 16 blocks was methodically reserved to present numerically such information as the "Activity Status" of the interceptor, its "Flight Status," and its "Hostile" status. The last would require a "Hostile Track-Slot Number" when the interceptor was assigned to a "hostile." A number of binary digits were assigned to tell the operator whether the interceptor in question was alone or one of several being assigned to the hostile aircraft. "Command Heading" was as important as "Interceptor Heading" and "Smoothed Interceptor heading." So were such items as "Command Altitude," "Indicated Airspeed," "Interception Possible," "Interception Impossible," "Parallel Course," "Bearing to Target," "Separation from Target," "Attack Heading," and "Time-To-Go" until intercept.[24] Such information, processed by appropriate programs and subprograms and compounded as many times as there were airplanes, became increasingly minute in one direction of machine operations committed to providing detailed information, as layer of program piled on layer. It also became increasingly broad in another direction, as the particulars of information were brought under program control in order to give the operators in the Direction Center the information they needed to understand and command the battle situation.

Such program control was provided, for example, by "Weapons Assignment" (WA) and "Intercept Direction" (IND) programs, which were frozen in their basic design by the middle of August 1954 in order to be incorporated into the 1954 Cape Cod System tests. The degree of sophistication of battle control being attempted was reflected in the major tasks these WA and IND programs were crafted to carry out:

> 1. Conduct the calculations necessary for the interceptor assignment and direction functions of the Weapons Direction Section.
> 2. Superimpose automatic restraints upon the results of these calculations so as to smooth and regulate the dynamic control of the interceptors.
> 3. Provide automatic communication of the relevant control data:
>   (a) to other sections of the Direction Center;
>   (b) to the operating personnel in the Weapons Direction Section; and
>   (c) when the automatic ground-to-air data link is used, directly to the airborne interceptors.
>
> These three functions are logically distinct. The interception calculation can and should be considered separately from the controls put upon its outputs. For the purpose of logical simplicity and ease of later modification, every effort will be made to separate these distinct functions within the computer programs. However, due mainly to timing considerations for data-link transmission and ease of indexing through the system capacity of 16 interceptors, the combining of the above three functions into a specific program (a separate read-in from the drum) which keeps these functions logically distinct may not be possible. Instead, some overlap within the computer program may be necessary, although this will not impose any serious restrictions on later modifications to the program.[25]

On a higher level of information processing and control of machine operations than that represented by the details of "Fighter Data Storage" or the broader "Weapons Assignment" and "Intercept Direction" programs were the three "Master Control Programs," which had been designed to control a number of essential functions in the operation of the computer. As was true of all these software programs applied first to Whirlwind in the 1954 Cape Cod System, if successful they would be rewritten for the more powerful XD-1, which would be used in ESS testing.

The first of the Master Control Programs to be called upon in the operation of the 1954 Cape Cod System would be the "Start Over Program," which would set the initial machine conditions of the system according to one of three selected modes: all tracks dropped and cleared from the system, or all track data maintained as they were when the computer was stopped, or friendly tracks dropped and all others extrapolated from the condition reported before the computer was stopped.

Equally complex and meticulous in operation were the other two master programs: the "Control Program," which established rational sequence and coherence among the multitude of subprograms that enabled Cape Cod to do its job, and the "Subframe Timing Program," which controlled the timing of the system.[26] Rather than descend to the machine-language level of examining how each binary digit was set in the appropriate registers in storage or in the arithmetic element or in the input-output sections of the computer, it is perhaps more enlightening to realize that experience with these programs and procedures in the 1953 and 1954 Cape Cod Systems was providing practical warrant that they would work as component features of the command-and-control system that would lie at the core of the SAGE approach to air defense. The SAGE System, unprecedented in military or civilian experience, would be composed of both hardware and software elements that had been tried, tested, modified, and improved in the light of test results piled upon test results in mock battle situations of increasing scope and complexity.

As the autumn of 1954 approached, the designers continued to suggest improvements for the system. Ground Controlled Interception had begun as a manual operation; however, as the techniques of gathering data and acting on them became more sophisticated, automatic GCI became more attractive. Indeed, when radar reception was relatively free of unwanted signals ("noise"), tracking procedures had become well defined in an automatic tracking program. But in other regions, the engineers admitted, "the function of the scope operator at a GCI Station remains more of an art than a science." So they developed a computer program to monitor questionable tracking data in such a way as to "combine the art of the scope operator (monitor), his imagination and ingenuity, with the science of the computer, its capacity and its speed of calculation." In a proposal that analyzed equipment, personnel, and operating procedures, they described a program that "senses the characteristics of the tracking function, decides upon the utility of manual monitoring for individual tracks, and allocates tracking trouble situations to humans according to a priority scheme."[27]

Our discussion of these representative efforts has focused on expanded tracking and intercept procedures, the outlines of which were developed while the improvements that would characterize the 1954 Cape Cod System were being established at the end of the winter of 1953–54. Ultimately these planned procedures became the formal "Operational Specifications" for the 1954 system. The circumstances of their genesis provide significant insight into the way Division 6 carried on research and development, and this genesis was rationalized and explained in the Weapons Direction specifications codified early in the autumn of 1954 in a detailed technical "working document" prepared by Herb Benington and Charles Zraket:

> Planning for the Weapons Direction Section of the 1954 Cape Cod System (CCS) was initiated in a series of meetings between members of the Weapons Direction Section and between representatives of this section and the Tracking Section. The overall philosophy was to design a system which would: a) change the operational layout, the displays, and the tasks of the operators in the Direction Center so that the system would more closely resemble the operation of an AN/FSQ-7 Direction Center; b) cause only slight modifications of the equipment used in the 1953 CCS; c) correct deficiencies which were observed in operation of the 1953 CCS; d) meet existing time schedules; e) incorporate improvements in the computer programs suggested by operation of the 1953 CCS; and f) prepare computer programs that would be amenable to modification at a later date.
>
> Results of this planning were noted in a series of inter-office memoranda between members of the groups. Operational specifications for the Weapons Direction Section were than prepared and issued in a single inter-office memorandum for comment and criticism by members of the group. Following agreement by the responsible people, this materiel was brought up to date and is now being issued as this M-Note which serves only as a working document. Programming specifications are also being issued in a series of memoranda which define program organization.

> Decisions on operational specifications outlined in this memorandum are final except that minor changes will be permitted if they aid program preparation and efficiency. It should also be noted that 1954 CCS equipment specifications for the switch panel wiring, console layout, scope display equipment, and telephone facilities have already been forwarded to Group 64 engineers and are frozen.[28]

In prospect, the Cape Cod tests had become an improvable stepping stone toward the higher-performance continental SAGE system. Nevertheless, the actual testing of the components of the revamped system took several months, and it was not until late 1954 that the final version of the 1954 Cape Cod System Air Defense Program took shape. Benington and Zraket's description required qualification. Zraket, as Chief of the System Operation Section of Group 61, explained in a November 1954 Biweekly Report that the 1954 Cape Cod System was "essentially completed and checked out except for the following features: weapons-assignment display, final-turn intercept computations, recording program, and minor parts of the simulation program and monitor-action program." He continued:

> These features will be added and checked out during the next biweekly period. The checkout of the 1954 Cape Cod program . . . in a period of about 2 months was possible because of the high reliability of the computer system and Direction Center equipment, the cooperation of Groups 22 and 64 in checking out radar inputs, and the large amount of overtime put in by members of Group 61. It is expected that live testing of the System will commence during the first half of December. A large amount of work remains to be done in the matter of setting up and documenting standard operating procedures for the System and coordinating the needs of the Test Program Section for test data.[29]

Once Forrester and Everett were satisfied the 1954 Cape Cod System was ready for "live testing," there followed almost immediately the inevitable institutional ritual of scheduling demonstrations for American, British, and Canadian military personnel and for representatives of selected government agencies and industrial and academic institutions. Selected Lincoln Lab personnel watched the first demonstration on December 16, 1954, and formal demonstrations for the Air Force and other interested organizations were scheduled for January.[30] Drawing a lesson from their previous experience, Forrester abandoned the earlier practice and arranged for demonstrations to be given in "groups of three days every few weeks instead of once a week." By March of 1955, this scheme was canceled under the pressure of testing that had to be done.[31]

While the performance of the many parts of the 1954 Cape Cod System was being checked out during the fall of 1954 preparatory to the System Operational Tests that lay ahead, the more elaborate live flight tests for the 1954 System also had to be planned to the last detail. This, too, was done, and preliminary descriptions of experimental flight-operation procedures that had been circulated to supporting aircraft units on an informal basis were codified so that all participants would know what was expected of them.

"In the future," P. O. Cioffi announced in an April 1955 technical memorandum, "as particular test planning specifications are defined," "modifications, additions or deletions" to existing specifications would be issued by Group 61 as they affected interceptor and target aircraft operations. Group 61 would generate crew manuals detailing the standard operating procedures that would provide "a sound basis for testing and evaluation." There was no question who would be in charge of the flight tests: "Liaison for all matters concerning aircraft participation with the exception of aircraft scheduling and visitations will be handled directly between the airbases and Group 61."[32]

"The distinguishing feature" of the 1954 Cape Cod System, said the crew manual that Cioffi drafted, would be "the extensive use and tie-in of electronics enabling the jobs of local air surveillance, weapons deployment and over-all air-situation defense evaluation and direction at higher levels to be made more comprehensively and efficiently. The focal point of such a complex electronics design is a large-scale digital computer."

While scanning a 200-mile-radius circle with South Truro at its center, the Cape Cod System would overlap the coverage of "a heavy radar at Montauk."[33] Both high and low coverage of the skies would be maintained by the FPS-3 long-range radars at South Truro and Montauk. Supporting these would be "several short range (about 30 miles) radars located in eastern Massachusetts, southern New Hampshire, and northeastern Connecticut," along with "several strategically located gap-filler radars," two height finders of long-range capacity at South Truro, and two height finders of shorter range, "one each at Pigeon Hill and Martha's Vineyard." The gap fillers would also give the needed radar coverage of the interceptor bases, while a Mark X APX-6 radar at South Truro would help locate interceptors more precisely in flight, and a Raydist navigational aid system would provide radar calibrations.

B-25 and B-29 aircraft flying as targets would soar no higher than "about 23,000' maximum altitude for maintenance reasons," and jet interceptors (F2H, F9f, F86F, F94C) would operate at altitudes consistent with target altitude, even though "25,000 to 40,000 are optimum for fuel consumption and range." Obviously, flight-safety requirements would supervene at all times, while "existing air defense tactical philosophy" and flight-system technical requirements would impose other standards.[34]

Ground-to-air communications with the interceptors would be of the standard sort; VHF and UHF frequencies would be used. But in addition, some F86F interceptors based at Hanscom Field would be equipped with a promising new automatic ground-to-air data-link system devised by the Air Force Cambridge Research Center in collaboration with Collins Radio (a small, independent supplier that AFCRC had found). This experimental data-link system, the engineers hoped, would allow control information to be transmitted automatically from the Barta Building to the aircraft. Radar data would be "received via multiple-input, electronic terminal equipment and stored temporarily on magnetic-storage drums (similar to wire and tape recorders)." The Whirlwind computer, under the control of the Direction Center, would periodically process the radar data.[35]

In order for the 1954 Cape Cod System to work properly, the shifting locations of targets and interceptors would have to be known accurately. To certify this type of information for intercepting aircraft, a preliminary series of special tracking-accuracy tests was developed in the spring of 1955. In these tests, F-2H aircraft carrying Raydist equipment would be allocated 8–16 hours of flight time in order to log the 5–10 aircraft-hours of test data needed. With these data the men would evaluate and compare the performance of the two tracking techniques they had employed, one traditional and one called "command tracking." The latter utilized computer calculations and was being considered as a desirable new feature of the 1954 system.

Familiar with established techniques of relying only upon reported radar returns to generate the required velocity components from successive reported positions of the aircraft, and aware of the distance errors that earlier Raydist measurements had exposed in this procedure, Group 61 proposed to track interceptors "on straight-line courses, both with and without command tracking." Using both Mark X radar and slowed-down video, they would track aircraft, flying at 20,000±1000 feet at an airspeed of 325±20 knots, in "three directions of flight across the quantizing grid at a range of 90 miles from the S. Truro FPS-3 and Mark X radars."

The Intercept Director in the Direction Center would have a "Command Tracking" switch that would enable him to call on the computer to predict the future position of the aircraft by generating velocity components, which it would calculate from the reported speeds and headings of the interceptor, instead of deriving these components simply from the aircraft's successive reported earlier positions. More sophisticated than the older technique, command tracking promised greater accuracy if it worked reliably. Hence the tests.[36]

Inevitably, there were delays caused by bad weather and equipment malfunctions. Bringing the 1954 system up to full operational test status was proving to be a more complicated job than the Lincoln Lab designers had expected, and it was taking more time than had been anticipated—partly because Division 6 and Division 2 had to continue planning the Experimental SAGE Sector testing programs and the "Operational SAGE Sector" that would follow on and overlap ESS. In short, there were organizational and staffing problems, workload problems, and unanticipated software and hardware problems. The only variables available to accommodate them all were man-hours and time.

The R&D gestation process that was involved in enlarging and refining the Cape Cod testing was not limited to producing the 1954 version of the Cape Cod System and bringing it into full operation so that further trials of the SAGE air defense concept might be conducted. It was time to be preparing also for tests that would employ Whirlwind's successor, the XD-1 prototype of the AN/FSQ-7. In consequence, the R&D program was elaborate. It required preliminary and extensive theoretical paperwork for each of the many subsidiary projects involved in readying the enlarged system to carry out the 1954 Cape Cod tests. At the same time, it involved detailed preparations to set up the tests that would employ the XD-1 in the Experimental SAGE Subsector. ESS would be not a fresh start but an extrapolation built on experience with the Cape Cod System, just as the XD-1 was a design extrapolation built on the engineers' experience with Whirlwind.

The R&D process that Forrester and Everett utilized in developing the XD-1 was a longer and more intricate version of the process they had used in developing Whirlwind. The basic pattern was this: As the underlying theoretical framework took shape, it became possible to go into detail and develop specifications regarding the design and performance of the apparatus required. And as the apparatus was then obtained or built, the performance of its components had to be checked out, first singly and then together, in order to provide the kind of detailed information that would tell the designers why a component, a circuit, or a subsystem worked or why it did not.

Deviations from acceptable ranges of performance had to be eliminated, of course, and the best way to do this, according to Forrester's and Everett's experience and habits of thought, was to understand thoroughly the principles of operation of any component or subsystem, to understand equally thoroughly the demonstrated performance characteristics, and to understand the cause of any difficulty encountered in design or performance. Obviously, if a device worked there was no need to fix it. If it did not work well enough, then the reason for the malfunction should be determined. Since this philosophy placed a premium on knowing why equipment misbehaved and why it behaved properly, the designers regarded thorough testing all along the way as a prerequisite to obtaining the high reliability that would be required under actual battle conditions. This was the sort of program they were carrying on with Cape Cod, and it was the sort of program on which they were embarking for the Experimental SAGE Subsector.

27

Problems with Manpower, Computer Programs, and Intercept Testing

In Division 6, at the working level, it was Group 61—Air Defense, under Bob Wieser—that was in the thick of the efforts to develop the 1954 Cape Cod System, then the Experimental SAGE Subsector, and finally the Operational SAGE Subsector. While the group was busy bringing the 1954 Cape Cod apparatus up to full status as a testing system, using live aircraft and radar reports tied in to Whirlwind I, it was also pressing forward with plans for the Experimental SAGE Subsector, which would make use of the new, larger capacity XD-1 that Division 6 and IBM were bringing into existence. And as the ESS test system was emerging in the period 1954–1956, design and development work on the first Operational SAGE Sector was in progress.

As the end of 1954 approached, it was increasingly clear to Lincoln Lab's managers that preparations to undertake the 1954 Cape Cod intercept tests were being slowed by lack of manpower, and Group 61 was not the only element in Lincoln Laboratory suffering from this blight. Less manpower meant that more time was needed to accomplish scheduled work. Forrester was well aware that, as the head of Division 6, he no longer had the freedom to move that he had enjoyed as Director of the Navy-supported Project Whirlwind, before the Office of Naval Research began to question his management of computer R&D. He therefore began, with characteristic thoroughness, to generate the information he needed to arrive at a solution and to justify it to his superiors. In the middle of November he issued a survey of the organization of Division 6 that listed tasks to be performed and detailed each group's specific needs for more trained workers.[1] The technical staff, according to Forrester's survey, needed 107 new members. At the December 7 meeting of the Lincoln Steering Committee, Albert Hill pointed out during a discussion of the budget for fiscal year 1955 that no restrictions were in force that would preclude the hiring of "good personnel," and that Divisions 2, 4, and 6 all required "substantial numbers" of new staffers.[2] A December 13 meeting of Division 6 Group Leaders found 107 to be "the number of new staff we would require if we were to do all of the work which has been planned. As a matter of fact, we have neither the recruiting effort, the space, nor the budget to obtain these additional staff members."[3]

By the middle of December, Wieser's assistant, R. J. Horn, was calling for help from division members in devising ways to "give more personalized attention to Division recruiting," and before the end of December he had placed Help Wanted ads in local papers. When these netted too many poorly qualified applicants, a different tack was taken. Dave Israel's and Walter Wells's recent trips to several graduate institutions in Los Angeles and the San Francisco Bay area while attending technical discussions at the Rand Corporation and Hughes Aircraft encouraged Horn and Israel before the end of the year to begin arranging for contacts to be made with professors and for recruiting talks to be given at student professional activity meetings on college and university campuses around the country.[4]

Through the winter and into the spring of 1955, Division 6 continued to give special emphasis to recruiting. But it did so at a price. In December and January, Dave Israel was calling attention in his biweekly reports to the fact that the Cape Cod test planning for which he and his group were responsible was being delayed by the multitude of tasks and the lack of time to carry them all out. He cited "a lack of sufficient manpower and the diversion of available manpower to other tasks, principally assisting in demonstrations" for visitors. "This state of affairs seems to be getting worse instead of better," he noted, perhaps unaware of the remedial policy action being initiated at higher levels. "Fortunately, this condition is balanced by similar delays in preparing the 1954 Cape Cod System for test purposes."[5]

By early February the first round of recruiting contacts had been accomplished and more than a hundred trips were being scheduled to colleges at which staff members would give technical talks and hold interviews. Forrester impressed on his staff both the nature of the problem they faced and its solution:

> The present tasks of the Lincoln Laboratory, even without very pressing demands to take on additional future responsibilities, exceed the load which our present staff can carry. For Division 6, in particular, the demands of the SAGE System are growing more rapidly than we are turning duties over to IBM and Western Electric.

> The Laboratory, therefore, is looking for unusually well qualified men to add to the staff. Several Lincoln teams are visiting colleges this month to interview candidates, and the most promising men will be invited to visit the Laboratory. The emphasis is on quality rather than quantity.[6]

In February and March the campus recruitment program began to show some success, and slowly the manpower situation began to improve.[7]

During the winter of 1954–55, other pressures were generating organizational changes. To Forrester and Everett, the most important of these pressures were approaching technical developments for which they had been preparing since 1950: the impending addition of the next-generation computer and the development of an experimental surveillance and intercept system to match its capacities. This prospect called for Division 6 to get ready to phase in ESS while finishing up the Cape Cod tests. So the first of these changes stemmed not from Cape Cod problems but from progress being made elsewhere in the SAGE program.

At the end of December, Forrester, Everett, Taylor, Dodd, and O'Brien had witnessed tests that IBM had carried out on the XD-1 computer in Poughkeepsie before disassembling the machine for shipment to Lincoln Lab. "The machine," Forrester reported to his division staff, "seems to work as well as could be expected at this stage in the program. We were much encouraged." It was time to give more attention to developing ESS, the Cape Cod System's successor, and Forrester and Everett had already taken managerial action.

"In order to better define tasks and assess manpower needs" of Group 61, Wieser reported in the final quarterly report of 1954, the group was reorganized into six major sections:

1954 Cape Cod System Operation

Experimental SAGE Subsector

Systems-Test Planning

Analysis and Simulation

SAGE System Planning

SAGE System Training

More emphasis was being given to planning the Experimental SAGE Subsector and to setting up the ESS system tests that would build on the Cape Cod System experience. The Systems-Test Planning Section would be responsible for "planning test programs for the 1954 Cape Cod System, the Experimental SAGE Subsector, and the SAGE System." It would coordinate closely with the Analysis and Simulation Section, which had as its charge to "analyze and predict the performance of the SAGE System" by examining and improving existing programs being used by the Cape Cod System or being developed for ESS.[8]

Near the end of March 1955, Everett announced the details of the second internal reorganization to Division 6. Pointing to recent studies they had made of Lincoln Lab's responsibility for the SAGE System and its computer program that would be needed to make SAGE a militarily operational system, he explained in some detail that Group 61 was being reorganized and enlarged to handle reoriented priorities:

> Changes in Group 61 include the combination of the former Test Planning and Cape Cod Operations Sections into a Test Program Section under D. R. Israel. Three sections under Arnow, Walquist, and Zraket have been formed to prepare the operational specifications for the SAGE computer program. A Special Studies Section, also under Arnow, has been formed. The Analysis and Simulation Section under Wells and the SAGE Training Section will continue as before.
>
> J. F. Jacobs has transferred from Group 62 to Group 61 to head a new section for coordination of formal Lincoln and Air Force concurrence on program specifications and liaison with equipment design groups. . . .
>
> S. H. Dodd, Leader of Group 64, will spend the next few weeks in a study of the SAGE test program now in progress in Lincoln and BTL. This program is of the utmost importance: a decision to increase the effort must await the completion of Mr. Dodd's study.[9]

In the new Cape Cod Test Program Planning Section under David Israel there was fresh determination to accelerate completion of the data-generation and the data-reduction programs needed to carry out the test-program activities. They would try to finish these tasks by the end of April. But Israel added a word of caution about the radar-accuracy tests, noting that "the work of the past several months indicates a number of unresolved problems in the processing and use of Raydist data." Although he was hopeful that recent troubles had been solved and that they could proceed with the tracking-accuracy tests they had scheduled, he admitted that "the problem requires a good deal more attention."[10]

By the middle of March 1955, Charles Zraket was able to report that each of the Cape Cod computer programs had "undergone a series of measurements under various conditions in addition to detailed analyses made by the responsible programmers. These results have been compared and extrapolated to the SAGE System program." Difficulties with radar and Mark X beacon data that had dogged testing since the previous autumn appeared to have been eliminated at the beginning of 1955, permitting more successful intercept and tracking-accuracy tests. All the data from these tests were recorded on magnetic tape for the detailed analysis that would follow.

But during the middle of March 1955 fresh difficulties surfaced while Group 61 was attempting two accuracy-training tests, two saturation tests, and three Raydist tests.[11] Though these problems were neither exceptional nor insoluble, they represented further delay, and they lent credence to any speculation that Division 6's management of the Cape Cod program was becoming less than competent.

The facts of the matter were that, although the Group 61 engineers were working to bring the Cape Cod 1954 System "on line" for the more ambitious tracking and intercept tests being planned, new equipment failures of the Mark X radar were among a variety of routine failures beyond the control of Group 61, and these were disrupting the limited flight testing that was going on.

In the first saturation training test, after five "enemy" aircraft were airborne, the interceptions had to be canceled because the Mark X radar failed again. On a second saturation test before the end of March 1955, "six of the interceptors scheduled were employed, but mapping operations, Mark X malfunction, and poor velocity computations resulted in poor intercepts."[12] As such familiar reports grew in frequency that spring, it became increasingly clear within Lincoln Laboratory that the 1954 Cape Cod System was in trouble in this area of its activities.

But in another area the Cape Cod System was laying elaborate, essential computer programming foundations for the ultimate continental SAGE system which Lincoln Lab was committed to build. This highly technical programming activity was neither conspicuous nor easy to describe, but to ignore it or remain ignorant of it would be to miss a major, basic, and strategically innovative dimension of the SAGE project's character as an R&D program. While those engaged in the computer programming on the expanding scale required for the Cape Cod intercept tests were breaking new ground, their activity was only a prelude to the programming that would be required for the Experimental SAGE Sector for the continental SAGE system. Regardless of the scale and complexity of the programming, if it were to achieve its intended results it had to be meticulously accurate, instruction by instruction and datum by datum.

To give here some assessment of the programming required for Whirlwind and Cape Cod, however brief, is to take further measure of the engineering sophistication that went into the SAGE project and the contributions to that project by Division 6. A small example is the Master Make-up and Display (MMD) program of the 1954 Cape Cod System. It was taking shape at the hands of six or eight staff members in the System Operation Section, of which Charles Zraket was chief. By Christmas 1954, the necessary documentation in the form of annotated program sheets, flow diagrams, and the like was nearly half completed. It still was lacking "certain minor features and some major sophistications," such as the transfer of interceptor control between Direction Stations in the Direction Center, computer-assisted control of final-turn interceptions, "split indication, tally-ho indication," and simulated dropping of tracks used in training and intercept-simulation sessions.[13]

Finished by March of 1955, the MMD program occupied "about 10,000 drum registers" in Whirlwind's auxiliary memory. Among the many programs that ran the computer, its task was to perform "most of the switch-interpretation, display, and bookkeeping functions required" when the computer was called on to support just the Weapons Direction Stations in the Direction Center.[14]

Long-standing habits of association with machines incline us to credit the machine with its important operations, whether it be a machine as simple as a hand loom or one as complex as an airplane. We credit the operator with making it work, of course, but it is most convenient to view the relationship between the operator and the "hardware" as simple and direct, whether it is or not. (In the case of an early Wright Flyer it is; in the case of a jet transport it is not.) These habits of thought do not serve us well when examining the computer, its relation to its operator, its relation to the programmer who writes the programs that make the computer useful, its relation to the programs themselves, and its mathematico-logical operations, the electromagnetic patterns of which make all the other relationships possible. Whirlwind (or any other stored-program electronic computer) could do nothing of consequence without its programs, and these operated at many different levels, some separate and some integrated.

To appreciate the R&D challenge that SAGE posed, it is useful to distinguish between a machine and a system of machines. This difference became important, and its implications began to be seriously explored, around the middle of the twentieth century. Among those active on the "cutting edge" of this advance in scientific technology were the engineers, mathematicians, and other experts working on SAGE at Lincoln Laboratory and in private industry. The Master Make-up and Display program that Division 6 produced is a small, representative example of the "systems R&D" that was becoming a hallmark of the emerging advanced technology which would, a generation later, give rise to the term "high tech." The MMD program itself was an integrated software composite of repetitive machine-language routines, subroutines, and many other small programs. Operating as a central interpreter and coordinator, it carried out "most of the WD [Weapons Direction] switch interpretation, situation-display make-up and display [on one screen], digital-display make-up and display [on another smaller screen], track bookkeeping [compiling a 'history' of any identified track in the event a human monitor might need the information], and communication [of the Weapons Director] with the Tracking Program."[15]

There was the machine, there were the programs that made it go, and there were the functions, or activities, controlled and carried out by the programs. In addition, there was the Weapons Director, who, in turn, had a number of other functions that he carried out with the aid of the computer and seven major "subprograms" composed under a common design philosophy. One of these seven was the MMD itself. The other six were Interception, Geography, Height, Identification, Antiaircraft, and Data Simulation.

The writers and debuggers of the MMD and the other six programs had followed a rather sophisticated principle when allocating individual operational functions to any of the seven programs, and its application reflects the character of their systems-R&D approach. This approach was born out of their several technical backgrounds, their programming experience, and their experience with the overall air defense problem. The covering technical memorandum put it as follows:

> . . . a one-to-one correspondence between Direction-Center organization and computer-program organization is not the best utilization of the computer for air defense; instead, computer functions should be organized in terms of switches, displays, bookkeeping, stable accessibility, and communications; only where tasks are long and isolable should they be performed by special, independent programs. This principle accounts for the fact that every WD function is performed partially by MMD, partially by a special-purpose program. This memorandum describes the operational functions performed by MMD and the programs which perform them.[16]

On the one hand, there were the "Functional Responsibilities," describing "in primarily operational terms the air defense functions which MMD performs." On the other hand, there were the "Program and Coding Specifications," describing "the organization of the MMD program and the responsibilities of each MMD subprogram." The technical understanding of the programmers was such that to approach the MMD program and its crafting from both of these directions—the one emphasizing air defense functions, the other computer functions—was to better understand and to dictate how the computer was to be led by hand to accomplish the syntheses of information it must perform in order to carry out its projected inconspicuous but essential air defense role.[17]

All these efforts—MMD, Raydist tests, reorganizations—characterized an R&D exercise that was racing against time while deliberately becoming more ambitious in scope and detail as the months and years passed. So it is not surprising to find that the MIT and Lincoln Laboratory managers, as 1954 came to its end and 1955 wore on, were undertaking what to them seemed appropriate reorganizations of their administrative framework in order to handle the already complicated, interlocking, and growing workloads of Divisions 2 and 6. In general, the performance of the 1953 Cape Cod System, along with the projections presented in TM-20, had established the technical feasibility of the ADSEC-Lincoln approach beyond serious challenge from the outside, and government financial and administrative support was ensured. But the next phase still had to be demonstrated in practice.

The Air Defense Command concluded by March of 1955, with the agreement of Division 6, that, given IBM's outline of its production schedule, several important target dates could be specified: "the first subsector could become operational on 1 March 1957," followed by the "first Combat Center by 1 July 1957" and the "entire SAGE System . . . by 1 November 1960."[18] But to state these goals in ADC's Operational Plan was not to achieve them, and the SAGE R&D program, for which Lincoln Laboratory was responsible, was by no means out of the woods.

28

Design Progress and Administrative Reorganizing

Although the major components and subsystems of the Cape Cod System were sufficiently developed for Forrester, Everett, and Wieser to report formally that the system was, in Wieser's words, "checked out and functioning as a complete unit" by the end of the first quarter of 1955, system-wide live tracking and intercept tests still lay ahead. So complex was the ADSEC–Lincoln Lab solution to the air defense problem that it was easy to confuse the resumption of limited live testing in November of 1954 with the tracking and intercept tests, which required integrated functioning of the complete Cape Cod System. The former was necessary in order to check the performance of the system and its components under test conditions; the latter would test the functioning of the system under increasingly rigorous mock battle conditions. "Field trials" still lay ahead. First, Division 6 would have to acquire enough manpower to finish the data-recording, data-generation, and data-reduction programs in April and May, as scheduled. Once that was done, Dave Israel predicted in February of 1955 from his vantage point as the man in charge of Test Program Planning, "a shortage of sufficient personnel to plan and analyze tests will become evident. The first milestone and first actual flight test of the program is expected to occur late in March and will be the first of a series of tracking-accuracy tests."[1]

Nevertheless, genuine design progress was taking place, and it had far-reaching consequences for the programming of the SAGE System that the engineers had in view. In the spring of 1955 the operating programs of the 1954 Cape Cod System comprised "about 25,000 computer instructions and about 10,000 additional registers of data storage," by one conservative reckoning. The system was making more efficient use of the computer's speed and capacity, and it possessed improved recording and documentation features crucial to the future SAGE program. These were crucial for two reasons: First, they preserved detailed records of the computer program and of system operation, and such records could be referred to when making future modifications and analyses. Second, they "demonstrated the necessity of using similar techniques and formats for the SAGE System." An indication of the value of these improvements was the fact that the 1954 Cape Cod computer programs were debugged and brought into full readiness more rapidly and easily than the 1953 Cape Cod programs, although the 1954 programs had almost twice as many instructions.[2]

It was in the midst of these efforts to cope with the growing workload and with a tight schedule framed to shift greater attention to designing the Cape Cod System's successor that a new "SAGE Test Committee" was created during the spring of 1955 to monitor the operation of the 1954 Cape Cod System and to prepare to monitor the Experimental Subsector System (as it was then called), which would succeed the 1954 Cape Cod System.[3]

In separate administrative actions that spring, George Valley was appointed Associate Director of Lincoln Lab and Marshall Holloway was selected to replace Albert Hill as Director. At the same time that Valley's change in status brought him increased responsibility and authority over the "technical direction of the Laboratory," there began a general overhaul of the two divisions primarily responsible for developing the air defense network. Division 2, under its new chief, Carl Overhage, was restructured to permit it to concentrate "more directly on problems relating to SAGE System planning, testing, and equipment." This division's Groups 23 and 24, which were involved primarily in radar and data transmission, were transferred to a reorganized Division 4 designated "Radars and Weapons" and made responsible for the "development of radars for aircraft detection and tracking."[4]

Late in 1954, the engineers of Division 6 began to reduce their efforts to adapt computer programs to the 1954 Cape Cod System and to give more attention to planning the Experimental SAGE Subsector tests that would follow the final Cape Cod series. Group 61, under Bob Wieser, was increasingly turning its attention to three overlapping assignments that winter and spring: developing and constructing a Direction Center for the Experimental SAGE Subsector, designing the Direction and Combat Centers for the finished SAGE System (the Operational SAGE System), and preparing the computer programs that the ESS and the Operational SAGE systems would use. Before the end of 1954, Group 61 received a new name that reflected its new duties: "SAGE System Test and Planning."[5]

By the end of the summer of 1955, Group 61, no longer officially reporting on the 1954 Cape Cod System, was preoccupied with tasks related to the Experimental and Operational SAGE Subsectors. The increasing responsibility for Division 6's role in making the 1954 Cape Cod System work fell to Group 64, which was renamed "Group 64 - Cape Cod System" in the spring of 1955.[6]

One of the aims of these reorganizations was to move the 1954 Cape Cod System into its final phase, System Operational Testing, as rapidly as possible. To achieve this goal on the working level during the spring of 1955, a SAGE Test Committee was created at Lincoln Lab and the efforts of Groups 22 and 61 were joined by those of a subdepartment of Bell Telephone Laboratories. Because construction of the ESS Direction Center was progressing rapidly, installation of the first operational subsector became more urgent. As a consequence, the engineers decided to schedule preliminary test operation of the Experimental SAGE Subsector to begin before the testing of the 1954 Cape Cod System would be completed.[7] The resulting overlap was fraught with hidden problems. Simultaneous operation of the Cape Cod System and the ESS was looming, and use of the same radar inputs and the same aircraft support threatened to cause confusion and conflicts in flight-test scheduling.[8] It was in anticipation of these difficulties that the Lincoln Lab administrators had created the SAGE Test Committee, giving it the "immediate objective of drawing up detailed plans and schedules for the entire analysis, test and evaluation required." A reasonable administrative solution in prospect, it proved to be a mistake that took many man-hours to rectify, and it provoked the kind of "spinning of wheels" in the conduct of R&D that Forrester and Everett had been able to avoid in the relatively autonomous days of the Digital Computer Laboratory, when they had developed Whirlwind and set up a unique collaborative arrangement with IBM to develop Whirlwind II (the AN/FSQ-7). Once the DCL became Division 6, it was exposed to the variety of constraints incurred by its being part of a larger and growing organization, Lincoln Laboratory.

The SAGE Test Committee, itself a fresh constraint, proved to be the prelude to an even stronger move to subordinate Division 6 to the managerial demands of higher authority at Lincoln Lab. (Ironically, while Lincoln Lab was trying to tame the former Project Whirlwind team under Forrester, military program and budget managers in the Air Research and Development Command under General Thomas S. Power were trying to tame Lincoln Lab.)

The members of the SAGE Test Committee included Steve Dodd of Group 64, Dave Israel and Walter Wells of Group 61, G. Harris and V. A. Nedzel of Group 22, A. Herckman of the Bell Telephone Laboratories, and F. Kline of the Air Force Cambridge Research Center. Since this committee was expected to meet only once every 3 weeks, a SAGE Test Office was established to assist the group in "administering the test program, maintaining schedules, monitoring the work, and preparing progress and status reports."[9]

The stage was set for disagreement and confusion.

The SAGE Test Committee set a due date of October 15, 1956 for the complete computer programs needed for the first Operational Subsector. This would give ample time, they felt, to meet the March 1, 1957 deadline sought by the Air Defense Command, and to achieve this deadline they gave "highest priority . . . to those tests and evaluations that [would] contribute most heavily to providing the best possible over-all system performance of the first operational subsector." Moreover, since time was precious, the committee's members agreed that demands imposed on the test schedule of the ESS should be limited to the "most important tests only, and to combining separate tests as much as possible." In this way they would achieve "maximum utilization of each period of system operation." When a choice was possible between the programs and procedures of the 1954 Cape Cod System and those of the ESS, the former system should be utilized. Under this arrangement, the Cape Cod System should remain in place through 1956, while the ESS was taking shape. It all looked quite reasonable on paper.[10]

In the spring of 1955 the SAGE Test Committee prepared a comprehensive outline of the 1954 Cape Cod System test program, describing in some detail "the work required for evaluation of the operation of this system, . . . the priorities assigned, and . . . the groups responsible for each work item listed." The committee then turned to preparing a preliminary outline of a test program for the Experimental SAGE Subsector. Since they were laying out ideal schedules, they found it easy to pattern the ESS test program after the Cape Cod program. Anticipating that they would obtain suitable information from the 1954 Cape Cod System, they assumed that they could pay more attention to the performance of the ESS than to how its components were functioning. In short, they assumed that the performance of the Cape Cod System would provide a basis for a complete evaluation of the ESS.[11]

As spring gave way to summer, it became apparent that the Cape Cod tests were not going well. At the end of June, Dave Israel was asked to give his appraisal of the results. "Actually," he said, "we have not yet obtained any quantitative results; however, during the past several months we have learned quite a bit. We have far greater experience now than in January of this year and are cognizant of areas to concentrate attention on now and in XD-1." Outlining six conclusions that he drew from the experience of planning and programming and conducting and evaluating the tests, Israel expressed his conviction that "there is a definite need for a single over-all direction of both the entire system and the test program effort. The systems viewpoint is essential."[12]

Whatever may have been the managerial reaction of Israel's superiors to his honest statement, it was quite obscured by the coincidental release late in June of a memorandum from the SAGE Test Committee. The committee voiced the same point of view that Israel had regarding the overall direction of the Cape Cod System and its test program, but the memorandum went on to criticize the current operations vigorously.

In the two divisions involved, the managerial reactions to the committee's criticisms were swift. Bob Wieser and Jack Jacobs of Division 6 and H. W. Boehmer of Division 2 were brought in to deal with the Cape Cod tests and the committee's criticisms. The three men spent most of the first two weeks in July reviewing the situation. They then went on record formally to the Director and the Associate Director of Lincoln Lab, Marshall Holloway and George Valley. It was no longer sufficient for them to sit down with Everett and Forrester and discuss what was to be done.

Wieser, Jacobs, and Boehmer admitted that the "test and evaluation effort, which is now principally a 1954 Cape Cod System activity[,] has not been satisfactory." But they rejected the conclusions of the committee, declaring that "the description of the problem and the recommended solution contained in the 29 June memorandum . . . (Subject: Coordinated Direction of 1954 Cape Cod System) are at best incomplete."[13] They promised "a detailed description of the problem areas and a recommended solution," which Boehmer and Wieser drafted in Division 6 memorandum L-212.

The SAGE Test Committee had considered all aspects of the test programs in an authoritative assertion of supervisory power that quite dismayed the leaders of Group 22 and Group 61. The committee had concluded that the "over-all program" was suffering from the lack of a central control, and they recommended that "planning, maintenance, operation and evaluation in the 1954 Cape Cod System be coordinated or directed by a single centralized source."[14] This proposal provoked a forcefully argued rejection from Boehmer and Wieser. The source of the problems noted by the committee lay, they argued, with the committee itself, and elimination of problems would "best be done within the . . . group structure," not by the establishment "of a completely separate SAGE test group." Extending their comments to the subject of the committee's responsibilities and authority, Wieser and Boehmer urged that its activities be confined to "planning and concurrence" and that "the responsibility for execution of tests rest fully with the operating groups." No test program, they asserted, should be considered final "until the operating groups have agreed that execution of the program is feasible with available resources."[15] The committee's chief functions, Wieser and Boehmer contended, were to review items that members of the contributing groups identified in order to "eliminate redundancies," to coordinate for groups cooperating on particular test projects the "joint use of equipment, programs, and operating personnel," and to "establish priorities among the work items."[16] Instead of serving the committee, the SAGE Test Office should be redesignated the Test Coordination Office and "should be reoriented to perform its proper function of serving the executors of the tests." Among its duties, the office should "convert the test plans approved by the SAGE Test Committee into scheduled tests, keep itself informed of the status of all system elements, make arrangements for scheduling each test, collect significant data from all observers, keep central files of test results . . . and summarize these on a monthly basis" for the groups involved and for the committee.[17]

Lincoln Lab's managerial disposition of the situation came early in August. It was orchestrated by George Valley, who as early as July 26 was asking members of the Lincoln Steering Committee whether they believed the SAGE program was on schedule. Carl Overhage, Valley's successor as head of Division 2, responded by voicing his reservations about the rate of progress of the Experimental SAGE Subsector testing program. He wondered what impact a delay might have on the whole SAGE program. Forrester noted that internal and external delays might well put the program several months behind schedule.[18]

Valley's question and the responses it elicited were followed 2 weeks later by an administrative order from the Lincoln Lab Director's office drastically revising the organizational structure. The SAGE Test Committee was abolished, and the authority it had sought to exercise in planning, supervising, and evaluating test operations and their results was given to Division 2. A "SAGE Experimental Test Policy Committee" appeared. Chaired by George Valley, it would be expected to "establish the management program, within the Lincoln Lab complex, of the tests to be executed and conducted upon the successor to the 1954 Cape Cod System, the Experimental SAGE Subsector." The Test Policy Committee was denied any authority over detailed implementation or execution of tests.[19]

Management has the power to give as well as to take away. When Lincoln Lab's directorate abolished the Test Committee and gave Division 2 the authority that committee had sought to exercise, Holloway and Valley also transferred to Valley, in his capacity as Associate Director, the power that Division 6 had exercised during the Cape Cod phase of the unfolding air defense program. This action deprived Division 6 of some of the administrative control of programs that Forrester, Everett, and their Group Leaders had been in the habit of exercising flexibly while coordinating their actions with those of Division 2. Indeed, Division 6 appeared to be left with only supporting functions to carry out for the ESS, but the fact of the matter was that the R&D and managerial realities were too massive and too complicated and had acquired too much momentum to warrant relegating the division to such a status.

This giving, taking, and redistributing of authority reflected more than internal stress and rivalries. The actions also reflected a longer-range shift of emphasis that was taking place, from a feasibility-and-experiment phase that had employed the Whirlwind computer in the Cape Cod System to the penultimate phase that would employ the XD-1 prototype of the AN/FSQ-7 computer in a more complex and sophisticated module of the continental SAGE system. Undoubtedly, as the program approached this operational status, greater centralization of authority over integration of all components was seen as an absolute necessity. Neither Holloway nor Valley would be inclined to leave this authority at the divisional level.

Since this redistribution of test-operation authority and responsibilities immediately affected Divisions 2 and 6, both again underwent reorganization, partly to reduce friction between them and partly to more efficiently discharge their duties under the new arrangement that Holloway and Valley imposed. In Division 6, Group 64 (Cape Cod) remained unaffected, but Group 61 lost Walter Wells and its Leader, Bob Wieser, to Division 2. There Wieser, taking responsibility for the Cape Cod tests with him, became Associate Head. He was also put in charge of SAGE operations, and in this capacity he wore another hat as Acting Leader of newly formed "Group 23 - Operations," which was given "primary responsibility to arrange for and conduct test missions to fulfill Cape Cod System and Experimental SAGE Subsector requirements."

Wells and his associates in the "Technical Evaluation Section" of Group 61 were incorporated en masse into Group 22 of Division 2, "Analysis and Evaluation." Operation of the Cape Cod tests had been removed from Forrester's and Everett's control. Only the program-support function was left behind.[20]

The rest of Group 61, under its new leader, Dave Israel, would concentrate on activities related to planning and developing the Experimental SAGE Subsector and the Operational SAGE Subsector. In this capacity, Group 61 would determine the operational and mathematical specifications without which the master computer program for SAGE could not be written. Group 62 would take care of the XD-1 and would cooperate with Group 311 in ESS subsystem testing. Group 64, already familiar with the Whirlwind computer, would also involve itself in the planning of tests of both the XD-1 prototype and the AN/FSQ-7 production-model computers, and to this end it would work closely with the Equipment Program Services Committee (EPSCOM), members of which also came from the RAND Corporation, from the Bell Telephone Laboratories, and from Western Electric. The managerial intricacy generated by this arrangement is evidenced by the role given EPSOM "to facilitate system testing of AN/FSQ-7," or (as explained at greater length in a Quarterly Progress Report) to "complement the activities of the subsystems test teams of Group 62 by providing the computer programs required for design shakedown, system debugging, systems test, maintenance, and equipment adjustment."[21]

In Division 2, Overhage, as Division Head, and his two Associate Heads, Bob Wieser in charge of SAGE operations and John Harrington in charge of research, continued during the autumn of 1955 to consolidate in the division the reorganization that Valley had begun. The major tasks associated with the intercept tests being carried out by the 1954 Cape Cod System and with the more ambitious tests being scheduled for the ESS were assigned to five different groups within the division. Wieser was placed in charge of three of these tasks: overseeing the "performance of the radar and data-transmitting devices that provide the input data in the systems under test," analyzing and evaluating the data obtained from system tests and recommending appropriate "revisions to equipment, program, and test plans," and "planning, scheduling, execution, and reporting of tests." Harrington was made responsible for tasks that two other groups would carry out: dealing with "problems encountered in conversion of radar signals to digital computer input data" and investigating "new methods for presenting computer output data to human observers."[22] One effect of this reallocation of responsibilities was that Group 21 would continue to manage the radar sites and the transmission of radar data. Group 23 would be in charge of scheduling the many tests and would also be responsible for training new personnel. Group 311 of Division 3 would handle radio communication and data-link facilities and would be partially responsible for ESS subsystem testing, while Group 38 would be concerned with "improvement in the human aspects of system design."[23]

Before the changes, Group 22 ("Cape Cod") had been the division's chief instrument for making the 1954 Cape Cod System work. Reinforced by the addition of the Test and Evaluation Section transferred from Division 6 to its own Analysis and Evaluation Section, it was renamed "Group 22 - Test and Evaluation." Its primary mission was to determine "the effectiveness of SAGE for air defense through test of the Cape Cod and Experimental Subsector Systems." Four activities were specified to accomplish this mission: formulating "the Cape Cod and Experimental Subsector evaluation test program according to the management directives established by the SAGE Experimental Test Policy Committee" (which Valley chaired); planning such "specific tests" and drawing up such test specifications as would be necessary for "coordination, scheduling, and execution" by Division 2's Group 23 - Operations; reducing, analyzing, and evaluating the data obtained from the tests; and recommending any "revisions to equipment, program, and test plans" that Group 22 deemed necessary.

When this new group was formed under its old Group Leader, H. W. Boehmer, it consisted of 37 members, all Lincoln Laboratory personnel. A year later, it had 72 members, including representatives from the Bell Telephone Laboratories, the Rand Corporation, and various Air Force agencies. (By that time V. A. Nedzel was its Group Leader and Walter Wells its Associate Group Leader.)[24]

In their vigorous rebuttal of July 28, 1955, Boehmer and Wieser had argued that the 1954 Cape Cod System's objective was to "support the SAGE System by investigating and evaluating the weaknesses of a SAGE type System so that the initial computer programs, equipment, and operational doctrine for the first subsector are as effective as possible in the fulfillment of their air defense mission."[25] Then they had proceeded to outline in 22 pages a suitable test program and to list the resources needed, after explaining that no previous test program had contained "1) a clearly stated objective; 2) a scheduled end date for accomplishing a stated series of tests; and 3) a statement of the resources needed to accomplish the program."[26]

It is a measure of how fast events moved that within 2 weeks the test program described by Boehmer and Wieser was endorsed and issued verbatim as a Division 6 M-Note by the soon-to-be replaced (and perhaps chastened) SAGE Test Committee.[27] The program scheduled more than a dozen kinds of tests, most of them designed to check various operations within the overall system. There were tests of radar performance, radar mapping to eliminate irrelevant signals, initiation of tracks by automatic and manual methods, guidance of interceptors to target, reduction and analysis of the data, and supporting utility programs.[28]

The next 2 years' tests of the Cape Cod System were being mapped out. They would also be the final years of the Cape Cod System's useful life, for the system had only these two more years to run as a tracking and intercept system using live aircraft. It was to be replaced by the Experimental SAGE Subsector, built around the Whirlwind machine's successor, the XD-1 (prototype of the AN/FSQ-7). During these two final years the Cape Cod System was used increasingly as a "training ground" for the military personnel who would one day operate SAGE and as a means of exploring new techniques and equipment that might improve the ESS and the continental system.

Of all the tests scheduled for the Cape Cod System in 1955, the most important and the most elaborate yet were the System Operational Tests (SOTs). These required all the functions to be tied together, and they monitored the performance of the total system "under realistic, complete, operating conditions." In an SOT all radars must supply data, all consoles in the Direction Center must be manned, and live aircraft interceptions must be carried out. This would provide the quantitative and qualitative information needed to indicate "the proper emphasis that should go into [the] remainder of the test program." It would also evaluate the overall performance of the system.[29]

The development philosophy being followed here assumed that the Cape Cod System was not a prototype of the proposed continental air defense system. That role would fall to the Experimental SAGE Subsector. Nevertheless, the Cape Cod SOTs were expected to provide the principal source of compiled information, to permit informed prediction of the SAGE system's performance, and to indicate which areas should receive special attention in order to guarantee the best operational performance of the SAGE system. Furthermore, the SOTs were expected to provide "valuable operating experience" for several groups, including those in charge of operating SAGE-type equipment, the Air Force operators in the direction center, and the Lincoln Lab groups responsible for running not only the Cape Cod tests but also evaluation tests of the Experimental SAGE Subsector.[30]

The SOTs were framed as a sequence of four series, and each of the last three was to be more complex and sophisticated than its predecessor. Three groups in Division 2 shared the task of carrying out these Cape Cod tests in 1956, and they coordinated these and their test results with the planning and development of the more elaborate ESS. Group 21 operated the equipment at the radar sites. Group 22 conducted the tests and evaluated the system's overall performance. Group 23 (which was formally organized under Group Leader J. C. Starks when Bob Wieser became absorbed in his duties as associate head of the division) planned, arranged, and conducted the flight-test missions of the aircraft participating in the tests and assisted the Air Force in training operating personnel to participate. Though all these tasks were associated with the Cape Cod System, the groups were expected to shoulder similar duties in planning the ESS.[31]

During the summer of 1956, as the second series of SOTs neared completion, it became increasingly apparent that the Cape Cod System's troubles were not over. All but 3 hours a week of system operating time were "used for experimentation with various parts of the system," and this left too little time to establish guaranteed system performance. The problem was compounded by operator inexperience, because they were spending too few hours at the consoles. An ingenious partial remedy was born in July 1956 out of the desperation of the moment: when an SOT could not be run for one reason or another, the Direction Center crew shifted to a "System Live Exercise" (SLE) in order to acquire further operating experience.

It became feasible to run two SLEs a week, the object being to match SLE activity as closely as possible to SOT procedures, using such live strike aircraft and defensive aircraft as were available. Since the aim of the SLE was to provide Direction Center operators with experience, no test data were recorded for future analysis; recording was saved for the SOTs.[32] A month after the SLEs were invented, they became officially System Operation Exercises (SOEs). When aircraft were unavailable or weather interfered, tests using simulated data fed to the computer were substituted in order to carry out further operator training.[33]

The SOTs were carried out between October 25, 1955 and May 29, 1957, ending 4 months before the 1954 Cape Cod System was permanently shut down to make way for evaluation testing of the Experimental SAGE Subsector.[34]

The problems the engineers encountered when running the System Operational Tests, though often exasperating, provided a wealth of information that was put to use to make the emerging SAGE system more reliable.

Eight SOTs spanning the period from October 25, 1955 to January 31, 1956 made up the first of the four series of tests called for. Plans called for "five simulated and five live strike aircraft (B-29) with an average velocity of 240 knots, and eight simulated and eight live interceptors." Half of the interceptors would be assumed to be carrying rockets, the other half missiles. The tactics of the fighters would "depend on the Intercept Directors," who would be "told about their assigned interceptors' weapons." All the interceptors would be equipped with Mark X IFF radar responders in order to assist in their identification by the operators at their consoles in the Direction Center. Two of the F2H fighters scrambled from South Weymouth would be Raydist-equipped, and the F-94s from Otis Air Force Base would possess "Airborne Intercept" (AI) radar, permitting them to close in on a preselected target after being guided to the vicinity by the Direction Center in Room 222 of the Barta Building. One antiaircraft battery would also be involved. According to the plans:

> Each of the live and simulated strike aircraft will participate in three raids. Several raids occurring almost simultaneously [i.e., all raids reaching the target center within a ten-minute interval] will constitute a wave. Each wave will consist of from three to five raids flown by either all live or all simulated strike aircraft. There will be three waves of simulated strikes alternating with three waves of real strikes. Each of the interceptors, both live and simulated, will fly at most three missions, with the live interceptors being assigned to the live hostiles and the simulated interceptors to the simulated hostiles
>
> All raids will be directed toward only one target area, Boston. There will be no altitude maneuvers, no crossing of tracks, and no jamming. The maximum number of maneuvers per track will be two, and there will not be more than one split per raid. The altitude range will be from 15,000 to 25,000 feet. Information between the Direction Center and the aircraft will be over radio.[35]

Since SOT No. 1 was carried out to make certain that instrumentation and operational procedures were adequate, no effort was made to run an extensive analysis of the results. Although the tests that followed in the first series "produced a large amount of data relative to system operation," according to the Quarterly Progress Report, their chief contributions appear to have been "the formulation of testing procedures and the considerable amount of system shakedown which was accomplished."[36]

Since Series I was the least complicated and ambitious of the four, SOT No. 1 employed a small number of propeller-driven B-29 bombers serving as strike aircraft and a limited number of interceptors. The other seven tests involved a total of 50 B-29s that penetrated the radar defensive screen "protecting" Boston. Sixty-two interceptors were "scrambled" from neighboring airfields to meet the "enemy," and 24 successfully sighted their targets. Voice telephony was the principal means of communication between the Direction Center and the interceptors, although occasional use was made of a ground-to-air data link with those aircraft equipped with AI radar.

The radars used in the first two series of SOTs included the long-range radar at South Truro, which was equipped with a Mark X beacon, four gap fillers, and three height finders. These were sited within the area of an arc that swung from Londonderry, New Hampshire, to Martha's Vineyard.[37] Series II, which followed Series I by 2 weeks, consisted of nineteen tests attempted between February 14 and June 20, 1956.[38] The objectives of these tests were similar to those of the earlier series but were more technically sophisticated. While using essentially the same physical system, they sought "more realistic system exercises" that would enable them "to collect and analyze data from more realistic tests for extrapolation to SAGE performance, to isolate areas of system performance that need improvement and to determine the nature of the changes required."

Multiple "air strikes" were carried out by newer, faster, jet-engined B-47 medium bombers, which were allowed to take evasive action. Portsmouth, Martha's Vineyard, and Boston were "attacked."[39] Of the scheduled 19 SOTs listed in the "box scores" published in the Quarterly Progress Reports covering the test period, only seven sorties proved to be available for Direction Center-guided intercept attempts, and these seven guided the pilot only to a point from which he could see the target and then presumably attack it. Cancellation of flights by the Strategic Air Command (which was contributing the bombers), adverse weather conditions, or equipment problems on the ground or in the air interfered with all the other attempts.[40]

As the tests proceeded, the 1954 Cape Cod System's designers and operators took steps to correct revealed deficiencies and to improve the system's performance. Computer programs were changed to provide better communications between the Weapons and Tracking sections, and a "new and improved tote board for the Weapons Director" was installed.[41] When the Collins Radio ground-to-air data link was replaced by a General Electric model and a new computer program was written for transmitting data-link messages, "considerable improvement" resulted.[42]

But not all problems were quickly solved. A wearisomely familiar malfunctioning of the Mark X radar identification system persisted. Of the 38 interceptors scrambled during the first series of tests, 21 failed to complete an interception because of the lack of an operative Mark X system. The same problem plagued the second series of SOTs. By the summer of 1956, engineers in Group 21 had concluded that the Fine Grain Data (FGD) equipment operating with the South Truro long-range radar, as designed, was not compatible with the Mark X radar beacon used to give more accurate readings of the momentary location of an interceptor. The performance of the Mark X appeared to be inhibited when it was coupled to the FGD-type processing equipment, so modification of the equipment was begun.[43] The failed tests obviously limited the amount of data available for analysis and evaluation of the Cape Cod System's performance, but by the end of the second series of SOTs preliminary technical judgments were being rendered and efforts were being made to provide what the engineers were willing to call "a realistic appraisal of the fundamental CCS capability."

Several general conclusions were offered during the summer of 1956 as Division 2 tried to put the best face it could on the situation:

> (1) The CCS will initiate and track the strike aircraft up to the point where there is a total of about 30 tracks in the system.
>
> (2) The strike aircraft will be identified correctly a high percentage of the time in these relatively simple air situations.
>
> (3) Interceptor aircraft can be scrambled and assigned to the strikes.
>
> (4) The interceptors will also be initiated and tracked; but the initiations will not be made as quickly nor will the tracking be as continuous as on the strike aircraft.
>
> (5) As long as the track capacity of the system is not exceeded, the capacity of the Intercept Directors is from 4 to 6 simultaneous voice-controlled interceptors.
>
> (6) As long as the Intercept-Direction capacity is not exceeded, the CCS can be expected to accomplish interceptor guidance such that

almost 100 per cent tallyho's [visual or AI radar contact with target by interceptor] occur.

(7) About 70 per cent of these tallyho's can be expected to permit successful firing passes (satisfying vectoring limits) for B-29 targets. Information is not yet available for the corresponding figure for B-47's, but it is not expected to be significantly different.[44]

Series III of the System Operation Tests began on July 6, 1956, after a lull of 2 weeks, and ended on December 19. Test specifications for the series had been published in June. At least 12 tests were wanted, and 19 were scheduled initially. By the time Series III was over, 25 SOTs had been run.[45] The physical system used was basically the same as before, but a fourth interceptor base was added and two of the three height-finder radars were silenced. The third radar, at South Truro, was replaced by an FPS-6 model but was left on standby for use in the event that its replacement became inoperable.[46]

This third series of SOTs was constructed to push the 1954 Cape Cod System to its limit. Many of the tests were framed to use 16 of the faster B-47 bombers, all of which would be carried by the system at one time. In addition, previously recorded track data were incorporated to supplement the 'live' data in many of the tests, in order to impose capacity loads on all system functions.[47]

The 16 B-47s supplied by the Strategic Air Command were opposed by 16 interceptors controlled by the CCS Direction Center. They were also opposed by antiaircraft in communication with the Direction Center. By November 7, 1956, the 19 initially scheduled tests had been carried out, but only four of them had been successful enough to yield "useful system operation data." Three were changed to System Operation Exercises when they could not be carried out as planned, and these were considered to be of some aid to the evaluation studies being made. Extensive "mapping" (blotting out, or masking, of incoming radar information) permitted only "limited system data" to be acquired in four other tests. The other eight either were canceled or returned insufficient data as a result of weather conditions or equipment malfunctions.[48]

While the 1954 Cape Cod System was performing usefully but not brilliantly in such tests, the Experimental SAGE Subsector system was being readied for evaluation and flight testing. At the time the first series of Cape Cod SOTs was being carried on, late in 1955, Group 22 had already set to work on detailed planning of the initial phase of the ESS

Evaluation Program. The first group of these tests was tentatively scheduled to begin on July 2, 1956 and continue to October 15. These dates were predicated on Division 6's ability to begin shakedown testing of the ESS on May 15, 1956.[49] But that was not to be. The computer programs that the XD-1-centered Subsector system required were not yet finished. Indeed, their preparation had fallen a whole year behind schedule. As a result, the Cape Cod System received both a stay of execution and a new lease on its life. While Group 22 had to reschedule the ESS evaluation tests to begin in the or early autumn of 1957 rather than 1956, it was freed to give further attention to analyzing and improving the performance of the 1954 Cape Cod System.[50] Because of the delay, not only was the Cape Cod System "the only SAGE-type system in existence"; in addition, the ESS, when it came on line later than had been planned, would have to be "so inflexible as to preclude its use as an experimental device for future improvement work."[51] Planning for the ESS would have to be continued on a "reduced basis" until the time grew closer for the prototype system to undergo appropriate testing.[52]

Walter Wells, Associate Leader of Group 22, anticipated in June of 1956 that evaluation of the Cape Cod System would be completed as planned when the Series III SOTs were finished and analyzed. Further evaluation of a SAGE-type system, he argued in a memorandum addressed to Carl Overhage, would have to await operation of the ESS or "be done on a substantially modified Cape Cod System." By the time the SOTs were completed and their results digested, a number of modifications of the Cape Cod System should be "well formulated." Some would be desirable, if not necessary, in order to make the Cape Cod System more closely anticipate SAGE. Others held promise of offering "significant improvements over . . . planned SAGE techniques [and] should be evaluated as soon as possible for inclusion in SAGE." And Wells saw another argument for continuing such tests: "Lincoln's desire to continue improvement of SAGE" made it "imperative" that they retain a "SAGE-like system" to continue the "experimentation and trial-and-error work needed to develop improvements."[53]

The third series of Cape Cod SOTs was completed later than Wells had predicted, but they indicated, as he had surmised, that "a number of system modifications would improve capability for handling heavy loads." The conclusion was inescapable: the Series IV tests would be the "proving ground for some of these modifications," because, as Israel and Wells noted in an August 1956 analysis of the Cape Cod System and SAGE, several operational problems were in need of serious investigation.[54]

Members of Groups 22 and 61 had met on July 20 to agree on the significant operational problems that they saw affecting the two systems:

> Group 22 reviewed the experience of their continuing test and evaluation of the Cape Cod System and suggested those areas requiring most urgent consideration; Group 61 discussed those features of SAGE which might bear on existing problems; and both groups attempted to indicate what future action was needed. The general objective of the meeting was to provide a guide to Group 22 in their revision of the Cape Cod System and to highlight for Group 61 those problems requiring future research and development.
>
> Discussion generally was limited only to those operational functions represented in the Cape Cod System; this excluded raid forming, crosstelling, Subsector Command Post, forward-telling, duplex, and standby, etc. In general, the meeting was focused on establishing the current situation and not providing specific suggestions for improvement.[55]

The two groups agreed on three major problems for which they had as yet no operational experience, they admitted, "either in Cape Cod or elsewhere." There would be the challenge of "initiating and tracking multiple aircraft flights in which the separations are large enough for radar resolution but too small for tracking resolution by existing computer programs." Also, it would be extremely useful to set up tracking programs that would employ "time tags." A revised Cape Cod program incorporating inputs from an FST-2 radar at Bath, Maine, would give them helpful information in this direction. Finally, there was the problem of "tracking aircraft in the overlap region of two or more radars." Since the Cape Cod System hitherto had had the use of only one heavy radar and had dealt with overlap problems using only gap fillers to cover one "very small area . . . in which operational missions are conducted and data taken," it was high time to be thinking of more ambitious testing that would provide needed information "within six months to a year."

Radar mapping, interceptor tracking, track trouble detection and monitoring, interceptor direction and displays, and weapons-assignment actions were all "major unresolved problems." In addition, radar inputs, identification, initiation of tracking, smoothing, and height finding could all stand improvement, nor did the engineers have enough information yet on overlap tracking and on tracking "closely-spaced aircraft formations."[56]

So intense was their preoccupation with the ever-unfinished business of advancing the state of the art that it was easy to overlook what sort of a system Cape Cod had become by the end of the summer of 1956: a miniature, experimental, increasingly sophisticated air defense subsystem centered around a Direction Center housing the Whirlwind computer. The computer received, over telephone lines, signals from one radar that could "see" about 200 nautical miles, as well as signals from four gap-filler radars, each of which could "see" about 32 nautical miles.[57]

A glance at the Cape Cod System's operation from the perspective of those responsible for blotting out spurious signals by means of "radar mapping" serves as a reminder of some of the things that the Cape Cod System could do. The antennas at the radar sites made a 360° sweep (a "scan") every 12 seconds, and their reports on the areas under surveillance were sorted into small cells ("boxes") in order to facilitate filtering out significant from superfluous information. If the radar antenna, as it rotated, picked up no echo from a box, that cell was empty and its information was discarded. The long-range radar was equipped with a Fine-Grain Data "trimming" device that called any cell empty and such information superfluous if it contained only one echo. These preliminary screening procedures thus provided data from "filled boxes" only—data worth converting to polar coordinates and sending on to the Direction Center, where the Whirlwind computer processed the data in order to present "the air situation to the AC & W officers and airmen for their subsequent action."

The Lincoln Lab designers had long known that "typically radars can supply much more data than can be handled by the computer" before the next antenna sweep begins, because of interference arising from random noise, jamming, and normal "clutter." Radar mapping technicians at the long-range radar site monitored and excluded much of this interference before the data went into the FGD machine, and similar Radar Mappers at the Direction Center excluded more clutter before the incoming data were presented to the computer. These necessary chores were better supervised by human technicians than by automatic machinery, and it was taken for granted that the Mapping Supervisor and the Tracking Officer in the Direction Center would be in constant telephonic communication. The former might report "large mapped-out areas, data interference, site shutdown" and the like; the latter could request, for example, that a certain area be mapped out or carefully not mapped out if a track of special interest was about to enter such an area.[58]

To free Whirlwind to process significant incoming data without interruption, a buffer drum was provided to receive and hold incoming "sweep" data until Whirlwind was ready to accept the data. In addition, data alarms were provided at the consoles to warn of "Sudden Increase, High data count, Excessive data count, and Data Rejected." The Cape Cod System was sophisticated enough to be used to break in new military trainees who one day would be expected to operate the Experimental SAGE Sector and then the fully operational, military SAGE System, which the 1955 Operational Plan had estimated should become available in the middle of 1957.[59]

But early in 1957 Cape Cod System Operation Tests were still being scheduled. When specifications for the fourth and last series of SOTs were issued in February of 1957, they included a number of changes in the system. Although these were not as extensive as Walter Wells had proposed, they were sufficient to remold the 1954 System into a "1957 System" should further modifications actually be undertaken.[60]

The eight modifications were divided into two groups: those to be done as soon as the Series III tests were finished and those to be done in the future. The first group comprised five modifications:

> Complete recoding of the 1954 CCS program for 1957 CCS.
>
> Addition of a heavy radar, Mark X and two height finders at Bath, Maine.
>
> Addition of one bank of core memory. . . .
>
> Operation of height finders in a semi-automatic mode, with improved height priority program.
>
> A new tracking logic similar to that planned for the ESS.

The second group comprised three future modifications:

> Improved display facilities.
>
> Improved manual input facilities.
>
> A Recovery Officer to aid INDs in returning interceptors to base.[61]

The improved 1957 Cape Cod System that was being proposed would operate as a fully manned, mock-operational system during each test, in order to ascertain "under certain controlled conditions . . . how well a semi-automatic system can perform the integral functions of air defense."[62] Unlike the previous series of tests on the 1954 Cape Cod System, which were "planned to provide data for one or more specific evaluation items per test," the 1957 System tests would "obtain data on all objectives throughout the first twenty-five scheduled tests."[63]

Furthermore, the testing plans stipulated that during a four-hour test period, between six and sixteen B-29s flying at altitudes from 10,000 to 25,000 feet, or a like number of B-47s flying between 23,000 and 39,000 feet, would penetrate the system's surveillance zone. The zone would be protected by 20 interceptors and the Boston-Providence Nike missile sites, all under the operational control of the Weapons Director in the Cambridge Direction Center.[64]

In each test the "number, location, and time of strike" would be changed in order to "increase the realism of the air defense problem."[65] Varying attack conditions would "prevent the tracking and weapons sections of the direction center operators from recognizing a pattern of attack."[66] However, they would restrict the number of "input variables" by requiring strike aircraft to "maintain a constant air speed and barometric altitude" while penetrating the system. Also, they would prohibit the use of radar jamming.[67]

Plans contemplated and plans carried out are rarely the same, and such was the situation for the 1957 version of the Cape Cod System. Since hardware and software had been altered, it was necessary to check the working of the parts of the system before realistic engagements with attacking "enemy" forces could occur. As they began to check the working of the system in preliminary flight and intercept tests, problems began to emerge. Beset by difficulties (especially with the computer programs prepared for the 1957 System), the Series IV tests began in late February of 1957, ended late the following May, and ran through only 13 of the 25 tests initially planned.[68]

The first eight tests appear to have set the pattern: only two appear to have actually occurred, four were canceled by the Strategic Air Command before the scheduled strike, and one was called off by Lincoln Laboratory to "provide more program checkout time."[69] It is not surprising that no detailed analysis of the first test was prepared.[70]

On May 1, 1957, Group 22 finally conducted a test that it was willing to call "the first test of the 1957 Cape Cod System." It consisted of "overlapping heavy radar and Mark X coverage from South Truro and Bath, three height finders operating in a semi-automatic mode, and a rewritten program for the Whirlwind I computer." The results were "fair," the engineers reported, "considering the fact that this was the first time the 1957 system had been tested in a real air situation." Once again, a full report was not prepared, since the test was regarded as more of a "shakedown" than an evaluation run. Nevertheless, the limited data obtained gave helpful information on "specific troublesome areas," most of which appeared to be caused by program errors.[71]

Five more SOTs were undertaken before the end of May. Only the last was completed. Program problems caused Lincoln Lab to cancel one and to convert another into a System Operation Exercise because "there were still too many program problems to allow worthwhile evaluation of the 1957 system." When the Strategic Air Command withdrew from SOT 12 because of bad weather near its Little Rock air base, that too became an Exercise. Even the thirteenth and final SOT was limited in its scope, conducted "primarily to get a complete picture of the errors still existing in the 1957 Cape Cod System program." Although these were discussed in the briefing session that followed, analysis appears to have gone no further.[72]

It was in this manner that the Cape Cod System tests languished and expired. At the end of July, Walter Wells officially informed Bob Wieser that "as of 1 August 1957, Group 22 will consider the Cape Cod System inactive. No further tests or exercises will be conducted for evaluation or experimentation purposes, until further notice."[73] Further notice came late in September and brought with it information that was no longer news to the Lincoln Laboratory staff. Whirlwind would no longer participate in SAGE intercept-test operations:

> Support of the Cape Cod System as an experimental breadboard air defense ground environment will be discontinued 27 September 1957. The Air Force operators in the Cape Cod (Barta Building) direction center will be released, and the service of maintaining the Cape Cod computer program will be discontinued. It is now planned to keep the system hardware intact, for experimental work, with the possible exception of some of the existing telephone and teletype facilities. This includes the Whirlwind I computer, which will continue to be maintained for experimental or scientific and engineering work.[74]

Beginning with George Valley's reorganization of 1955, Division 2 increasingly had taken over responsibility for the tactical, live-intercept flight testing carried out during the last two years of the Cape Cod System. These tests simultaneously had provided needed information on the performance of the air defense command-and-control system under development, had given both Lincoln Lab and military personnel further training and experience in operating the system, and had made it possible for them to apply the lessons they learned to the emerging

Figure 28.1
The Whirlwind computer in 1952.

Experimental SAGE Subsector. The Cape Cod System had played an essential empirical role in the testing and the development of the ADSEC–Project Charles–Lincoln Lab concept of continental air defense. If it appeared to come to an end with a whimper, that was because the R&D process had pushed it out of the center of the stage. On October 2, 1957, five days after all testing and evaluation of the Cape Cod System had come to a formal close, Group 22 conducted the first evaluation test of the Experimental SAGE Subsector, using the XD-1 computer. Whirlwind no longer would be sending interceptors out to answer live radar reports.[75]

29

Checking Out the ESS

The period 1954–1957 was both successful and exasperating for the SAGE project in general and for Division 6 in particular. Not only did the men of Division 6 carry more than their share of crises; they clearly demonstrated the vigor and the growing complexity of Lincoln Laboratory's R&D effort as its emphasis shifted from designing and developing the SAGE AN/FSQ-7 computer to hooking up the prototype XD-1 to several regional radars and integrating the performance of the resulting network to produce the Experimental SAGE Subsector. Ideally, the ESS would serve as an exemplar of how each of the sectors of the continent-wide air defense system should work.

As application of the new computer to its national defense role loomed closer, the story of the SAGE air defense system and the affairs of Lincoln Laboratory began to eclipse the story of the computer itself and the affairs of Division 6. In view of the nature of the enterprise, the transition was inevitable, and the pressure of technical and managerial events did not bode well for the established independence and freedom to carry on its own R&D affairs that the MIT Digital Computer Laboratory had enjoyed at the time it became Division 6. In retrospect it is clear that the creative crisis and the impressive innovation of 1950 were gone by 1956. Perhaps the clearest harbinger of the approaching institutional routinizing of engineering R&D at Lincoln Lab was the departure of Jay Forrester at the end of June 1956. Bob Everett took his place, and the transition was remarkable for being unremarkable.

What was happening? More and more of the same, and perhaps this eventually became the issue. For example, in 1954 Division 6 had fewer qualified engineers and computer programmers than it needed, and, because the SAGE R&D enterprise was expanding, the shortage persisted month after month while the managers were taking special measures to hire and break in more men and women. There were managerial crises and reorganizations. The continual jockeying for power and control did little to speed up the effort, whether it was going on within Lincoln Lab or within the Air Force, between Lincoln and the Air Force, or between either Lincoln or the Air Force and industrial firms in the private sector.

Amid schedule slips, unanticipated technical problems, and unanticipated solutions, the major problems were being addressed and solved even while the stakes were growing higher. Net progress was occurring, but it was not occurring spectacularly, it was not occurring as rapidly as they had scheduled, and because it had to do with national security it was all going on quietly behind a veil of secrecy. Finally, its success was not a foregone conclusion.

"We had all grossly underestimated the overall programming job for SAGE," John F. Jacobs recalled 30 years later, referring to the pioneering, unanticipated software design challenges the engineers encountered as a consequence of being committed to the extravagant and close coordination of detail required if the SAGE air defense concept was to be converted from theory to practice. Indeed, so new were these challenges that the term "software" had not come into common use yet.

"During the early part of 1954 it had become clear that the programming job would be much larger and much more expensive than we had predicted," remarked Jacobs. "The operational program would be over 100,000 instructions and the utility and instrumentation programs would be almost as large."[1]

Software problems were not the only ones they encountered along the way, of course. And in 1954 the first order of business for Division 6 was to help IBM finish building and testing the XD-1 and get it ready for the move to Lincoln Lab. At the same time, they had to work with Division 2 to continue the Cape Cod intercept tests. Furthermore, they had to begin getting the Experimental SAGE Subsector up and running so that, in turn, the first operational SAGE sector, at McGuire Air Force Base in New Jersey, would immediately follow the ESS. These activities were to become major goals of Division 6 between 1954 and 1958, and the newest of these was the ESS—until the technical problem of integrating new weapons systems into the SAGE air defense network arose.

Consequently, in November of 1954, while Jay Forrester was persuading upper-level IBM administrators that ineffective management of Project High was delaying the delivery of the XD-1 machine to Lincoln Laboratory, he and Bob Everett were consolidating a modest internal reorganization of Division 6 they had begun in September. It would, they thought, enable them to proceed to phase in the XD-1 and Experimental

SAGE Subsector R&D activities on top of the activities already devoted to the 1954 Cape Cod System tests.[2]

Forrester put Bob Wieser's Group 61 in charge of SAGE System test and planning in addition to continuing with 1954 Cape Cod System testing. As he and Everett saw it, the men would be handling computer programming, training personnel, and operating, testing, and evaluating three R&D systems both simultaneously and in overlapping succession. Building these systems was part of the way that Division 6 had chosen to proceed through "the 'breadboard,' prototype, and production stages," as Forrester called them—that is, the Cape Cod System, the ESS, and the continent-wide SAGE system.[3]

Norm Taylor's Group 62 remained in charge of the XD-1's design and was consequently getting ready to install the computer in Building F at Lincoln Laboratory so they could check its internal performance. The group already had been establishing the "specs" for the AN/FSQ-7 production-model computer, guiding IBM's Project High "in the design, construction, improvement, and maintenance of prototype FSQ-7 (XD-1)." In consequence, the men spent most of their time on "logical design, the basic circuitry design, memory development, major computer component test" and on developing the display system that the human operators in the Direction Center and the Command Center would require.

Containing about 25,000 vacuum tubes, the computer was being assembled in Poughkeepsie during the autumn of 1954 by IBM engineers, assisted by six Division 6 experts on the test floor at Project High, before being checked prior to its delivery to Lincoln Lab. Once moved to Building F, there to be prepared for the ESS tests, the machine would require, Forrester estimated, some 22 Group 62 engineers—"a strong group of the most capable people we have." Some of these engineers would be transferred from other sections; at least nine new ones would be brought in. For the next 2 years they could expect to be preparing the XD-1 to operate in the Experimental SAGE System Subsector, and in this way they would identify any major errors in either hardware or program design, correct them, and make sure the corrections were incorporated as IBM built the FSQ-7 production machines. Also, they would determine the maintenance procedures for field use.[4]

Steve Dodd's Group 64 ("now badly undermanned," Forrester reported in the autumn of 1954 to the Lincoln Steering committee) was responsible for planning and modifying the Direction Centers for the Cape Cod, XD-1, and SAGE systems and should expect much heavier demands as the XD-1 and production SAGE systems took shape and underwent testing, which almost surely would call for further design modifications.[5]

The inconspicuous but essential and quietly effective Systems Office, which Jack Jacobs had set up, would continue to expedite formal coordination with IBM on design details of the FSQ-7. As the development schedule permitted, the Systems Office would expedite completion of the first XD-1 prototype. "It is through Systems Office efforts that Lincoln Laboratory guides the 262 engineers working on the FSQ-7 at IBM," Forrester explained. The Systems Office, too, should expect to expand.[6]

In spite of ongoing efforts to add staff members, the division was understaffed and overloaded. Forrester proposed to the Lincoln Steering Committee in November that the number of engineers, mathematicians, and programmers in Division 6 be increased from 190 to 297.[7] This provided even more of a reason to accelerate recruitment.

In September of 1954, Forrester and Everett timed their release within Division 6 of the news of the pending reorganization so that their engineers would have time to anticipate the changes they would have to make in their daily activities at the personal working level. In Wieser's Group 61, Charlie Zraket would be in charge of 1954 Cape Cod System Operation, Bob Walquist would be in charge of XD-1 Programming, Jack Arnow would be in charge of Transition System Planning (which henceforth would become "SAGE Planning"), Dave Israel would be in charge of Cape Cod Test Program Planning, Walter Wells would be in charge of Cape Cod Analysis and Simulation, and the division would undertake a new "SAGE Training" task to provide the large numbers of Air Force personnel that would be needed to operate the Direction Centers in Building F and subsequently at Air Force bases.[8] Releasing such information was a part of Forrester's and Everett's usual pattern of ensuring as instrumentally as they knew how that progress would be made and schedules adhered to, although it wasn't always obvious at the time to observers outside Lincoln Laboratory whether any particular reorganization was coping with or adding to the technical, staffing, and personnel problems customarily being encountered. Air Force commanders occasionally voiced their disquiet over the way Lincoln Lab was conducting its affairs. Part of the problem lay in the fact that the lab's style of carrying on advanced R&D could not readily be reduced to a reassuring routine.

On December 28 and 29, 1954, Jay Forrester, Bob Everett, Norm Taylor, Steve Dodd, and "Gus" O'Brien were in Poughkeepsie watching closely as IBM engineers ran through a final series of tests before dismantling the XD-1 for shipment to Lincoln Lab. "Electrical margins were very good," Forrester reported to his Division 6 engineers. "Except for troubles from a defective lot of tubes with intermittent connections at the base pins, the machine seems to work as well as could be expected at this stage in the program. We were much encouraged."[9] The XD-1 was ready to be moved. This unquestionably was progress.

Yet progress was rarely as straightforward as Forrester or Everett or Valley or Taylor or Wieser in their optimism planned it to be, for several reasons. Mention has been made, in the story of the 1954 Cape Cod System, of the internal reorganization of Lincoln Laboratory that occurred in the summer of 1955 amid mounting objections to the recent actions of the SAGE Test Committee. Here was another factor that prevented the sort of straightforward managerial and engineering solutions the engineers of Division 6 had become accustomed to invoking ever since the days when they were still the Digital Computer Laboratory on the MIT campus, developing the Whirlwind computer.

As the remaining months of 1955 passed, it became increasingly clear to rank-and-file engineers of Division 6 that the division was no longer so firmly in charge of its own destiny, and friction began to develop as they realized that the relatively inert hand of a programmatic and undiscretionary administrative routine imposed from without and above was supposed to displace the bracing challenge they had always enjoyed of being themselves innovatively in charge of the task of deciding how to reach assigned technical goals. Or so it seemed to them.[10]

"The Valley reorganization which had led to all the personnel changes," Jacobs recalled, "did not, as had been intended, close the rift between Divisions 2 and 6." The fact that one purpose of the reorganizing had been "to ensure that the technical results turned up by the Cape Cod System would influence the design of the various parts of the SAGE system" still did not heal the growing breach.[11]

More important than these growing tensions, however, was the R&D momentum accumulated by the unfinished business of completing, with the ESS, the R&D enterprise that had begun several years earlier with the Bedford tests. The fundamental situation was simple, stark, and known to all: the Experimental SAGE Subsector must be brought into being, and by its nature the job required the continuing extensive collaboration of Division 2 and Division 6 personnel.

This collaboration on the engineering level was forthcoming. While the intercept-testing operations of the 1954 Cape Cod System were still going on, Division 2 was bringing more radar stations into the projected subsector network and at the same time developing Group 22's charted managerial arrangement. The latter was intended to control the elaborate testing programs that would be associated with bringing the ESS on line.[12]

Concurrently, Division 6 engineers began to ready the XD-1 for the functions it would perform in the Experimental SAGE Subsector and to identify and assign the R&D activities that would bring the ESS into being. They did not need to be reminded that, compared to the ESS, the Cape Cod System was "an elementary system of air defense performing only basic tracking and intercept-direction functions." The ESS, the engineers' creature as the prototype exemplar of future SAGE sectors, would add overlapping radars and such functions as height finding, automatic flow of information from one center to another ("crosstelling"), and up-to-date information on such diverse factors as the weather, scheduled airline flights across the sector, and the status of the weapons systems being sent into battle. The ESS, and right behind it the SAGE system, as the engineers learned the hard way, introduced computer-programming complications of an entirely different order of magnitude. Bringing these complications under control took longer than they anticipated.[13]

So R&D realities, rather than reorganization charts, caused a new and larger equipment-testing regime to be launched during the summer of 1955. The new regime focused on the XD-1 and the problem of integrating it with the ESS expanded radar network. Six inter-divisional test teams were organized to carry out testing on the equipment of eight subsystems identified as essential input or output channels to the XD-1.[14] By the end of 1955 the subsystem tests were being scheduled in accordance with the policies of the new SAGE Experimental Test Policy Committee, which had been made responsible for scheduling ESS system operation tests to be carried out under a formal ESS Evaluation Program.[15]

Operating under the auspices of this new and apparently formidable managerial machinery during the summer and autumn of 1955, the Lincoln Lab engineers nevertheless went ahead largely as before. "Much effort," Overhage, Wieser, and Harrington reported, "has been directed toward a joint Division 2–Division 6 test activity, both in plans and execution."[16] In this way the engineers prepared to initiate ESS equipment tests while moving on into preliminary testing of ESS subsystems into which the pieces of equipment were integrated. The eight subsystems they needed to test were gap-filler inputs, long-range radar inputs, height-finder inputs, ground-to-air outputs, automatic teletype outputs, crosstelling, ground-to-air voice radio, and wire communications. As these were proved out for the ESS, they would become available for application to SAGE sectors as the latter were phased into operation. The test teams were composed of individuals from Division 6, Division 3, Division 2, Bell Labs, Western Electric-ADES, IBM, and the RAND Corporation, and their membership grew as the weeks passed.[17]

One measure of the sophistication of their design-and-test philosophy is to be seen in the way they went about equipping the Whirlwind computer to accomplish crosstelling with the XD-1. Crosstelling would test the crucial capacity of the future SAGE system to pass intercept and battle information from one Direction Center to another and from one sector to another, giving ground commanders unprecedented opportunity to influence if not control the course of battle. Aware in May of 1955 that the schedule called for XD-1 and SAGE hardware and software specifications to be ready by October, 14 Division 6 engineers met on May 18 and May 31 to determine how crosstelling equipment could be installed in the Barta Building by September. It would "occupy about 5 racks of the standard Whirlwind two-tube plug-in unit and would give Whirlwind the full complement of SAGE input and output test capabilities." Furthermore, the engineers would seize the incidental opportunity to build output equipment first, in order to test the feasibility of placing the height-finder subsystem under semi-automatic computer control.[18]

They recognized that the overall arrangement would permit them to carry out simulated systems tests of guiding the BOMARC and TALOS intercept missiles and to investigate the possibilities of linking the Direction Center to antiaircraft and Nike missile batteries. At one stroke, then, Whirlwind could be equipped to "(1) supplement and complement the testing of the Experimental subsector, (2) test the future input and output systems for the SAGE system, and (3) help evaluate in the future the new data gathering devices and weapons to be used with SAGE."[19]

By October of 1955 the equipment had been installed in the Barta Building and had been checked. It would be waiting "over the next four months" while crosstelling tests were written and the output system of XD-1 would become available for the scheduled tests.[20] But before the end of the year the complexities involved in setting up the ESS tracking and intercept system while also completing the Cape Cod System intercept tests suggested to the men that they insert another R&D step to explore crosstelling: "a small-scale air defense system, Whirlwind I SAGE Evaluation (WISE), which is much simpler than the 1954 Cape Cod System now undergoing evaluation." "As integration proceeds," Forrester and Everett explained in a Division 6 summary report issued in May of 1956, "WISE will be modified for crosstelling to XD-1."[21]

While the design and experimental testing programs for the ESS and for the SAGE system and its components were being organized, an ever more pressing planning problem arose. In addition to obtaining additional qualified engineering personnel for Lincoln Lab's divisions, they needed to begin large-scale training of the Air Force enlisted personnel and officers who one day would staff the SAGE system for the Air Defense Command. Already such men had been brought in to learn to operate the Cape Cod System, but more would have to be brought in and taught how to operate the Experimental SAGE Subsector, and many, many more would be needed to operate the Air Force's 20-odd SAGE subsectors across the continent.

The Air Defense Command and Lincoln Lab turned to the RAND Corporation, which had been training scope operators for the manual air defense system. RAND possessed the competence to take over responsibility for the operational programming. In July of 1955, Division 6 was getting ready to receive a small number of RAND people, who would learn what was involved and then take over the two tasks. "They plan to build up a staff of some 117 people over the next two years," reported Jacobs to his fellow engineers. They would be trained on the job by Lincoln Lab so that after the installation of the third Direction Center the RAND Corporation could assume responsibility for system training and operational programming as Direction Centers and Combat Centers were sequentially brought on line in the added SAGE Sectors.

To determine how much training over how much time would be required to prepare a sufficient number of Air Force personnel to operate the Cape Cod System (using Whirlwind) and the ESS (using the XD-1), S. B. Hibbard of Division 2 and K. E. McVicar of Division 6 undertook a detailed planning study in October of 1955. In this way the expanding effort devoted to the job of training military personnel began to generate a need for a new organization. Before the job was finished, the contingent from RAND would become the Systems Development Corporation.[22]

In September of 1955, Lincoln Lab engineers gave a "Planning and Progress Report" on the status of the ESS to representatives from the Air Force and from the Bell System, explaining that their tentative schedule called for them to finish integrating the Cape Cod radar network with the XD-1 computer by that December, to check the components and subsystems of this emerging ESS system until April of 1956, and then to swing into operational testing (which would involve at least two heavy radars and eight gap fillers). The ESS Direction Center in Building F should already have been completed by the middle of February 1956, they estimated, so they should be able to finish the "shakedown testing" part of the operational phase during the first two weeks in April. Further "system operational checkout" would then last from June through September of 1956. The computer program for the first military sector, at McGuire Air Force Base in New Jersey, should be ready to be shipped in October of 1956. Testing, evaluating, training, and preparation of the Combat Center Program would continue under the collaborative efforts of Lincoln Laboratory, ARDC, and ADC.[23] Unfortunately, this happy schedule would be disturbingly stretched out by bleak and unforeseen realities.

But problems were not the only order of the day. On December 9, 1955, the XD-1 "first successfully tracked live aircraft," using a simple, single-track program and a computer-generated map presented on a display console. By the end of 1955, all the major elements of the ESS network except the Jug Handle Hill radar station were electronically interconnected.[24]

Meanwhile, by the middle of December 1955—and representative of the extraordinary detail that hardware and software planning required—three sets of operational and mathematical specifications had been defined, "in order to expedite preparation of an initial XD-1 program" to be used "during early shakedown phases" of ESS operations.[25]

It was not always possible, unless one was busily at work on the inside, to distinguish readily between those reorganizations that turned out to be administratively cosmetic and those that were genuinely instrumental. During the first quarter of 1956, for example, Forrester and Everett set up Group 67, "Program Production," under Jack Arnow. Its assignment was to "translate system requirements as established by the operational and mathematical specifications" developed by Group 61 "into coded programs ready for the computer." In this way they planned to put together coded programs ideally so honed and polished as to be usable not only in the XD-1 prototype but also in the subsequent FSQ-7 computers that would be installed in the militarily operational SAGE direction centers. This experience would enable members of Group 67 to act as consultants to IBM when that corporation became involved in preparing the appropriate software programs for the standby computers that would be part of the "Duplex" arrangement for the SAGE Direction Centers.[26]

This course of action required elaborate attention to details in order that the men might develop the basic master program that would energize and control the digital operations of the air defense computers to be installed in every sector. First, Group 61, formerly headed by Wieser, had to provide Group 67 with the appropriate operational and mathematical specifications. During 1956, guided by these, Group 67's staff of more than 23 programmers plunged into a four-phase course of action that called for them first to generate the coding specifications they would need and then to write the actual codes for each component subprogram. Each subprogram, as it was finished, then underwent "parameter testing." Finally, as the parameter tests were completed, the programmers and engineers began to assemble the subprograms and the larger programs they composed in order to test them together. Such an assembly test, when successful, would indicate "that the subassembly, or the complete system of programs, meets the operational and program specifications." And when all these coded subprograms had been assembled, forming "the complete system of programs that meets the requirements of the operational specifications" for the ESS and the modular SAGE sectors, it was time to turn them over to those who were "responsible for shakedown testing and air defense evaluation."[27]

The complexity of this master program, of its preparation, and of integrating it with the other features of the Experimental SAGE Subsector had no precedent in the history of computing. At first, the men did not realize the immensity of the thicket they were walking into. It defied their efforts to set up a realistic schedule and completion date.

They had been following the same exhaustive R&D routine that Division 6 had brought with it to Lincoln Lab from the MIT campus. It continued to be imposed at the operating level of getting things done: design, build, test, modify, retest, and continue to modify and retest until the equipment was operating up to specification; then link the equipment into subsystems for further testing and modification before linking subsystems together, step by step, to build the overall, prototype system that would constitute the Experimental SAGE Subsector. The underlying theme was the same, over and over again, as the electronic hardware and the operating software were developed and brought into use in an ever more complex arrangement: design, build, test, modify in the light of the test results, retest, modify further . . . and continue to carefully elaborate the system as specified performance was achieved.

This was the way they set about readying the master operational program the computer would need—the Direction Center Active (DCA) program, as they called it. Their failure to anticipate how elaborate this constellation of computer programs would be was partly responsible for a schedule slip of a year.

The prospect that they beheld was one in which the DCA would operate each of the sectors of the prospective SAGE system, starting with the sample ESS module. It comprised "about 60,000 instructions and over 1,000,000 bits of internal data storage" which accommodated "20 tables of instructions containing another 35,000 instructions."[28] This program would be exhaustively checked on the XD-1 in Building F and then tailored in each of the subsequent SAGE sectors to whichever computer would *not* be on standby—the "active" computer—in the Duplex Central installations of the final SAGE system (figure 29.1).

In 1956 it became evident that the overall R&D schedule had been way too ambitious. By that time "enough was known about the computer programming job requirements," Jacobs remarked many years later, "so that we could state with certainty that we were too ambitious with respect to the capacity requirements for tracking and interception that we had originally set, and that it would take considerably more time to train the programmers and create the program than we had anticipated." "This was a very delicate matter," Jacobs continued, "because the whole SAGE schedule was keyed to the availability not only of the FSQ-7 machines, but also of the operational programming."[29] When they surveyed the situation, they came to realize they would have to add a year. No alternatives and no "quick fixes" could be found.

"Lincoln Laboratory now forecasts a very substantial delay in preparation of the Master Computer Program for the Experimental Subsector," reported S. P. Schwartz of Western Electric/ADES to the Air Force on June 12, 1954. Schwartz was co-chairman of the SAGE System Phasing Group in the Air Defense Engineering Service Project Office in New York City. Air Force officers in the Project Office immediately saw the implications: "This program is to be later adapted to the first two SAGE DCs [Direction Centers] and the first CC [Combat Center]. While the effect of this slippage can not immediately be related to slippage in [military] operational dates it could well be of serious proportions."[30]

There was an immediate institutional effort at damage containment. "All agencies at both the ADES Status and Progress Meeting and Phasing Group Meeting joined in emphasizing that the ESS program slippage does not provide reason for relaxation of effort . . . to minimize the impact of the . . . problem on SAGE schedules." Now they had to await preliminary reevaluation of the situation while facing the bleak fact that

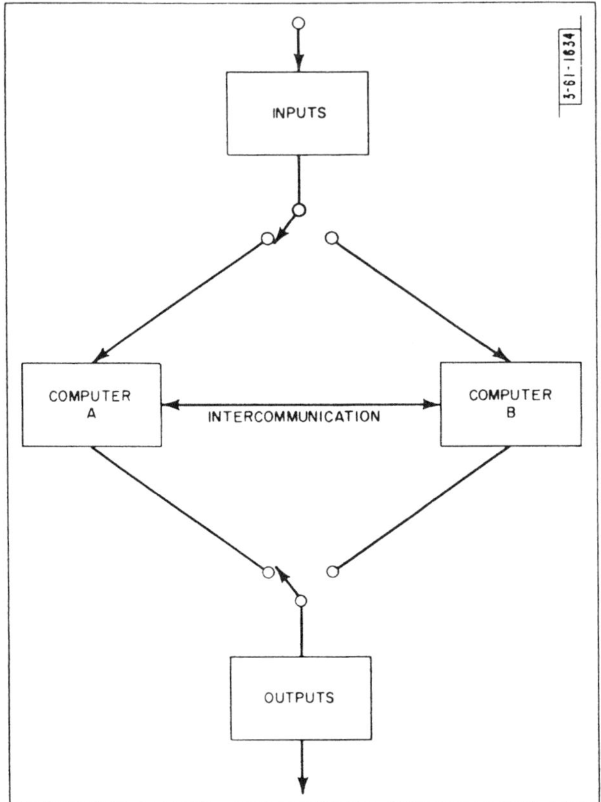

Figure 29.1
"AN/FSQ-7 Duplex Operation. The AN/FSQ-7 is a duplex machine, i.e., two central computers (A and B, above) are available although normally only one computer is connected to simplexed inputs and outputs. Over-all system reliability is increased by the fact that either computer A or computer B can perform the air defense function. (In the example, A is performing the air defense program and is called the active computer; B is called the standby computer.) Equipment whose failure would cause system failure is duplexed and includes drum and core memory, arithmetic elements, etc. Consoles, individual radar inputs and similar equipment are simplexed."

"the problem of obtaining a computer program is most serious as now presented."[31]

In September an "Interim SAGE Status Report" presented to the Air Force the grim news that, apparently "due primarily to slippage in delivery of the computer program (instructions)," the two Direction Centers and the Combat Center would be late "by one year and forty-five days." A "delayed SAGE Implementation Program" schedule would have to be invoked, with the not unhappy consequence that (in classic bureaucratic phraseology) "the primary change in the manning requirements is an extension of the period during which the initial complement will be on site."[32] They would have more time to train more men and train them better. It was a mix of bad and good news not uncharacteristic of the course of R&D.

The bad news of June 1956 was followed in August by a request from the USAF Air Staff for "cost information and recommendations" on a possible alternative delivery schedule. Two days after formal issuance of the request, three alternative schedules were presented by the ADES Project Office. The most attractive of these, from the ADES point of view, called for a "reduction in computer production from one every two months to one every three months, with procurement reaccelerated in early 1958, deliveries reaching a maximum rate of one computer per month in January 1961."[33]

Western Electric, "based on their belief that it would be good engineering and management judgment to avoid additional major commitments for SAGE until more conclusive tests can be completed about 1 January 1958," recommended adoption. Not surprisingly, "Lincoln Laboratory emphatically rejected adoption," and the issue was referred to "the top managements (Vice Presidential level)" of Western Electric and Lincoln Lab/MIT, while arrangements were made for a joint "task force . . . to examine these discrepancies." The fact that total costs would be reduced, if Western Electric's recommendation were followed, "by an estimated amount of $288,154,000" was counterbalanced by the perception that no simple saving would result, "inasmuch as fewer subsectors would be in operation with consequent forfeiting of potential air defense capability."[34]

Six months after the fact, Forrester and Everett acknowledged for the record in their contractual Quarterly Report that "an estimated slippage of one year in the programming schedule" had occurred. "Schedules set up approximately two years ago," they admitted, "significantly underestimated the magnitude and complexity of the effort required to produce the computer programs for the air defense functions and supporting activities of the SAGE System." The report continued as follows:

> The SAGE programming task was scheduled on the basis of the fundamentals established by the Cape Cod System, an elementary system of air defense performing only basic tracking and intercept-direction functions. The SAGE System introduced the complication of overlapping radars and added many new functions, such as automatic crosstelling, height-finding, command post, and weather and weapons status totes. In addition to the greatly increased size and complexity of the SAGE System, the inexperience of available programmers and excessive demands on computer time have also contributed to the delay.[35]

But the intricacies of the SAGE system and the momentum of its R&D effort were too massive to change quickly. For example, the SAGE Master Program, a suitably modified version of which would control each subsector, was based in large part on 22 operational specifications, "one for each air defense function. e.g., identification, air surveillance, weapons direction, etc." Of these, 14 required mathematical specifications too.

By the spring of 1956, all the detailed mathematical and operational specifications had been completed, and Group 66 had begun adapting special versions for "the McGuire, Stewart, and Syracuse centers." The "major job" of writing the coded programs had just been started. They needed to be completed and then checked in individual and system tests. As they were completed, they would be adapted first to XD-1 and the ESS, which by that spring consisted of "the direction center, four long-range radars, eight gap-filler radars, and five interceptor bases."[36]

None of these events eased the pressure on Group 64 ("ESS Test Planning—WWI, MTC Operation," under Edwin S. Rich) to lay out detailed plans for the next phase, "the integration and shakedown testing of the components which make up ESS." They went ahead with plans to set up a "preliminary period of combined program and equipment testing," which would be followed by "two phases of integration and testing of air defense functions." They would use what they found out during the first phase to shape the second phase, parts of which would call for them to join the air defense evaluation activities of Group 22.

At the start, already assured that the XD-1 and the radars would perform electronically up to expectations, they would begin with "favorable,

Figure 29.2
XD-1 control consoles.

limited, tactical testing conditions" before proceeding to "full-scale evaluation under a wide range of air defense conditions such as poor weather, saturated manpower, and insufficient weapons." These tests would check all major portions of the system and tell them how effective were the operational and mathematical "specs" under which, by that time, the engineers would have constructed the coded programs.[37]

The drastic slippage in the computer software schedule did not prevent Division 6 from carrying out a variety of subsystem tests at the same time, including successful early crosstelling operations between the Whirlwind and the XD-1 computers that confirmed the logical validity of the crosstelling program they had constructed. There was close enough design resemblance among Whirlwind, the next-generation XD-1 prototype, and the production-model FSQ-7s to indicate, they felt, that the crosstelling program should also work among FSQ-7 machines without radical alteration when the SAGE sectors they controlled would be activated by the Air Force. Meanwhile, in view of Division 6's habitual commitment to the idea of never taking performance for granted, it made compelling engineering sense to continue with equipment reliability tests.[38]

"Normal subsector capability" was reached by July of 1956. This brought appreciably nearer the moment when, completely integrated, the defense-sector radar screen centered around the long-range radar site at South Truro would extend to the south of Montauk, northward beyond Bath, and eastward beyond an offshore "Texas Tower" erected near Georges Shoal in the Atlantic.

Plans called for each long-range radar site to have one or perhaps two height finders, along with a Mark X radar that would serve as a beacon-assist in the tracking of aircraft equipped with IFF equipment. Radar data originating at each of these sites were converted by an "AN/FST-2(XD-1) coordinate-data transmitting set" into "fine-grain data suitable for transmission over telephone lines to the computer at the Direction Center."[39] Plans also called for six gap fillers. These short-range radars would provide low-altitude coverage and would be located to fill the gaps between the long-range sites. Slowed-down video (SDV) equipment would transmit the data from these radars to the Direction Center.[40]

Communication between radar sites and the Direction Center would be accomplished with "data grade telephone circuits." Similar circuits were to link the Center to already existing ADC Air Defense Direction Centers (which lacked computers) and also to the Cape Cod Direction Center in the Barta Building, "to simulate crosstelling and intercommunication between adjacent SAGE Subsectors."

In addition, there were Ground Observer Corps Filter Centers, the Fort Banks Antiaircraft Operational Center, the Civil Aeronautics Authority's Air Route Traffic Control Centers in Boston and New York, and enough other sites so that planners had estimated during the preceding autumn that "139 circuits between the Direction Center and ancillary sites" would be necessary. Finally, ground-to-air communication links would be needed. The Intercept Director in the Direction Center, for example, would be "expected to handle up to ten aircraft equipped with data link, or five by radio."[41]

To bring the ESS into demonstrated running condition, a sequence of three categories of tests was envisioned: equipment testing would check and guarantee the adequacy of the electronic equipment, operational testing would ensure that various segments of the system met both electronic and operational specifications, and system operation testing would evaluate SAGE's capacity to carry out its air defense functions.[42]

In practice, these three categories of tests were multiplied. Major testing of the Experimental SAGE Subsector under Lincoln Lab's direction began formally with the first "shakedown test" in December of 1956 and continued until January of 1959, when responsibility for operating the ESS passed to MITRE, a new corporation that had been created to meet the challenges of systems engineering for the SAGE system.[43] During that period, Division 2 and Division 6 ran 276 tests on the system. Of these, 182 were shakedown tests that checked both the initial ESS system and subsequent changes in equipment, programs, or operating techniques.

On October 2, 1957, around the time when Phase 2 of the shakedown tests was getting underway, the first of 19 evaluation tests was undertaken. These tests measured the prototype system's compliance with Air Defense Command specifications framed to define an effective semi-automatic air defense sector. Then there were 64 operator-training tests—exercises designed to improve operator proficiency and to detect and correct problems arising from the interaction between men and machines. Finally, there were several miscellaneous special demonstrations and tests to check individual functions within the system.[44]

The "shakedown" tests were sorted into three phases to examine "all major portions of the Experimental SAGE Subsector . . . in their operational environments." These tests were intended to "isolate and correct any malfunctioning programs or equipments" and to obtain "error-free system operation under limited tactical conditions."[45] The tests called for scrutiny to a seemingly infinitesimal degree, in keeping with the Division 6 philosophy of never taking the actual performance of any component, any integrated group of components, or any team of human operators for granted. This meant, Group 64's leader Ed Rich explained to Lincoln Lab's Director, that major components had to be integrated, and any incompatibilities among them had to be identified and removed. Any defects that stood in the way of the system's successful operation had to be diagnosed and eliminated, the proficiency of the operators had to be developed, and appropriate operating procedures had to be established. In this way they would proceed to provide a truly operational sector ready for System Evaluation testing.

In order to confirm, said Rich, that "the ESS system, as assembled, conformed to its design specifications," they would also have to verify "proper implementation of all major aspects of the operational and mathematical specifications for the system." Then the data obtained from the evaluation tests would indeed be "representative of performance of SAGE Sectors." In consequence, the performance of the "various system components . . .

particularly the Direction Center Active Program" would be in accord with design specifications.[46]

Plans and subsequent reality rarely maintain one-to-one correspondence, and such was the case with the Phase I tests. Three rather than four long-range radar sites formed the defensive screen, and of these only two, one at Truro and one at Bath, proved to be fully operational during the tests. Of the four semiautomatic height finders and the four gap fillers, the latter had not been "fully shaken down with the system," Rich acknowledged, although their data were "accepted and correctly processed by the first model of the DCA program."[47]

In spite of such less-than-perfect test conditions, the major objectives of the first phase of shakedown tests were accomplished by the end of September 1957. Ten months of Phase I shakedown testing had brought the test sector to an integrated performance level acceptable for evaluation testing. So Group 64 promptly moved on to the Phase II tests at the beginning of October and, with Group 22, began the first evaluation of the operating test sector. Also in early October, the RAND Corporation assumed formal responsibility for preparing the "production-site program" for future SAGE sectors, and Group 66 turned over SAGE installation of the program to RAND's Implementation Group. Two weeks later, the new System Development Corporation, which had been formed out of RAND's System Development Division, took charge of the preparation and installation duties, freeing Group 66 to study "air traffic control in its relationship to SAGE."[48]

The tests of the second phase continued until July 1, 1958, when Phase III began. The engineers had to contend with limited testing periods, electronic and man-machine systems that were continually changing in innumerable small respects, and extremely variable weather and operating conditions. They routinely faced such difficulties, reported C. W. Uskavitch, as "malfunctions of the experimental system components, the anomalous propagation of radar signals, . . . the delaying of interceptor take-offs by bad weather conditions," and the general circumstance that there were on occasion "uncontrollable variations in performance [to be] expected because of the experimental nature of ESS and its operational dependence on the varying atmospheric conditions."[49]

They recognized, too, that their efforts to set performance standards and plan tests were complicated by unrealistic test conditions that they could not always avoid in practice. If they did not already know it, they came to recognize that the number of test aircraft employed could not match "the number of aircraft expected in any real air battle." Ideally, the defense system should be in constant operation, but in actuality there were only "intermittent test operations of short duration." They had to resort to loading the detection system with electronically simulated aircraft, and they had to conduct the tests in a peacetime environment, identifying non-military aircraft that would not have been airborne in time of war. Normal commercial air lanes had to be avoided by interceptors scrambling to meet the "enemy." Hostile aircraft that were supposed to have been shot down remained in the system after being "intercepted." And strike aircraft had to be identified from the ground, since there were no military air patrols maintaining surveillance.[50]

A proper operational exercise to test the performance of a sector air defense system required, as the Group 22 engineers came to realize, a scenario of "artificial realism," approximating as authentically as possible the conditions of a wartime attack. All sector equipment had to be operational. Any B-47 medium bombers or B-52 heavy bombers sent by the Strategic Air Command on maneuvers to penetrate the sector had to be detected, identified, and tracked. Counter-defensive action had to be mounted: interceptors were scrambled from cooperating military air bases in the region and guided to interception. When tactically feasible, Army antiaircraft batteries had to be alerted.

The Strategic Air Command compounded the difficulties. For example, on February 5, 1958, it "withdrew permission from all agencies for the test interception of SAC aircraft." "Consequently," the Lincoln Lab test team reported, "obtaining a further measure of overall Sector air defense capability under realistic conditions appeared impossible." Thereupon, they modified Evaluation Test 18, which they had scheduled to run on February 12, and conducted it "merely to obtain additional data samples from some of the system functions." The next test, scheduled for February 19, was canceled.[51]

Before the end of February, the Strategic Air Command modified its directive. Without permitting a return to the freedom allowed in earlier tests, SAC stipulated that each "interception mission . . . had to be broken off at an interceptor-target separation of at least five miles," and the interceptor aircraft themselves were allowed no closer than 2 miles to any SAC aircraft.[52] Shortly thereafter, SAC prohibited "night interceptions of SAC aircraft even under the interim regulation," adding, as Group 22 reported, "some unrealism to weapons operation."[53] Realistic night tests had become virtually impossible.

Evaluating sector operations became an increasingly intricate process as more radars, more strike aircraft, and more interceptors were added. The first test (October 2, 1957) had involved nine SAC B-47 bombers flying "straight-line penetrating paths" at altitudes between 29,000 and 39,000 feet. Eleven of the 14 available interceptors were scrambled to meet them.[54] Two and a half months later, after the Montauk radar was integrated into the system, thus bringing the long-range and height-finder radars to their planned strength, 20 bombers "attacked" the sector. Twelve of these were B-52s; eight were simulated bombers electronically fed into the computer. They penetrated the system at nearly the same altitudes, 27,000 to 39,000 feet, but they resorted to evasive tactics, flying "crossing courses, dog-leg turns, and split penetrating paths." Of the 24 defensive aircraft scrambled, 12 carried data-link equipment that permitted the computer to send guidance instructions directly, without voice communication. Six of the original 24 "aborted without accomplishing the mission."[55] This was the way the intercept tests went in practice.

The incorporation of gap-filler radars in February of 1958 permitted Test 21 to be scheduled as the first of a series "with the specific objective of observing system performance in the presence of low-altitude attacks." It was run on March 5, when 29 live and simulated strike aircraft were met by 22 interceptors. Thirty-one had been expected to scramble, but nine aborted. This test was less than successful, for "a cloud layer with tops of 9000 feet" prevented the planned low-altitude strikes. A second test, scheduled for March 18, encountered even greater difficulties. Planned to be conducted at night (to "minimize the effects of local air traffic, especially as detected by gap-filler radars"), it was thwarted by a ban on night interceptions that SAC instituted just before the test was to be run. "Simulated interceptors had to be used as weapons," and so this exercise too was less than realistic.[56]

Other tests were carried out to examine the effect of "chaff" and similar electronic countermeasures (ECM) on the air defense sector's system. In January of 1958, SAC gave Lincoln Lab permission to join an ECM exercise in which SAC aircraft mounted a series of "surprise attacks" against Air Defense Command installations in the general area of the Experimental SAGE Sector. (By this time the Air Force had officially changed the name from "Subsector" to "Sector.")

Two months later and with SAC's cooperation, Lincoln Lab began its own ECM tests. The first, carried out on March 28, 1958, was a Group 64 Shakedown Test; the two that followed it on April 16 and 30 were

Evaluation Tests using the fully operational screen of radars. Heavy jamming equipment and Lincoln-designed chaff were used. The effect of the jamming equipment was "slight" and created no major difficulties. The effects of the chaff were variable, ranging from slight to troublesome. In the final test, both initiation of tracking and automatic tracking were "seriously affected," although only in certain dense areas.[57]

Carrying out the ESS testing program was an enormous and complex R&D challenge in itself. Time—or rather, the shortage of it—was a factor of major importance. "Consequently," remarked C. W. Uskavitch of Group 22 to an audience in September of 1958 at the Willow Run Symposium on the Prediction of Performance of Large-Scale Systems, "planning of the whole testing process including the selection of performance measures and the specifications of data reduction programs, had to be completed prior to observation of full system operation."[58]

The testing process was not simply an affair in which Group 64 operated the equipment while Group 22 determined "over-all system effectiveness . . . resulting from the coordinated activity of the various system functions of data inputs, track initiation, tracking, height finding, identification, and weapons direction."[59] It was much more than that, because it depended on rigorously orderly methods of collecting the great "variety and quantity of data required from each system test." Procedures had to be established for test analysis and reporting, and there was elaborate communication and interaction going on among many organizations, each with its own needs and priorities.

Uskavitch estimated that some 2000 people participated, drawn from Lincoln Lab, IBM, the Bell Telephone Laboratories, Western Electric, the RAND Corporation, the System Development Corporation, ARDC, ADC, AMC, and SAC.[60] Group 22 alone reflected this mixture, for its staff included personnel from Lincoln Lab, IBM, Bell Labs, and RAND. Controlling and coordinating the reporting of test results was as difficult as controlling and coordinating the testing process itself, and the volume of raw information obtained required elaborate reduction techniques to render it susceptible to analysis.[61]

Once Group 22 and BTL programmers had prepared more than 20 programs, containing some 30,000 instructions in all, an IBM 704 computer was pressed into service. It took that computer 4 hours to process the data from one test hour, and the procedure "produced over 5000 pages of printed data per test."[62] Though quantities of data could not guarantee quality of performance, quantity was nonetheless essential, as were organization, accessibility, and control of the data in order to interpret the test results.

The size and complexity of the ESS made a comprehensive and unified analysis of all its testing operations impossible. The task of analysis was therefore divided into manageable parts, each of which corresponded to one of the system's major functions, such as identification, tracking, or height finding. A bonus of this approach was that the evaluators became well-informed specialists who were better able to identify problems and their causes and come up with appropriate solutions. A drawback that emerged (reminding the engineers of the truism "you never get something for nothing") was that communication became difficult among the evaluation groups. This situation produced two results, commented Uskavitch: "The selected measures tended to describe function performance without necessarily relating it to system performance, and the measures of different functions could produce apparently incompatible conclusions." The pragmatic solution taken was to appoint an individual who was expected to possess the necessary comprehensive "understanding of all interrelations of the system functions and of the test conditions and their effects." He would control and edit the reports.

All this took time, of course. On average, a report took 12 weeks to prepare (from test to publication) and ran to 105 pages. Not only were there individual test-mission reports, but also there were other categories of reports. These included cumulative quarterly reports based on the mission reports, special reports on "particular system deficiencies discovered during test analyses" accompanied by "suggested remedies," and "summary reports prepared after the termination of testing and containing many performance measures not included in the other reports."[63]

On September 26, 1958, just two days short of a year from the first test, Lincoln Lab engineers ran the last evaluation test on the Experimental SAGE Sector. But work on the continental SAGE system itself had not awaited completion of the ESS evaluation tests. About 3 months earlier, on July 1, the first SAGE facility, at McGuire Air Force Base, which at the time had only a limited operational capability, had been turned over to the Air Defense Command.

Formal contractual time for the ESS and Lincoln Lab was running out: the Shakedown Test of December 31, 1958 was the last system test of the ESS that was performed by Lincoln Laboratory personnel. Subsequent responsibility for the ESS passed to The MITRE Corporation, which had been incorporated the preceding July.

When The MITRE Corporation took over the ESS in 1958, the ESS became the "Evaluation SAGE Sector," to serve thenceforth as "a test vehicle for research, development, weapons integration, and evaluation of new additions to the overall Aero-Space Defense Mission." Formal responsibility for operating and maintaining the new ESS was assigned to MITRE as part of its activity as technical adviser to the Air Defense Systems Integration Division of the Air Research and Development Command.

In some respects the Experimental SAGE Sector had largely carried out its intended informative function, but critics could call the results equivocal. Sometimes there was a shortage of data, sometimes an excess, and interpreting the data was often difficult. Meanwhile, in the years between 1950 and 1958 the task of upgrading the nation's aerial defenses had been passing, year by year, slowly at first and then with increasing rapidity, out of the hands of the engineers and into the hands of the appropriate fighting forces.

At the same time, in the realm of research and development (in theory, at least), bombers began to give way to missiles as the delivery system of choice. In this respect, SAGE, with its radars and its computers, would itself become a prototype for later computer applications designed to cope with intercontinental missiles.

Meanwhile, during the 1950s there emerged a large piece of crucial unfinished business that, even before the SAGE effort reached the ESS phase of its R&D story, had begun to tax the patience, the ingenuity, and the resourcefulness of Lincoln Laboratory's engineers, administrators, and policy makers. The question was how to go about incorporating new weapons systems into the SAGE continental air and space defense network, which itself was destined to go on changing. The solution, curiously enough, lay in the character of the R&D organization that Forrester and Everett had started to put together at MIT during World War II. This circumstance was not understood at first, even by Forrester and Everett.

30

The Problem: Weapons Integration and Systems Engineering

During 1954, 1955, and 1956, Jay Forrester, Bob Everett, and their engineers and computer programmers became more and more aware that their conduct of R&D was controversial inside and outside Lincoln Laboratory, apparently both in its style and in its substance. It was characterized by an independent, single-minded commitment to selecting, anticipating, analyzing, and solving technical engineering challenges, often before anyone else had considered them and sometimes with flair. The men of Division 6 had developed their approach in the MIT Servomechanisms Laboratory during World War II and brought it with them when they moved to Lincoln Lab from the MIT campus. The approach was out of the ordinary, and it was not regarded as a paradigm of how R&D should be carried forward, especially in the more budget-minded postwar environment. And this situation persisted after Forrester left the division in June of 1956, for Everett had been trained in the same school by the same mentors.

So Division 6 found itself becoming vulnerable sometimes to administrative intrusions by managers committed to more conventional and orthodox procedures, vulnerable sometimes to administrative intrusions of a political nature, and vulnerable sometimes to unexpected slippages and complications in completion and delivery schedules. These vulnerabilities increased the division's exposure to external administrative "assistance" and interference. The issues revolved not so much around naked grabs for power as around legitimate concerns arising out of different management philosophies.

Everett remarked reflectively several decades later that had upper-level managers understood, in the way that they would a generation later, the nature of the unprecedented technical challenges facing Division 6 and Division 2, they would not have permitted them the options and the freedom of movement that then prevailed, for they would have had too much knowledge and experience at their disposal.[1]

Hindsight suggests that unexpected slippage in hardware and training schedules occurred because the schedules had overoptimistically been made too tight. It could be argued, less charitably, that perhaps the men's management style skirted too close to the ragged edge between expertly forcing the state of the art and unwittingly overextending themselves. On the other hand, and viewed more charitably, the technical knowledge they were acquiring as the SAGE program went forward produced a mixture of computer software problems and challenges so new and complicated that no one could be expected to anticipate their cumulative capacity to swallow time.

There was also the argument that had a more prudent course been followed—one such as Bell Telephone Laboratories and Western Electric jointly had advocated in the summer of 1954 and again in the summer of 1956—nobody would have been caught off guard when several of the R&D efforts in the elaborate SAGE program fell behind schedule. Least of all should Forrester's Division 6 team have been surprised, in view of their habitual commitment to planning, to exploratory and confirmatory testing, and to riding close herd on the need to correlate work done with milestones scheduled. Yet when had so ambitious and elaborate a technical program ever before been attempted—to devise an arrangement that would link electronic digital computers and radars of advanced design in an integrated man-machine system of such unprecedented speed, scope, and complexity?

Or perhaps the root problem was an endemic, inconspicuous feature of the kind of organization that Division 6, formerly Project Whirlwind, had become, in some subtle way causing the slippages and the stretchouts to be everybody's fault and no one's fault in particular. Simply not a proper way to run a railroad! Did their familiarity with the R&D process, along with their acquaintance with the associated scheduling techniques they adopted to give direction and to measure rate of progress, perhaps lull them into overconfidence? Were they victims of their own hubris, born of their engineering experience? Heedless of the mounting difficulties, did they go on stubbornly believing, as the enterprise grew larger and larger, the scope of their knowledge grew, and they seized new opportunities to innovate successfully, that they could nevertheless continue to measure accurately the proportion of future effort required and meet the promises they had made to themselves and to high-level policy makers about meeting completion dates?

Perhaps, because they were attempting so much in such unorthodox ways, their combined efforts and accomplishments and mistakes, their actions, and their style almost *had* to be suspect and controversial. And perhaps in their eagerness they were willing to risk too much. In any event, Lincoln Laboratory's and the Air Force's ways of doing business in general and managing the ESS in particular were an increasingly forceful reminder to the engineers and computer programmers in Division 6 that their way of managing affairs was once again being challenged. As Marshall Holloway's and George Valley's actions appeared to indicate, the policy makers of MIT and Lincoln Laboratory were increasingly drawing the conclusion, for a variety of reasons, that the way Division 6 preferred to conduct its affairs was not a pattern that Lincoln Lab or any of its divisions should follow.

Indeed, it could be argued that telling evidence against Division 6's way of carrying forward R&D had been presented for any thoughtful administrator to see when ADES and its affiliated Joint Project Office had attempted, in June of 1954, to take over R&D management of the "Transition System." Although that attempt had failed, it had impelled MIT and Air Force administrators to seek a "new look," to adopt a new name and concept (SAGE), and to issue a new and more elaborately descriptive "Operational Plan" in order to continue on the course that Project Charles and Air Force Headquarters had, years earlier, decided in principle to adopt.

These developments brought Division 6 again under closer scrutiny by Lincoln Lab's Director and Associate Director. Signs of slippage in the schedule for completion of both the XD-1 and the unprecedentedly elaborate computer software could only make Division 6 more vulnerable to remedial administrative assistance and intrusion, nor did events at the time demonstrate that this intrusion was not justified.

So the intelligent, self-directed, traditionally unorthodox, forward-looking independence that underlay the technical pride and élan Division 6 had inherited from the days of Project Whirlwind in the Servomechanism Laboratory on the MIT campus came under sustained, quiet attack. Division 6 was being compelled to submerge its personality and its ways of carrying forward its affairs into those of Lincoln Laboratory.[2]

But one particular technical issue was of more immediate and persistent engineering and managerial concern to Forrester and Everett and their Group Leaders. This was the problem of incorporating new defensive weapons systems, whether unmanned missiles or manned interceptors, into the SAGE air defense system. It was not until rather late in the game that many members of the Division 6 staff would come to regard solving the problem of weapons integration as their unexpected path to salvation and renewed independence as an R&D laboratory, nor does this consideration appear to have been present in the minds of Forrester and Everett when they began to grapple seriously with the challenge of weapons R&D and the intricacies of advanced systems engineering.

From the very start, Forrester and Everett had taken it for granted that integration of new weapons into a warning and battle control system was essential to the success of such a scheme. They had incorporated it in their detailed 1947 scenario for waging antisubmarine warfare. Soon after George Valley first encountered them (in January of 1950), he learned they were ahead of him in their thinking. They were already investigating how the computer might be used in peacetime to help pilots and ground-based air traffic controllers to avert collisions.

In the first Quarterly Progress Report that Division 6 issued in June of 1952, after Lincoln Laboratory was organized, David Israel, Jack Arnow, and Bill Linvill addressed technical details of how weapons and the computer might be integrated. After discussing tracking and control in general terms, they went into the subject of tracking and control. Looking ahead, they anticipated the operations of air defense centers and a number of associated matters, such as radar data "smoothing" and using the computer to calculate aircraft throttle settings. They selected this last example, they said, "because it was a simple and significant illustration of how a computer could be used in a feedback control system."[3]

In view of the nature of advanced electronic technology and of military challenges to obtain an advantage over the enemy, it was inevitable that weapons integration would pose more and more complex technical problems as the years passed while the continental air defense system was being brought into being. Even before the spring of 1953, when the Air Force selected Lincoln Lab's Transition System, the commanding general of the Air Research and Development Command had warned that any air defense system, whether designed by the University of Michigan's Willow Run Research Center or by MIT's Lincoln Laboratory, had to accommodate not only contemporary weapons systems but also those planned for the future—especially the F-99 BOMARC, a pilotless interceptor being developed and tested at Boeing.[4]

BOMARC proved indeed to be a case in point. After the Air Force adopted the Transition System described in Technical Memorandum No. 20, ARDC initiated action to make the future air defense system compatible with the F-102 fighter and with BOMARC.[5]

The authors of TM-20 had been well aware that the defense system they were proposing would be required to employ data gatherers and weapons not yet in the inventory or even conceived. Consequently, they had recommended that the design of the proposed system incorporate features facilitating the introduction of new equipment.[6] The computer and its associated ground system should provide "all possible aid to airborne weapons," because "any computing or data processing on the ground [would] simplify weapon design." Unavoidable limitations in aircraft size and weight also made the "ground environment" concept superior for purposes of equipment operation and maintenance. It was simply good engineering sense to have the computer center on the ground "provide . . . any reasonable services . . . needed by weapons designers."[7]

Furthermore, the flexibility built into such a system would render any changes in the electronic equipment structure unnecessary when integrating new and advanced weapons, for the designers could simply modify the programmed instructions that they fed into the central computer.[8] Appealing though this conception was, it proved to be an oversimplification that systems engineering later would lay bare in practice.

Looking ahead, the writers of TM-20 took care to make it clear that Lincoln Lab would have no responsibility for *guaranteeing* compatibility between the proposed ground-environment control system and future weapons systems. By default, then, if not by design, that task would be left to the various weapons system contractors, who would be expected to coordinate with Lincoln Lab.

The Lincoln Lab engineers agreed that, as an interim solution, they should initiate action to accommodate weapons already in the inventory and those that would be available at the time the air defense system became operational. So, while it was tacitly assumed that the manufacturers of such latecomers as BOMARC and the F-102 should be the ones to bear primary responsibility for ensuring compatibility, the Lincoln Lab engineers expected to be of all the help they could in introducing the manufacturers to the special problems of integrating the weapons with the looming SAGE system.[9]

Joint discussions on technically integrating BOMARC and the F-102 into the air defense system, begun in June 1953, continued into autumn before tapering off rapidly as Lincoln Lab and the manufacturers became preoccupied with developing their respective systems.[10]

Their involvement with integrating BOMARC helped the Lincoln Lab engineers to further crystallize their thinking in 1953 regarding Lincoln Lab's responsibility for the integration of new weapons into the Transition System (as the SAGE system was still being called).

Although the experience did not provide them with a considered policy position on how weapons integration should be carried out (that was still a couple of years down the road), it proved to be both a significant and a typical episode in "educating" those involved. Viewed in retrospect, it reveals once again how engineering problems and their solution guided Forrester, Everett, and their colleagues in Division 6 in initiating courses of R&D action and establishing R&D policy. It subtly set their R&D approach apart from that of, say, Air Force managers, whose custom it was to approach the problem from the other direction: by identifying and defining a military mission, then calling for engineering solutions.

At the first joint meeting on BOMARC (June 15–18, 1953), representatives from Boeing, ARDC's Wright Air Development Center (WADC), ARDC's Cambridge Research Center, and Lincoln Laboratory gathered to discuss how the Transition System might be used to control the interceptor.[11] The question arose, reported Forrester, "whether or not the Transition System can do the required control job for BOMARC. This was discussed at many times during the three days, with not enough information to reach firm conclusions." Boeing and Lincoln Lab engineers agreed they should explore the problem further, and they scheduled a second meeting for June 26. The purpose of the second conference would be to "determine whether or not the Transition System can and *will* absorb the guidance requirements of BOMARC."[12]

The Lincoln Lab engineers had taken care to note at the first meeting that the initiative for carrying out the studies on the BOMARC ground control system rested with the manufacturer, while Lincoln would assist "where possible" and would conduct those particular tests to which it had agreed.[13]

But before the conference of June 15–18 had ended, the representatives from WADC announced a decision that the engineering effort to provide ground control of the BOMARC missile should be provided by Lincoln Lab's system rather than by the Westinghouse G-20 Ground-Control System, which had already been contracted for. They recommended that the G-20 contract be canceled, thereby saving the Air Force an estimated "$13 million during the coming fiscal year." When Forrester reported this turn of events to the Lincoln Steering Committee at its June 22 meeting, Valley became "disturbed that Westinghouse will think that Lincoln supported this conclusion which is not so." Lincoln Director Al Hill asked Forrester to write a report of the action, copies of which were to go to James Killian.[14]

WADC's recommendation to cancel was subsequently overturned by higher authority in the military chain of command, and Boeing received permission in the autumn to "go ahead with the G-20," at least for the test phase of BOMARC's development program. This turn of events, to Valley's and Forrester's relief, removed Lincoln Lab from the center of the picture. Forrester later informed the Lincoln Steering Committee that Boeing had consequently "lost most of their interest in the proposed use of WWI for Florida tests."[15]

Meanwhile, the joint technical discussions begun in June were resumed in August, when Lincoln Lab engineers spent a week at Boeing in Seattle. George Valley reported to the Lincoln Steering Committee that they had spent their time "re-evaluating the data accuracy requirements of the BOMARC System and determining whether the proposed data coding, the transmission system, and the computer smoothing of these data were adequate for BOMARC."[16]

Further meetings during the next 2 years were only occasional, and references to BOMARC in Division 6's quarterly reports grew less frequent. It was not until the autumn of 1955 that the issue of responsibility for the integration of weapons into SAGE resurfaced in Division 6. As often occurs with complex problems, the focus was blurred at the start and the field was wide. At the October 24 meeting of Division 6's Group Leaders, Steve Dodd offered a proposal intended to describe "Lincoln's position in acknowledging SAGE System responsibility." After some discussion, Dodd was asked to clarify Lincoln Lab's position further and to come up with better examples of what was at stake. Weapons integration as a formal Lincoln Lab R&D responsibility, although implied here, was not yet a front-and-center issue.[17]

But at the next Division 6 Group Leaders' meeting, on October 31, Forrester opened a general discussion of the future of the engineering of the SAGE System, and David Israel "inquired about the integration of future weapons such as BOMARC." Forrester replied that nothing had been firmly decided, but it "would likely be a job for Rand if the problem arises *after* the completion of the initial phase of the ESS program."[18]

Weapons integration was still a representative example of future engineering aspects of SAGE research and development; it was not yet the exclusive object of focus it would become. At the October 31 meeting (attended by Forrester, Everett, and thirteen Group Leaders and Associate Group Leaders),

> Taylor and Dodd spoke of the need for a positive program for Lincoln both for the purpose of giving Lincoln personnel evidence of continuing challenging tasks and for the purpose of improving relations with the SAGE contractors upon whom we are continually thrusting responsibilities as SAGE System emerges from the

> development to the production and operating phases. Forrester mentioned the AICBM program to which Lincoln is committed and cited the air traffic control problem and other problems in the offing. He pointed out that Lincoln's preoccupation with SAGE System will continue two years into the future and that it is not feasible at present to predict the optimum direction of Lincoln effort that far in advance. He pointed out the need for action to avoid Lincoln involvement in future ESS activities clearly not within our province and mentioned recent LPO [Lincoln Project Office] letters raising this question. He suggested that Lincoln might relax concerning long-range plans, meanwhile keeping alert to challenging problems arising in the future for which Lincoln is suited; alternately, he said we might make a tentative program based on presently known problems and allow the program to be modified in accordance with priorities as they develop. He suggested that we clarify our thinking concerning the SAGE System responsibilities which do *not* belong to Lincoln. . . . Further discussion brought out the opinion that Lincoln is unique in its experience in systems design and in the management problems that arise in connection with a large complex systems problem. . . . Everett closed the discussion by pointing out that our future program is likely to emerge from our own thinking. He, therefore, urged that we make a concentrated effort to prepare realistic proposals for presentation. . . .[19]

Two weeks later, at the next Group Leaders' meeting, Jack Jacobs brought up the question of "Lincoln responsibility for integrating missiles (not just BOMARC) into the SAGE System." He was asked to prepare a proposal, to be turned over to Everett, indicating how Lincoln Lab might handle such responsibilities in the future.[20] A week after Jacobs had asked his question, he proposed at a Group Leaders' meeting that a special group be formed to start studying the problem of weapons integration without further delay, "immediately listing the jobs and outlining their scope as well as suggesting which tasks Lincoln should accept." Dodd requested a more ambitious examination "of the future of Lincoln, pointing out that some day Lincoln participation in SAGE will diminish." Israel remarked that "a survey was started last week to develop the programming load [on Lincoln] represented by new weapons systems," whereupon Jacobs urged that some policy decisions be made about "overlapping responsibilities of Rand, ABC, ARDC, APG, and Lincoln."[21]

It was in part as a consequence of these growing concerns and in part as a result of policy questions being raised in upper echelons of the Air

Force that Forrester, late in January of 1956, offered a formal, detailed policy statement on the issue. By this time weapons integration was emerging as a major R&D management problem, and it brought in its wake important larger questions of systems engineering that were to catch up with it and engulf it. At the center lay the issue of Lincoln Lab's future role in the continental air defense program, now 6 years old and moving toward completion.

Division 6 was not the only organization involved. Before Forrester laid out his analysis of the weapons-integration problem and described what should be Lincoln Lab's role in solving it, the tripartite Joint Services Advisory Committee became involved. They had become aware of the fact that the position Lincoln Laboratory was informally expressing during that January was rapidly hardening. The lab's position, intended to simplify its problems, only complicated the Air Force's problems.

When Lincoln Director Marshall Holloway requested on January 10 or 11, at a meeting of the Joint Services Committee, that the lab be allowed to "phase out of certain areas of the SAGE program at an early date," the members of the committee could see no reason why, in principle, it should not be permitted to begin doing so, once it was clear that the Air Force was capable of handling the system. But who would assume the responsibilities that Lincoln was giving up? As the ensuing discussion revealed, no plan existed that "showed a phase-out of either Western Electric Company or Lincoln activities, with a designated Air Force agency to take over these responsibilities as Lincoln and the Western Electric Company were scheduled to phase out."[22]

Plainly, the Air Force needed to take action. Early in February of 1956, Lieutenant General Donald L. Putt, Deputy Chief of Staff for Development at Air Force Headquarters, recommended to Lieutenant General Thomas S. Power, the commander of ARDC, that the ADES Project Office in New York be instructed to meet with representatives of Lincoln Lab, Western Electric, and the affected military Commands to consider four questions: What tasks being performed by the civilian contractors "normally would be accomplished by an Air Force agency"? "What Air Force agency should assume these responsibilities?" Could the designated agency "take over these functions when Lincoln and the Western Electric Company are scheduled to phase out"? On what dates should the two contractors "phase out of each responsibility"?[23]

Colonel Albert R. Shiely Jr., the chief of ARDC's Air Defense Systems Operating Division in New York, relayed Putt's questions and his request for action to Holloway and requested that Lincoln prepare a tentative plan for its eventual withdrawal from SAGE. This plan would then become a basis for further consideration.

Shiely also asked for Lincoln Lab's "thoughts or recommendations" as to whether the Air Force or a contractor should be designated to take over from Lincoln, and he pointed out that the Lincoln Project Office had been established at the Cambridge Research Center to develop with the Air Force a capability to "take over development responsibilities in the SAGE Program when the phasing-out of Lincoln Laboratory became desirable."[24]

Lincoln Lab's administration pondered the Air Force's request for 6 weeks before a response went off to Shiely over the signature of Associate Director George Valley. Central to the position enunciated in Valley's letter was the formal policy statement on weapons integration that Forrester had written and that Lincoln Laboratory had issued on February 2, 1956, the day before Putt (presumably informally aware of what Lincoln was about to do) wrote to Power.

To understand what was going on, it is necessary to look at Forrester's new description of Lincoln Lab's old policy, and at a number of organizational responses that were being developed inside and outside the lab. Then Lincoln's reply to Shiely's request, along with its delay, begins to make historical sense.

Lincoln Lab's policy statement was Forrester's analysis of the weapons-integration problem. It went into considerable technical detail, and it set about explaining to the Air Force, to manufacturers, and to anyone else involved why Lincoln should *not* undertake the R&D task of integrating the new weapons that lay ahead. This memorandum went forward on February 2 to the Air Force, endorsed by Lincoln Laboratory Director Holloway, and it set forth Lincoln's policy on responsibility for weapons integration that would apply after the initial SAGE sites at McGuire, Stewart, and Syracuse had been activated. Its message was clear. Direction Centers would be able to handle contemporary, "all-weather, manned interceptors of the F-86, F-89, F-94, and F-102A types," as well as the Army's Nike ground-to-air missile batteries. But thereafter weapons manufacturers must be the ones responsible for ensuring the compatibility of their weapons with the ground environment.[25]

Forrester pointed out that the demands already placed on Lincoln Laboratory's resources by SAGE and other programs prevented the laboratory from taking "the initiative in executing the integration of new weapons, nor [could it] supply manpower to the solution of problems arising from a particular weapon nor carry the burden of experimental tests and the equipment and equipment-modifications necessary for them."[26]

Until other suitable arrangements were made, Lincoln Lab would be quite willing to "make a small group available for broad coordination of weapons system integration, to provide necessary information on SAGE arising from Lincoln's work, and to work on generalized solutions to problems which are common to all weapons systems." The group would also offer an occasional "familiarization course to which weapons contractors personnel can come for SAGE System Orientation."[27]

Forrester made the point that the work that would be required to integrate a future weapons system would prepare the way to resolving "the specific problems imposed by the characteristics and desired operational use of the weapons with SAGE." Weapons integration, he insisted, "should begin early in the weapons development program, to insure the earliest integration and the best possible exploitation of the combined characteristics of the new weapon and the SAGE System."[28] He could not recommend any detailed plan that would apply to the integrating of any and all new systems. That would be impossible. But a sequence of four general steps could and should be followed:

> 1. The special characteristics and desired operational employment of the weapons system must be considered with respect to the existing SAGE computer program and equipment so that a broad plan for integration can be prepared.
>
> 2. The necessary analysis, test, or simulation must be performed to support the broad plan for integration and to permit preparation of detailed plans and specifications for new or modified equipment and SAGE master computer programs required for operation of the weapon system.
>
> 3. Tests must be conducted to evaluate the joint operation of weapon and ground environment.
>
> 4. SAGE computer programs and equipment must be modified and checked out.[29]

Forrester regarded this preparatory work as "logically an extension of the normal research and development effort on a weapons system that is to be used with SAGE." The greater part of this work, he emphasized, must fall upon the weapons contractor, since "only a small part of the total effort is spent in revising the computer program."[30]

Forrester then laid out a general program of four phases covering an estimated 30 months. In addition to detailing it over the next four pages—the remaining half of his memorandum—he presented it in a bar graph that showed how the phases would relate to one another over time (figure 30.1).[31] Here was a blueprint, then, to guide the Air Force and those manufacturers interested in integrating their weapons systems into the SAGE defense network. But Lincoln Laboratory would be too busy with other R&D problems to accept the responsibility for accomplishing the integration itself.

Forrester's memorandum of January 30, 1956 reflected no basic shift in policy. It was simply a restatement of Lincoln Lab's position on weapons integration, which had been expressed in January of 1953 when the proposal for the Transition System had been submitted to the Air Force. But significant organizational changes within Division 6 had already been set in motion in January of 1956, without waiting for Forrester's policy memorandum, and they reflected both a reorienting of the work of some of the groups to new tasks that had developed over the preceding year and a sharpened appreciation of the weapons-integration problem. Dave Israel's Group 61 was made the "contact point within Lincoln Laboratory for technical discussions with weapons systems contractors and planners for the integration with SAGE of new weapons and control systems." This arrangement was a natural addition to Group 61's responsibilities to prepare the operational, mathematical, and adaptation specifications for SAGE and ESS computer programs and to review the operational results. "Weapons Integration" began to appear as a formal entry in the Quarterly Progress Reports of Division 6 with the March 1956 issue.[32]

Other steps were being taken. Out of discussions with Boeing engineers that had begun in late autumn of 1955, a joint Boeing-Lincoln committee (which also drew upon IBM personnel) emerged to explore testing of BOMARC with SAGE. Theirs became the initial study undertaken under a new policy outlined by the end of January 1956: "General Procedure for Integrating New Weapons with SAGE."[33] Of peripheral interest to Division 6 was the earlier news that, at the request of the Assistant Secretary of the Air Force for Research and Development, Trevor Gardner, a SAGE Improvement Committee had been appointed to recommend areas for improvement.[34]

In February of 1956, when Putt asked the commander of ARDC to accelerate the search for a solution, he was worried by the dissatisfaction that Lincoln Lab and weapons contractors were expressing, and he was concerned that the problems of weapons integration and programming might still be unresolved by the time the Air Force took over operation of the first continental subsectors.[35]

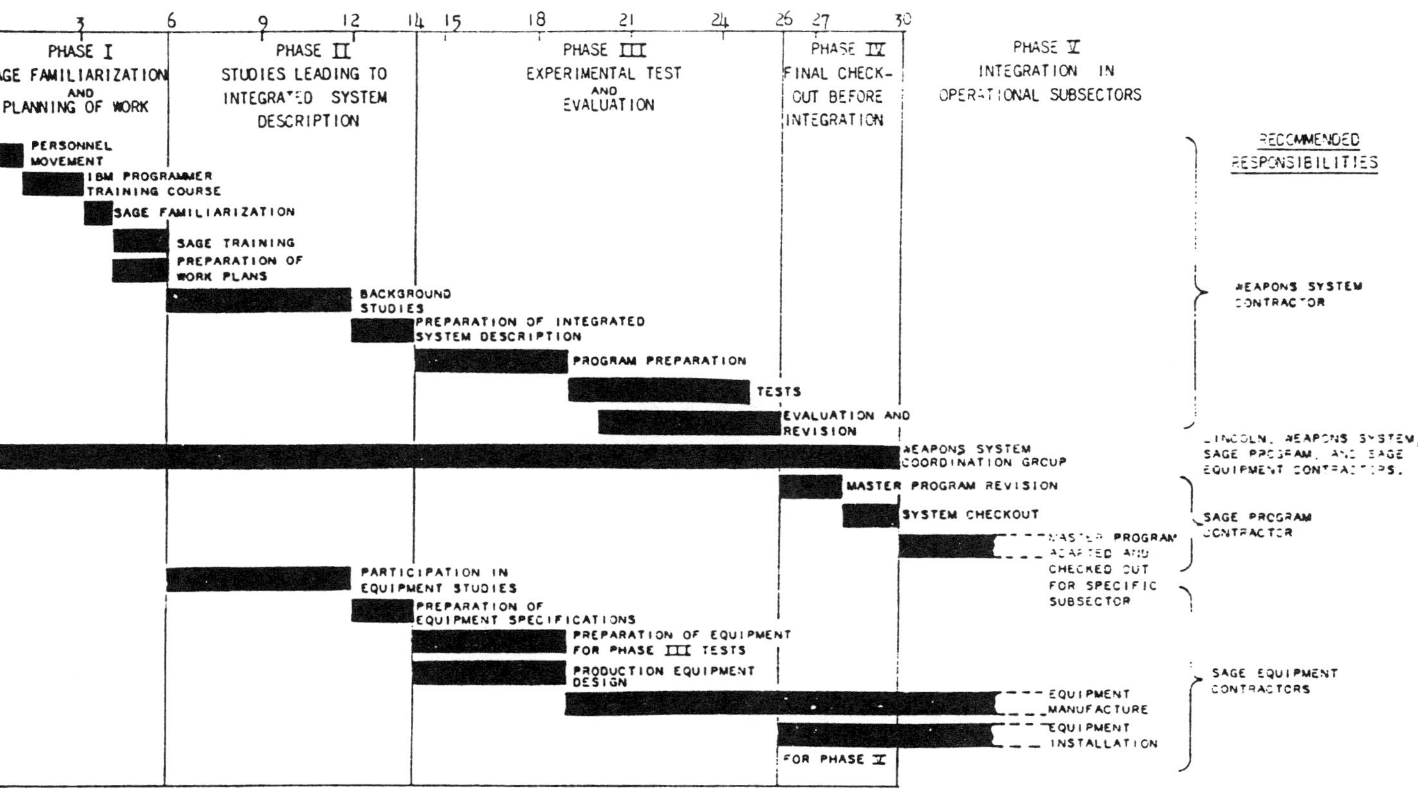

Figure 30.1
"Proposed Integration Procedure."

ARDC contemplated designating a separate agent to coordinate and oversee the integration of weapons with SAGE—perhaps the RAND Corporation or its offshoot, the Systems Development Corporation. But the Air Defense Command raised strong objections. "No other contractor could possibly satisfy the requirements that the work be accomplished by a contractor basically familiar with the air defense problem, the computer, and air defense weapon systems," reported an ARDC officer after a mid-April visit to ADC Vice Commander General F. H. Smith Jr. Instead, Smith recommended they should consider holding Lincoln Lab's feet to the fire and threaten to "take away the XD-1 computer and give the whole program to someone willing to do a complete R&D job and not selected parts." Selecting RAND might create more problems than it solved, since such an assignment might well overburden the corporation and keep it from meeting its existing responsibilities.[36]

The Vice Commander of the Air Materiel Command, General William F. McKee, was no less outspoken when he responded to General Putt's letter of February 3 to General Power. McKee assumed that the first air defense subsectors would be "tested and operationally suitable" by the November 1957 deadline, and he urged that "no substantial phasing-out of Lincoln Laboratory effort should occur before 1958." Lincoln's responsibilities were "so unique in the design and development of electronic systems, equipments, and projects," he insisted, "as to have no comparable parallel in the Air Force at this time."[37]

Valley's letter expressing Lincoln Laboratory's response to the Air Force request went out in mid April. As Shiely read it, he realized that Lincoln appeared to be "waffling" over how to proceed toward a solution, for Valley advised that his letter's contents were not to be taken "as a formal statement of policy on the part of Lincoln but only as friendly advice."

There were two areas of activity, Valley explained, that, by reducing the anticipated rate of tapering off, could affect "considerably" the planning of Lincoln Lab's "orderly" withdrawal from the SAGE program. One was "a joint ADES-Lincoln detailed study of the problems of getting McGuire Subsector into operation." It would supply a great deal of information not yet known "about the integration of SAGE into the Air Defense System." The other was a possibility that "some Lincoln work will be desired on the problem of integrating SAGE into the Air Traffic Control System." Of course Lincoln would meet its obligations to finish SAGE, but once these were met it would "restrict itself . . . to basic R&D problems associated with SAGE." Valley foresaw a worrisome possibility, however: non-R&D problems might arise that only Lincoln Lab's experience and knowledge could resolve, yet Lincoln would be unable to attack them "because of limitations of manpower, supervisory staff, and material resources."[38]

Valley saw three directions in which Air Force action might "significantly relieve Lincoln of major continuing responsibilities for the SAGE Program." As Shiely read on, it became clearer what sort of fine distinctions Valley was interested in drawing as Lincoln Lab's informal spokesman. If the Air Force were to mount a production-improvement program on the AN/FST-2,[39] if it were to persuade weapons contractors to accept the task of integrating their weapons with SAGE, if it were to establish "an agency to *manage* integration of SAGE" with the existing air defense system, and if it were to focus its attention on these three things, then it could "let decisions relative to Lincoln's future role in SAGE-oriented R&D work await more information." Lincoln should first be extricated from the "non-R&D installation and production improvement programs."[40]

Colonel Shiely replied on April 20 to Valley's letter of "friendly advice," and the ideas he presented were presumably not his alone. He minced few words. Lincoln Lab was "the only presently existing source of such a [system engineering] capability." Weapons integration and Lincoln Lab's future association with SAGE were both only parts of the greater problem of obtaining a systems engineering contractor that would be responsible for integrating "individual weapons or equipments . . . , simulation studies of new weapons and new equipment . . . . , active participation in preparation of system employment plans for individual weapon systems or new equipments . . . , and provision of detailed computer programming guidance." All these responsibilities, Shiely wrote, were being incorporated into a "contract work statement," which upon completion would be forwarded to Lincoln for review and "a definite formal commitment." The only way to postpone a major commitment, as Valley had preferred, would be for "Lincoln Laboratory to continue (at least as an interim measure) its responsibility for Air Defense System Engineering, including integration of the immediate urgent weapons and equipment such as BOMARC, TALOS, Nike, and SIF."[41]

Shiely was forcing the issue. He assumed that Lincoln Lab was sincere in its intention to withdraw from the program and probably would reject the role of systems engineer. Also, he had his reservations, which he expressed privately and bluntly to higher-echelon officers in the Air Force. He wondered if the laboratory might not be avoiding a firm commitment in order to "retain for itself complete authority to engage in such system engineering activities as seemed attractive but not to accept responsibility for any part or all of the system engineering problem."[42]

A systems engineering capability was essential to the success of the program, and so was Lincoln Lab's cooperation. To secure the latter, a definite work statement was necessary. Lincoln must be given the "formal opportunity to turn down the job and concur in [the] proposal for a contractor." Lincoln might, Shiely thought, be more willing to accept the task after reviewing the statement of work. If so, then his letter to Valley might prepare the laboratory to discuss the matter further.[43]

A few days earlier in April of 1956, Shiely had pointed out to George Valley that SAGE, as the nucleus of an elaborate air defense system, was still only one of many interconnected elements in a large and complex system. Consequently, weapons integration in SAGE was "only part of a much larger problem, namely, overall Air Defense system engineering." Incorporating SAGE into the national air defense system meant, said Shiely, that "the development and integration of any single weapon or piece of equipment must be continually analyzed and controlled on a system basis."[44] Although neither man mentioned it (neither had to, to the other), Shiely and Valley clearly realized that the practice of systems engineering raised further questions of Lincoln Lab's future role in the air defense development program that the Air Force was expected to fund and manage as an ongoing responsibility.

The work statement that ARDC delivered to Lincoln Lab assigned to a single contractor overall responsibility for "successful integration of new weapon systems, supporting systems, and allied equipment with the SAGE-equipped Air Defense Electronic Environment."[45] Did this appear to Lincoln to be a practical solution to the problem of weapons integration? ARDC waited for Lincoln Lab's reply.

Lincoln Laboratory's response, submitted almost 2 months later, was still less than positive, suggesting that Shiely's reservations concerning Lincoln's real intent were quite accurate. A long period of difficult and unsatisfactory negotiations had begun, as the correspondence demonstrated.[46] In receipt of further advice from Lincoln but no commitment, ARDC officers continued to review the situation internally and with Lincoln in July and August, but neither side could find a solution that satisfied them both.[47]

ARDC and Lincoln Lab administrators were not the only ones disagreeing about the future role of Lincoln during that spring and summer. Within Division 6 in March of 1956 a regular meeting of Group Leaders and Associate Group Leaders fell into a lengthy discussion of "the relationship of Division 6 employee morale to the long range Division program." The discussion was provoked by Dave Israel's observation that the weapons-integration problem was "snowballing." Norm Taylor commented that what they needed was a "positive program," for his interviews with departing staff members during the normal turnover of personnel had persuaded him that many were "unhappy about sitting around administering a program of *not* doing various things." Forrester's response was not surprising to the 21 men at the meeting. He urged them to give the matter serious thought and to assign within their respective groups "challenging problems" that would "push forward the research program and build morale without serious detriment" to existing responsibilities. He recalled how "in the past, for example, we developed magnetic storage while we were under the pressure of starting up Whirlwind." The discussion moved from weapons integration to computer program preparation for the ESS and McGuire, then to "various interesting assignments . . . including ICBM, high speed tracking, assistance on the SIC Working Group and assistance to the RAND penetration study."[48]

Although Forrester suggested that they keep the matter open for further discussion and exploration from week to week, other routine events continued to dominate future Group Leaders' meetings. When Israel reminded everyone a week later that "the weapons integration task is growing faster than contemplated," it evoked no resumption of discussion of challenging goals.[49] And when Jacobs reported late in April the efforts being made "to define in detail the RAND-Lincoln responsibilities in the SAGE program," Forrester simply reiterated Lincoln Lab's standing policy that it "must concentrate on ESS and confine its activity at McGuire to the programming responsibility which we have specifically accepted." In his own thinking, he distinguished sharply between general thoughts about the future and unfinished specific programs to which they were morally and contractually committed.[50]

Whatever current problems were demanding attention at whatever levels in the laboratory, there was among the working engineers a general restlessness, caused in part by the lack of consensus about the future at the upper levels of Lincoln Lab management. Early in April, Marshall Holloway put a number of questions about plans for the future to the thirteen other members of the Lincoln Steering Committee. Were the AICBM and the continuing SAGE programs suitable for Lincoln Lab to undertake? What should be the balance between research and development? What should the laboratory be like in 1960, a year far enough away to allow for orderly change? George Valley, Henry Fitzpatrick, and Jay Forrester were among those who replied to these questions. As the report of the meeting describes it, Valley called attention to staff morale and the importance of identifying "categories of satisfaction" above and beyond salary, especially "financial security, personal satisfaction, and power or the ability to control or influence people and events." Fitzpatrick pointed out that Lincoln Lab was "becoming a court of last resort to solve problems that no one else can solve." Forrester observed that "this was indicative of the change in the Laboratory which started out doing things that were new and unaccepted." Now it was turning "more toward programs which are extensions of known techniques and which have great public support." They should expect "changes in the atmosphere and kind of people in the Laboratory."[51] It should not have surprised Forrester that other members of the committee took issue with his analysis and began to set up ideal distinctions, such as the relative merit of working on problems suggested by someone else compared to working exclusively on problems of one's own, or "the relation of nonprogrammatic research and development to component and systems work and the percentage desired of each," or the difference between carrying on research and carrying on development work. Carl Overhage thought a separate research division should be formed. Others inquired "how far into development the Laboratory should go . . . in the light of present and past programs." Taking his cue from Lincoln Lab's experience, Holloway suggested that Lincoln "should carry ideas far enough to know as much or more than the contractor." When J. Freedman proposed that the laboratory might create a special planning group of its own, Valley objected: "It is extremely difficult to have a continuing planning group separate from those who carry out the program." Others disagreed, noting that "those working directly on a project are usually too busy to do long-range planning."[52]

At that time and for the rest of 1956, Lincoln Lab was in no position to forge its own fate. Although no one appeared to be aware of it, this circumstance gave Division 6 the opportunity to work out its own salvation.

31

The Solution: MITRE

A paradoxical feature of the history of science pure and applied is the overlapping identity of endings with new beginnings. The story of Division 6 (formerly Project Whirlwind) is no exception. Perhaps it is a consequence of the very nature of inquiry, of which research and development is a part. When Jay Forrester, looking backward and looking ahead as was his habit, realized that the creative effort that had produced the SAGE computer as the successor to Whirlwind I was coming to a close, he concluded that it was time for him to move on.

The politics and the personnel problems that inevitably accompany the wielding of administrative power attracted Forrester less than the creative play of ideas. As Jack Jacobs (subsequently an Associate Director under Bob Everett) remarked, Forrester was an inner-directed, idea-directed engineer who "never seemed to have doubts about anything." At least, so he appeared to Jacobs when Jacobs joined the group in the middle of 1951.[1]

Jay Forrester's decision to leave Lincoln Lab for an MIT professorship in industrial dynamics appears to have been due in part to the endemic and persistent "hassle" of administrative and political difficulties and rivalries he encountered in his relations with Lincoln Lab's administrators. Also, as Forrester explained years later in an interview with Emerson Pugh of IBM, he "had come to the conclusion that great technological successes depended more on the managerial environment than they did on the underlying science. If you had the right environment, you would get the science, but you could easily have the science in an environment that did not make it effective."[2]

Although Project Whirlwind had run into difficulties with the Office of Naval Research, Forrester had found the administrative independence and the innovative atmosphere of Gordon Brown's Servomechanisms Laboratory and the Barta Building more favorable to relatively unfettered creative technical activity than the increasingly institutionalized environment of Lincoln Laboratory.

Of course, Forrester's departure at the end of June 1956 was not the end of the Division 6 R&D team. The natural result was that Bob Everett succeeded him as director of the division. At the time Forrester left, both men, each for his own carefully thought out reasons, were convinced that involving Lincoln Lab in weapons integration ran contrary to the R&D policies and philosophy that had brought the lab into existence. Everett would change his mind in the months ahead.

To understand how Everett's change of mind came to pass, it is helpful to go back to the week before Forrester formally severed his ties with Lincoln Laboratory. There we begin to encounter some clues in reflections Everett entertained about some aspects of the R&D process that he was communicating to Lincoln Laboratory's Associate Director, George Valley. Everett was considering how Lincoln should respond to the Work Statement that Colonel Albert Shiely of the Air Research and Development Command recently had sent to Lincoln Lab in the effort to resolve the issue of who should carry on weapons integration with SAGE in the future.

Everett offered a number of detailed recommendations concerning Lincoln Lab's role in the future of SAGE. ARDC's proposed Work Statement, which Shiely had sent to Lincoln to prod a decision, had elicited from Valley a response which argued, as Everett put it, that ARDC had "misunderstood the problem which lies in the non-R&D parts of SAGE." Most of the items listed in the proposed Work Statement would eventually, Everett felt, be "done by Lincoln or others in the normal course of events." But what the Air Force really needed was "a systems engineering contractor who will be responsible for SAGE installation, modification, and operation."[3]

"There should be no half measures," said Everett. "Lincoln should either accept full responsibility for SAGE or else should accept no new responsibilities and withdraw as rapidly as possible from those we now have." Favoring the latter course, he called for Lincoln to commit itself to a statement of "specific responsibilities for SAGE R&D, since the definition of this term is so open to misconstruction." "We will, of course, continue to do R&D on SAGE as on other areas of air defense," he noted; "the point is that this R&D will concentrate on areas of our own choosing." He was not thinking of abandoning the Lincoln engineers' commitment to SAGE; they would complete the programming that was called for, and they would continue to operate and evaluate the ESS as they had promised. But looking farther ahead, he felt they should reject ARDC's proposed Work Statement and make a clear policy statement of their own. Perhaps the way to persuade the Air Force to choose a systems engineering contractor for SAGE was to make it clear, he said, "that we are not going to do the job."[4]

In Everett's view, SAGE had brought Lincoln Laboratory to a fork in the road in the summer of 1956. Down one path lay the systems engineering job that in the years ahead would maintain and update SAGE as technical progress and enemy threats would dictate—an ever new and expanding job that would generate responsibilities and patterns of action quite different from those that had brought Lincoln into being. A single organization was what was needed, he said—one capable of carrying on an open-ended job. But the SAGE R&D challenge had produced no such organization when it had brought Lincoln Lab into being. Instead, to design and test the SAGE concept while building it, all the engineering groups and the manufacturers had been given specific jobs to do, "with the result that there have been unfilled gaps and intolerable pressures on all concerned, especially Lincoln." To carry forward the installation, operation, and modification of SAGE by continuing to apply the familiar approach of defining anew "a complete set of specific SAGE tasks" would simply perpetuate "the old misunderstandings and . . . pressure us to do more than we can and want to do."

Down the other path lay the opportunity for the lab to be "left free to conduct R&D on those aspects of SAGE that appear to Lincoln to be most in need of improvement, to attack problems, make inventions, give advice, and make suggestions as it is able in support of the Air Force." This, after all, was the philosophy under which the lab had been founded.

It was not immediately clear to Everett which of these two widely separate courses of action Lincoln should pursue, "either for its own good or for the good of national defense."

Were Lincoln Lab to undertake the systems engineering job that ARDC was pressuring it to assume, the effect on the lab—"probably undesirable," he remarked—could well be so profound as to change Lincoln Lab's character and its conduct of R&D. To get on top of the job and stay on top might well double the size of the laboratory and call for "two or three times the number now working directly on SAGE." And in addition to the technical challenges there would have to be a lot of time spent on "nontechnical struggles with the Air Force and with the contractors. The "transient" that would occur in establishing this organization," Everett continued, invoking the electrical engineering metaphor of a temporary disturbance that would run its course, "would take a considerable time while we hire and train the people, including the necessary hard-headed supervisory types, and also while we gradually retrieve various items of control that we have been abdicating over the last year." And it would be altogether satisfying, no matter what the cost, to really finish the job that Lincoln Lab had invented. But Everett's real reason for bringing up these considerations, he said, was "not because I feel we should undertake the job but because it is a solution that should be discussed."[5]

Should the Lincoln engineers decide to pursue their past policy and turn down the systems engineering job, they must recognize that "we would lose control of SAGE; we would be forced to work by indirection; we would be forced to watch while our creation was warped, perhaps beyond recognition, by strange hands." He foresaw another unwanted "transient" that would try their patience "while we extricate ourselves bit by bit from our present position.

"After much consideration," Everett concluded, "I feel we should get out and pay the price."[6]

At a meeting between high-level Lincoln Lab and ARDC administrators in July of 1956, the former remained steadfast in their resolve to get out. They realized that, if ARDC were to accede, a period of transition would follow, and that Lincoln should expect to help smooth the transition. Accordingly, they indicated the laboratory's "willingness to 'pull on the rope' to expedite the air defense program," as Everett described it to his Group Leaders. But at the same time, in order to resolve the problem of weapons integration, they "pointed out the need for Air Force long-term arrangements for contracting responsibility outside MIT."[7]

But ARDC was not yet willing to agree, and its response was to insist on a more definitive reply. Late in August, persuaded that there was no other real option, ARDC rewrote the Statement of Work, named Lincoln as the implementing agent, and presented the proposal in this form to Lincoln.[8]

In the meantime, the Lincoln Steering Committee (of which Everett had long been a member) was deeply involved in recurring discussions of the issue. Many of the twelve members attending the July 2 meeting had severe reservations about accepting the proposal of ARDC's Colonel Forrest Allen to make Lincoln Lab "Captain of the SAGE Team" that would direct an expanded SAGE effort.[9] On August 6 a straw vote of the nine members attending opposed Lincoln Lab's adopting any greater responsibility than it already had for weapons integration, and the informal vote ran seven to two against accepting the "captain of the team" role. When they tried to cut the deck another way, supposing that "a real price tag" might be attached to the costs involved and that Lincoln Lab would be given "commensurate authority," the straw vote swung to five in favor and four opposed. After Allen put the request directly later in August, a vote taken on the assumption that the Air Force would provide the necessary support in the "way of money, manpower priority, authority, etc." swung to five in favor and seven opposed. The "general sentiment" of the Lincoln Steering Committee was sufficiently doubtful and controversial to add up to a negative verdict, and so the committee formally concluded. They had been gnawing at the problem like a dog worrying a bone, dropping it and then returning to it time and again.[10]

During the summer and the autumn of 1956, the question Lincoln's administrators appeared unable to answer—a question underlying the integration that ARDC desired of weapons into the SAGE system—was this: What should be Lincoln Laboratory's future role in the air defense program once SAGE was up and running?

In July, the Lincoln policy makers expected "to maintain . . . responsibility and authority in the SAGE program" until 1958 or 1959, when the first module would become operational. Directly related work would then begin to taper off over the next 4 or 5 years. By 1963, "the major Lincoln research efforts relative to SAGE would be on improvements either wholly self-generated by Lincoln or for which only the operational desirability had been stated to Lincoln by the Air Force."[11]

Defending their reluctance to assume responsibilities that they felt exceeded Lincoln Lab's R&D mission, the Lincoln administrators argued that the SAGE program did not fit into the usual procedural pattern followed by the Air Force when it developed new equipment, a pattern that entailed research, development, and testing under the direction of ARDC, procurement by the Air Materiel Command, and use by the appropriate operational Command. Under this arrangement, when modification was required, the equipment was temporarily withdrawn from use, modified, then returned to operational status. However, in the case of SAGE this would not be possible, since the air defense system had so many subsystems and components that it would always be undergoing modification. As Lincoln Lab's managers saw the situation, it was "no longer possible to provide a time separation between the activities of the various Commands working on the job; all must work at once," and this necessitated the kind of overarching coordination and control that only a single systems engineering contractor could provide. But that contractor should not be Lincoln, for this was a job that Lincoln had not been created to do.[12]

While these inconclusive skirmishes were taking place at higher administrative levels, Division 6 came to take, in the words of one of the division engineers, "an active but limited part" as 1956 progressed. It began to mount a course of engineering action that would both "plan and accomplish integration of . . . new weapons for Air Defense" under the general but limiting procedures that Forrester had set forth in January of 1956 in Memorandum 6M-4048.[13] Thus, Division 6 engineer Clare Farr's September internal memorandum on weapons integration, which echoed Shiely's earlier comments to Valley, argued that "unified air defense demands that the entire system be treated as a whole and that all additions and modifications be centrally coordinated."[14]

In September, as a result of the pressure of engineering events, weapons integration became Group 61's primary area of concentration. Work was done on problems associated with the F-89, F-102, and F-104 manned interceptors, the Army's Nike surface-to-air missile, the Air Force's BOMARC, and the Navy's TALOS.[15] By September the Lincoln engineers, in conjunction with the manufacturers, were carrying on in-depth studies of BOMARC, TALOS, and the F-102A, "to define the scope of the work involved in the actual integration effort." Similar studies for other weapons systems were being set up by the Air Research and Development Command.[16]

Meanwhile, ARDC had put to Lincoln Lab the basic policy question of who should handle the technical engineering details of weapons systems integration. In September, facing ARDC's unequivocal Work Statement, the Lincoln administrators had to return an unequivocal answer. After a month of deliberating, they did: "No."

Holloway provided a lengthy explanation, noting in part that the responsibilities to be assumed for weapons integration would exceed those Lincoln Lab had assumed for SAGE. The lab was in no position to set about increasing its staff by "something like 50–100 percent." Echoing sentiments Everett had earlier expressed to Valley, Holloway also pointed out that accepting such a task would change the very nature of the laboratory as it shifted "from emphasis on technical matters to emphasis on materiel coordination." Weapons integration would require competence to be developed in a new group of personnel, and with Lincoln Lab's assistance this could be accomplished just as easily by another organization. In sum, their decision to decline was, they felt, the only one that MIT at the level of Lincoln management could make.[17]

But at the working level within Lincoln Lab the problem of weapons integration could not be eliminated so neatly. In a draft memorandum surveying weapons integration at the end of 1956, David Israel reported that staff members, "with the assistance of IBM and Rand," had concentrated their efforts, within the limits imposed by Forrester's January 1956 memorandum, on those weapons "expected to be operational with the SAGE System prior to 1960." They had given their primary attention to BOMARC and the F-102A, paying less attention to the Nike and TALOS systems.[18]

Also during 1956, the Air Defense Command had arranged a series of meetings to consider operational deployment of the weapons systems that were to be integrated with SAGE. Representatives from Lincoln Lab, RAND, the Department of Defense's WSEG (Weapons Systems Evaluation Group), affected Air Force units, and the contractor for the particular weapons system under review all met to examine such details as the role each weapon should play in a future air battle, the manner in which it should play that role, the effect its presence should have on the general conduct of the battle, and the support and restrictions that SAGE control should provide. Each meeting provided basic information that Lincoln Lab and the weapons system's contractor would use to determine the most appropriate means of integrating the weapon with SAGE. By autumn's end, "employment meetings" had been held on the BOMARC, TALOS, and Nike missiles and on the F-89H and J, F-101B, F-102A, F-104A, and F-106A manned interceptors.[19]

Although in January of 1956 Forrester had recommended a limited advisory and coordinating role for Lincoln Laboratory, and although the lab's administrators had endorsed his recommendation at that time and later, in September (when Lincoln rejected ARDC's proposed Work Statement) engineering events continued to take a contrary course. In the closing months of 1956, the Division 6 engineers became more heavily involved as further details and problems came to light. By December, as members of Group 61 prepared to draw up a year's-end report, they found themselves making a soul-searching analysis of what was transpiring and seriously entertaining a number of reservations about what was going on.

The Group 61 Biweekly Report for November 30–December 14, 1956 noted their concern: Members of the larger joint group engaged in the integration studies "had come to feel that the integration of manned interceptors on an individual basis was ridiculous." Their own investigations, as well as those carried on in the "employment meetings" called by

ADC, had made it clear that the integration of the "century-series" interceptors would have to be carried out as a single integrated task and not as separate activities, for these interceptors had many problems in common.

Israel pointed out in his December memorandum that treating each interceptor as "a separate integration problem" would "result in excessive duplication of effort and lack of coordination." Furthermore, the future addition of technologically superior weapons systems, "with their greater speeds, maneuverability, and new airborne weapons," raised "a host of unsolved problems not earlier apparent with the F-102." For a number of reasons, then, Israel concluded mildly, "the desirability of building up separate groups of weapons contractor personnel to integrate each interceptor . . . seems to be lessened."[20]

On the one hand, these problems at the working level were indicating that Lincoln Lab's official policy to not become responsible for weapons integration was not effectively keeping Lincoln engineers from becoming more and more heavily involved, insidiously creating a slowly gathering momentum of technical activity in what appeared to be the wrong direction.

On the other hand, Lincoln Lab and Air Force policy makers were not having much better luck striking a bargain or resolving the basic issues on their level. In view of these circumstances, it is perhaps not surprising that by the summer of 1957 Everett was recommending a reversal of the policy of limited participation that Forrester's January 1956 analysis had recommended. In a curious, slow turn of events that neither he nor others had anticipated, Everett had come to the conclusion that Lincoln Lab should, after all, "undertake the task of systems engineering for the air defense environment," that it should do so "not for a limited time, but indefinitely," and that it should "undertake whatever expansion in size or budget may be required to perform the task successfully."[21]

The fact of the matter was that in 1956 and 1957 weapons integration as a policy issue simply would not go away, partly because of the belief inside and outside Lincoln that the lab was best qualified to handle it, partly because Lincoln's policy from the start was to phase it out rather than chop it off, and partly because it called for systems engineering of the very kind that Forrester, Everett, and their Project Whirlwind had become involved in a decade earlier, although in a less complex way, as soon as they had decided to develop the general-purpose digital computer that became Whirlwind I.

As Lincoln Lab engineer Clare Farr pointed out in September of 1956 when reviewing Division 6's limited participation in weapons integration, a unified continental air defense made absolute the requirement "that the entire system be treated as a whole and that all additions and modifications be centrally coordinated."[22] If any group was qualified to implement this systems view, it was Division 6, which had the experience and the expertise.

Systems engineering and its application in weapons integration were prominently on the minds of six of the Division 6 engineers who, in September of 1956, authored a "Six-Point Proposal" that went beyond current considerations of weapons integration to discuss in some detail "the first major SAGE revision." They described the revised system they were proposing as "not a new system" but "a redesigned and rewritten computer program" entailing "incorporation of new or improved operational techniques, additions to the AN/FSQ-7 terminal equipment, possible small-scale modifications to the AN/FSQ-7, and modified system capacities and capabilities required to meet revised operational requirements or changes in the air defense environment."[23]

Both Farr's analysis and the six-point proposal agreed that the Air Force would have to establish a central point of managerial control, and that Lincoln Lab had a definite role to play, if only a limited one. Acting promptly on the recommendations offered in the six-point proposal, Everett moved on October 5 to divide the problem of weapons integration into two parts: long-term and interim. In a memorandum to Valley, he set forth his recommended approach to each. In regard to the interim aspect of the problem, Lincoln Lab would "take the initiative" and, except for BOMARC, would provide "an interim capability for all weapons in the 1959 time period." BOMARC would be too large a task for Lincoln Lab to take on, but Lincoln could help by appointing a small group, under "a senior man," to "plan, guide, and monitor the effort and to define tasks that the Air Force must contract to other organizations."[24] For the long term, Everett adopted a recommendation from the six-point proposal calling on the Air Force to organize a Weapons Integration Project Office with "sufficient power to assign tasks, let contracts and see that the job is done." Lincoln Lab would act as the proposed office's technical staff, enabling the office to assign tasks to selected contractors as a consequence of Lincoln's capacity to advise on various technical matters and carry out "research and development on related items."[25]

Everett's recommendations were not the only response to the problem of weapons integration. Indeed, assigning technical responsibility for weapons integration had become so thorny a problem between ARDC and Lincoln Lab that it elicited action at higher policy levels. In

October of 1956, Generals Don Putt (from Air Force Headquarters), Earle Partridge (of the Air Defense Command), and "Tommy" Power (of ARDC) met with MIT's President Killian and Vice-President Cochrane and Lincoln Lab's Director Holloway and Associate Director Valley. But they, too, could find no mutually agreeable solution, and in consequence Killian and Cochrane were meeting in December with the Lincoln Steering Committee to learn its members' views on what should be Lincoln Laboratory's future responsibility with regard to SAGE.[26] The managerial issue had become so critical that it was involving top-level administrators.

The Air Force made more than one attempt to fill the relative vacuum of technical competence that Lincoln Laboratory's intransigence was creating. Colonel Albert Shiely, in his capacity as Chief of ARDC's Air Defense Systems Operating Division, and Colonel O. M. Scott, AMC's representative to the ADES Project Office, drew up a detailed proposal based on actions that had been recommended in October of 1956 at a meeting—convened by General Power of ARDC—at which representatives from involved Air Force organizations had explored the problems besetting the air defense effort and had recommended two courses of action: combining the ARDC and AMC elements of the ADES Project Office into a single agency that would be responsible for the electronic ground environment, and creating a new agency to manage the overall air defense program.[27]

The Scott-Shiely plan, duly coordinated at various command levels of the Air Force, was presented to the Air Staff at Air Force Headquarters in December of 1956. It proposed to combine the ADES Project Office with the Electronic Defense Systems Division and form a single agency in charge of the ground electronics part of the overall air defense system. It called also for a new air defense systems management office to oversee the entire air defense program. And it recommended that a systems engineering contractor be selected to provide the technical resources the Air Force lacked.[28] Months passed as a decision was deferred.

In June of 1957, after Air Force Headquarters had carried out its own investigations into the air defense problem, the Air Force was ready once more to undertake formal action, this time by creating the Air Defense Systems Management Office (ADSMO) that Shiely and Scott had recommended. This upper-level managerial action, too, implementing the findings of the conference that General Power had convened, offered prospects of a solution that subsequent events failed to confirm.

But during the intervening months Brigadier General Ivan L. Farman of ARDC had been encouraged to form "an Air Force-industry advisory group" to "correlate the performance of individual air defense fighters with the capability of the SAGE system" and to tackle immediate difficulties associated with the larger, worrisome problem of weapons integration. Farman had set up an informal SAGE Weapons Integration Group (SWIG) under Colonel R. S. Carter to act as "an *interim* contact point" for weapons systems manufacturers, for Lincoln Lab, and for concerned Air Force agencies. Carter's team operated for several months at Hanscom Field.[29]

Until its duties were taken over by ADSMO, SWIG worked with air defense fighter contractors, Lincoln Lab, the Systems Development Corporation, and the Air Defense Command to "establish some pattern toward which the SAGE people could program their requirements, and around which the aircraft contractors could arrange performance envelopes."[30] The success of SWIG in this 1957 endeavor proved mildly controversial. Colonel Carter enthusiastically endorsed his group's work, but Jack Jacobs of Division 6 concluded after several months that SWIG was unable to "establish itself or . . . get the needed technical support from the weapons contractors."[31]

While SWIG and then ADSMO were trying to bring systems engineering under Air Force managerial control, efforts to achieve effective action on the technical engineering level continued to sputter. It was in this general situation that Everett, before he had been head of Division 6 for a year, acted on several pieces of knowledge that had shaken into place in his mind. It was these that appear to have caused him to rethink and abandon the position on weapons integration he had held since the days of the Transition System. "It had become clear," he said some 30 years later, "that Division 6 was the only one really qualified to do the job."[32]

On June 3, 1957, Everett, using a disarmingly direct and simple line of reasoning, explained to Lincoln Lab's new Director, Carl Overhage, why Lincoln should reverse its position and accept the responsibility of weapons integration, which had become but one aspect of the larger problem of systems engineering and the complex integration of increasingly elaborate electronic systems. Everett was aware that, should Overhage be receptive to his proposal, he could so notify Lincoln's Joint Services Advisory Committee at its next meeting in July.[33]

Everett began his analysis of the situation with a bit of history. Lincoln Lab had been created in 1951 to develop SAGE and the data-processing machines SAGE would need. Clearly excluded from Lincoln Lab's R&D responsibilities were the radars, the data links, and the interdicting weapons to be put at SAGE's disposal. But in the matter of tying the weapons to SAGE, an efficient course of action had been less than clear, for Lincoln was not explicitly charged with the responsibility of making the ground environment apparatus of radars, computers, and data-link telephone lines work with new weapons in the years to come. Instead, serious questions had arisen over who should integrate future interceptors, anti-aircraft weapons, and guided missiles into SAGE.

"A little over a year ago," Everett continued, Lincoln had issued Forrester's analysis as a policy statement "pointing out the problem, stating that Lincoln did not propose to undertake this work and recommending that the work be done for each weapon by teams of men from the corresponding weapons contractors." But this suggestion had not been followed, so the Air Force had proposed that a third contractor be hired and made responsible for carrying out the "technical work necessary to integrate new weapons."[34] Everett related how in 1956, when asked to be this contractor, Lincoln had formally declined, and Director Marshall Holloway had given two reasons, "one internal and one external":

> Internally, the management of the Laboratory concluded that half-way measures would not suffice for this situation. The Laboratory would either have to go wholeheartedly into the job with all this implied in the way of increased size and budget and reduction of freedom of action or else withdraw completely from a controlling position, concentrating instead on research and development to be carried out primarily on the Laboratory's own terms. The sentiment of the Laboratory was in favor of the latter course.
>
> Externally, the Laboratory management was concerned about the way in which the Air Force proposed to manage the entire air defense development picture, spread as it was over all parts of the Air Force with no semblance of overall coordination except at the very top. In view of this organizational situation, the Laboratory management had considerable doubts about the success of the proposed third contractor arrangement. It was clear that at best much frustration and difficulty were involved.[35]

While the Air Force was continuing to search for a workable option, attitudes within Lincoln Lab toward the problem began to shift, Everett noted. One major cause was the fact that in 1956 Division 6 engineers had discovered they would have to shrink the tracking capacity of the SAGE system as it became militarily operational in 1958, making it a less effective defensive instrument. Another was that BOMARC's integration into the air defense system was proving to be more complicated than they had expected. Consequently, in 1956 Lincoln had agreed to tackle these systems engineering problems by embarking on "the so-called 1959 SAGE revision."[36]

So far, Lincoln Lab's commitment had always been limited in scope and time where SAGE was concerned. Should this policy be continued? There were persuasive traditional reasons for doing so, Everett realized. And there was the attractive prospect that "Lincoln should turn primarily to unfettered research and development." But this might diminish the laboratory's role to a supporting one, and in consequence the lab's influence on air defense planning might become so minimal that it might not be able to maintain its budget and its physical plant.

Yet this prospect disturbed Everett less than the disaster that might well result when R&D was no longer vigorously affecting systems-design decisions: systems design would deteriorate. R&D and systems design needed each other, Everett was convinced, for he had seen the healthy interaction at work "within the narrowly defined SAGE concept."[37]

By June of 1957 the entire situation was changing for the better, Everett felt: "While the Laboratory is now confronted with an Air Force proposal not substantially different from that made to it last year, the Air Force has revised its internal organization for systems management." Everett thought he saw signs that the old stalemate was being broken. In his opinion, "the sentiment within Lincoln" was "now toward taking on the job." With that in mind, he offered a formal proposal.

Everett's proposal revealed how extravagantly his thinking had changed, for he was now supporting a position he had earlier considered in corresponding detail and rejected:

> It is proposed that Lincoln Laboratory undertake the task of systems engineering for the air defense ground environment; that this be undertaken not for a limited time but indefinitely; that Lincoln undertake to support the Air Force in this effort; that Lincoln undertake whatever expansion in size or budget may be required to perform the task successfully.[38]

Since Division 6 was already the division most heavily involved in systems engineering, it should take on the job, Everett concluded, and the head of the division should be Lincoln Lab's "official point of contact" with the head of the Air Defense Systems Management Office that the Air Force was then preparing to set up.[39] The balance of his proposal went into detail regarding how it might be implemented within the lab.

But if Everett's proposal was Division 6's solution to the problem of weapons integration, it was not automatically Lincoln Lab's, nor could it automatically qualify as the Air Force's solution. They were by no means out of the woods yet, as is evident from subsequent events. On the same day on which Everett happened to send his proposal to Overhage, June 3, 1957, the Air Force Chief of Staff announced the results of the Air Force's 6-month study.

This latest formal attempt by senior Air Force officers to reorganize their managerial efforts to bring the problem of weapons integration under control and at the same time take care of any future updating of SAGE consisted of three administrative actions to be undertaken: an Assistant Chief of Staff for Air Defense Systems would be appointed to manage and direct systems under development; ARDC would form an Air Defense Systems Management Office (ADSMO), with the participation of the Air Materiel Command and the Air Defense Command; and existing SAGE-related agencies within ARDC and AMC would include within their areas of responsibility the total electronic ground environment.

ARDC accordingly activated ADSMO in July, placed it under the Directorate of Systems Management, and gave it a dual mission: to complete the air defense system already being developed (SAGE) and to manage the design and development of future systems.[40]

But subsequent events demonstrated that more than these administrative actions would be required. Although ADSMO replaced SWIG, within a few months it became clear that ADSMO lacked the status and the authority it needed to accomplish its mission. Though as an executive office it was able to coordinate and recommend, it lacked the power to direct, and the result was an instructive but not very effective exercise in the voluntary coordination of three independent Commands. Consequently, in March of 1958 the Air Force tried again: ADSMO was reorganized and renamed, becoming the Air Defense Systems Integration Division (ADSID) under the command of not a colonel but a general officer, Major General Kenneth P. Bergquist, who, as a Deputy Commander within ARDC, possessed greater authority and autonomy.[41]

While these efforts were supposed to take care of the Air Force's role as the funding agency participating in managing these R&D programs, they did not themselves directly address the unsolved problem of who would wield the technical engineering authority to get the weapons-integration job done and update the SAGE system to take advantage of oncoming technical developments. This issue was now being handled carefully with a combination of formal and informal actions at a higher administrative level, with the Secretary of the Air Force and the president of MIT involved.

Formal sanction was provided early in March of 1958, when Secretary of the Air Force James H. Douglas proposed to MIT that what he was careful to call an interim solution to the problem of providing the Air Force with technical assistance be worked out. Douglas asked MIT to play a mediating role in providing the new ADSID with a "technical systems management team."[42]

One of the aforementioned informal actions occurred in April of 1958 when the heads of Lincoln Laboratory and the System Development Corporation became involved in exploratory discussions regarding what it would take for either or both of them to provide the technical systems engineering services that the Air Force required.[43]

Their efforts, too, came to naught. Lincoln Lab and SDC were unable to find the common ground that would permit them to collaborate, as the directors of the two organizations explained to MIT's Vice-President for Governmental and Industrial Relations, James McCormack Jr. Two days of intensive discussion, in which Bob Everett and Jack Jacobs of Division 6 and W. S. Melahn of SDC were also involved, demonstrated that, while the two firms were "in qualitative agreement on the nature of the ADSID support-contractor's task," they disagreed fundamentally in their "quantitative estimates of the manpower required to perform the task." SDC estimated a professional staff of 300, while Lincoln felt the number should be more than twice as large: 690, plus support from nearly 1300 non-staff employees. Everett and Jacobs began to feel that SDC did not realistically understand what such a commitment would entail.[44] The issue was further complicated by the fact that Lincoln insisted on more freedom of technical activity than ARDC, mindful of its past difficulties with the laboratory, was unwilling to grant.[45]

MIT was not the only organization to which Secretary of the Air Force Douglas turned for help in obtaining technical engineering support for weapons integration. He had also approached RCA, Bell Telephone Laboratories, and Western Electric. All declined to take on the burden, even though BTL and Western Electric remained committed to finishing the job they had started with ADES. "They felt," Jack Jacobs recalled, "that the central design and system engineering role ought to remain with Lincoln."[46]

Jacobs reflected on the general situation nearly three decades later: "As the machine and the computer program were assembled, the enormity of the computer system integration task stood out." In the beginning there had been a handful of radars and one interceptor to integrate, but by 1957 there were "at least three new interceptors, four or five different kinds of radars, 'Texas Towers' [far offshore in the North Atlantic], picket ships, AEW&C aircraft [forerunners of AWACS], the B-version of BOMARC, the Nike Hercules ground-to-air missile—all to be controlled by the SAGE direction centers. In addition, there was cross-tell and forward-tell integration among the centers themselves."[47]

Nearly a year had passed since Everett had proposed to Overhage that "Lincoln Laboratory undertake the task of systems engineering for the air defense ground environment . . . not for a limited time but indefinitely." He had suggested at the time that "Division 6 which is presently responsible for SAGE systems engineering, be given responsibility for the new work," but the top managers of Lincoln and of MIT had not been ready to move.

Now, 11 months later, during the last week of May 1958, Everett was at work on a draft of a "General Plan," the opening sentence of which was "The new corporation should be formed as rapidly as possible and its sponsorship and officers chosen and announced."[48] And before another week had passed, Stratton was writing Douglas: "This is in reply to your letter of March 3. . . . As I am sure you know, the delay in making formal reply to your letter does not reflect any actual delay in operations responsive to your request. . . . MIT personnel have worked with General Bergquist and other Air Force representatives in clarifying the details of the task and in formulating longer term plans for its accomplishment. We are now in a position to inform you of our concept of the job."[49]

Stratton then went into detail, "taking up the major points generally in the order in which your letter deals with them" and declaring that MIT proposed to "sponsor the formation of a new non-profit corporation."[50] This corporation should incorporate policies and operating practices already employed in successful industry. "We therefore propose that high level business and industrial management experience be brought into the new corporation initially through temporary loan of operating personnel and membership on the board of directors. . . ." These people should understand that they were performing a public service to their country and should be drawn from "sources which minimize the possibility of conflict of corporate interests." From this beginning the new corporation could then proceed to build "a more permanent management team."

Because the technical side of the new organization would "have to be built from various sources," MIT would "encourage" it to hire "those Lincoln staff members who are qualified for the task and who are moved to shape their professional careers in the direction of systems engineering." To this end, Lincoln Laboratory's "personnel recruitment facilities" would be available to the new corporation until it got on its feet. MIT would not be responsible for hiring the personnel and shaping their organization; that would be the duty of the new corporation. Furthermore, once the new corporation was sufficiently established, it should take over, "as a systems engineering tool," the operation of the Experimental SAGE Sector.[51]

Starting out as a subcontractor to MIT, the new corporation should in due course switch over to a prime contract with the Air Force, allowing MIT to end its interim responsibility once the new systems engineering organization had gotten off to a successful start. Stratton attached to his letter to the Secretary of the Air Force five detailed pages of "Some Initial Actions in Organizing ADSID Support Operations." He closed the letter with a courteous request "that we be informed if your staff sees any important omissions or points of potentially serious divergence of views."[52]

It was as a consequence of these events outside Lincoln Laboratory, as well as of the creative R&D activities of Division 6, that a majority of the professional staff of Division 6 left Lincoln Laboratory in 1959 to join a new R&D venture in systems engineering and SAGE weapons integration: The MITRE Corporation, to be located in nearby Bedford.

The wheel had turned full circle since George Valley had discovered Project Whirlwind, thereby setting in motion the events that caused the Whirlwind engineers to leave the Servomechanisms Laboratory on the MIT campus and join Lincoln Laboratory. Eight years after Valley had discovered Whirlwind, the "Division 6, Lincoln Laboratory" phase of the story would in turn draw to a close. And for a while after 1958 the computer-centered SAGE continental air defense system would come into full military operation and would draw on the continuing technical assistance of MITRE.

Still the wheel would continue to turn. While ahead lay the continuing adventures of many of these same Division 6 experts in systems engineering, theirs would become not only a part of the story of the emergence and growth of the new MITRE Corporation but also a part of a larger story in which missiles would supplement manned aircraft, these would give way to intercontinental ballistic missiles equipped with atomic warheads, and SAGE would be replaced by ever more complex and diversified military command-and-control systems in a restless world inhabited, for a while, by only one national superpower curiously devoted to exploring the research and development process.

32

Epilogue

On June 27, 1958, Lincoln Laboratory formally turned over control of the first link in the SAGE continental air defense system, the largest and most complex data-processing system yet attempted, to the Air Defense Command. The product of nearly a decade of cooperative innovative efforts by electrical engineers, mathematicians, scientists, and administrative policy makers in government and academe, SAGE required the combined contributions of tens, then hundreds, then thousands of individuals working within the constraints imposed by traditional academic, industrial, and military institutions. Subsequent events would show it to be a major national accomplishment and a technical innovation of such consequence as to make it one of the major human accomplishments of the twentieth century—claims that might appear extravagant to those not acquainted with the historical details.[1] On the immediate, practical level of military affairs, a digital computer was, for the first time, incorporated into a military command-and-control system to receive, process, and display data essential for prompt and effective countermeasures against an attack. It was "by any man's standards, a tremendous advance in the air defense art—certainly in the use of data-handling devices."[2]

Of course, SAGE was never used under wartime conditions, and its main impact was on the technical state of the art. It proved to be an application that would expand into all areas of the national life of the United States and then all over the world, producing a subtle, conceptual, philosophical revolution and also a revolution in the daily conduct of human affairs. The full consequences of these revolutions in peace and in war still lie ahead.

Balanced estimates of revolutionary technical developments are not easy to come by until the years have passed. If one is a gradualist at heart about the progress of human affairs and a believer in the primacy of uninterrupted sequences, sharp discontinuities and shifts in direction

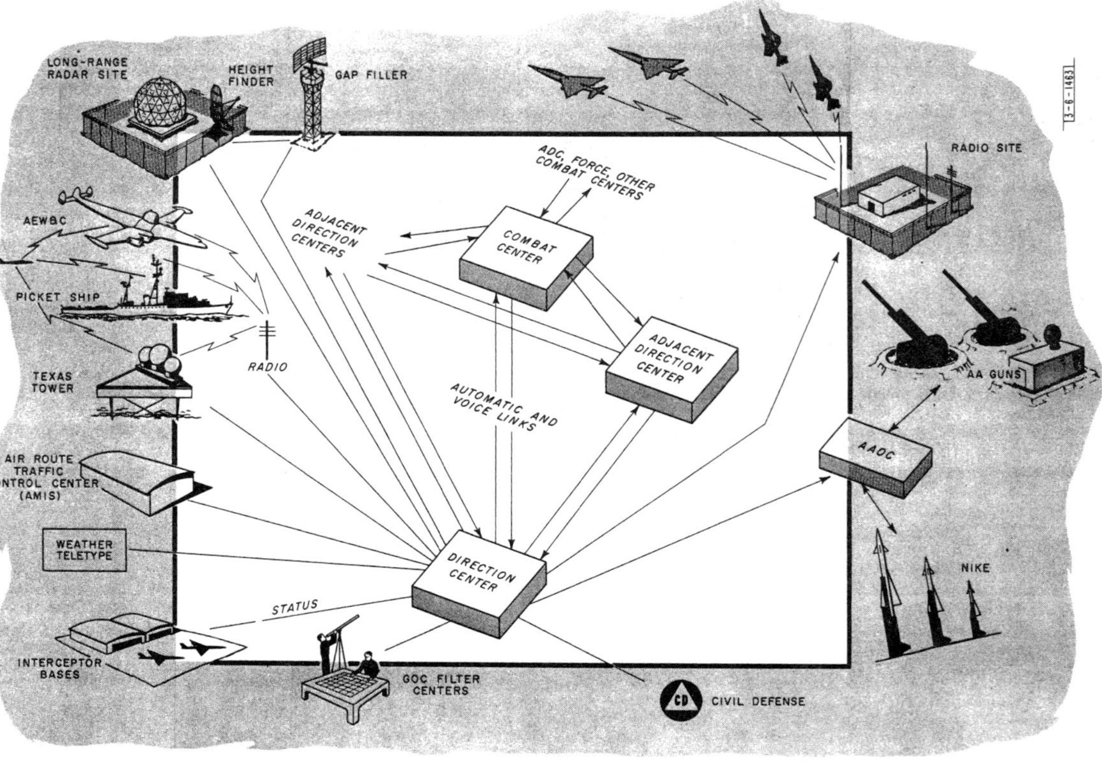

Figure 32.1
"Relationship of SAGE to the Air Defense System. The unshaded area of the figure, including portions of other air defense elements in which SAGE equipment has been installed, represents the SAGE System."

are difficult to recognize, appraise, and account for. If one is a revolutionary at heart and committed to seeking out and identifying the strategic discontinuities, evolutionary tides are tedious to trace and process is relatively drab. Having found much evidence to support both of these views, we shall embrace both in this final chapter, recognizing that the ways in which the past produces the present can range from the mundane to the heroic, with the mundane dominating most of the time.

June 27, 1958 may be proposed as a ceremonial date that marked the inconspicuous dawn of a new era, an era to which most people at the time were oblivious. Viewed narrowly, it was a new era in the practical command and control of military operations. Viewed more widely, it was a new era in communication: massive volumes of information now could be transmitted, virtually instantaneously, anywhere that suitable equipment was installed. Viewed more widely still, it would fundamentally transform the ways in which the business of government, industry, agriculture, medicine, education, and most activities in the adjoining realms of the arts and the sciences would be carried on.

Furthermore, the story of the SAGE computer is essentially a contingent one, nested in the circumstances of the 1940s and the 1950s. And those circumstances were contingent on events that occurred before and during World War II and in the course of the Cold War.

In 1958, when the first sector of the SAGE system became operational, the Air Force's Vice Chief of Staff, General Curtis LeMay, noted that what had been "little more than a concept when the Air Force decided to go ahead with the project" had become "a completely new system constituting an improvement of great magnitude." Praising those who had foreseen the danger and who "with vision for the future" had cast aside "outmoded practices," LeMay observed that their "imaginative thinking" had made "the fundamental concept of a coordinated air battle and a defense in depth a practical reality."[3]

On December 15, 1961, 3½ years after the activation of the McGuire site, the last American link in the SAGE air defense system was turned over to the Air Defense Command at Sioux City, Iowa. Twenty-one sectors stretched along the east and west coasts, along the northern tier of states, and into Canada. Each sector, with its own direction center housing the computer and associated equipment essential to its operation, reported to one of three regional combat centers, which in turn reported to the North American Air Defense Combat Operations Center under Cheyenne Mountain in Colorado.[4] A year later, the SAGE installation at North Bay, Ontario was tied into the system.[5] For the next quarter-century, SAGE

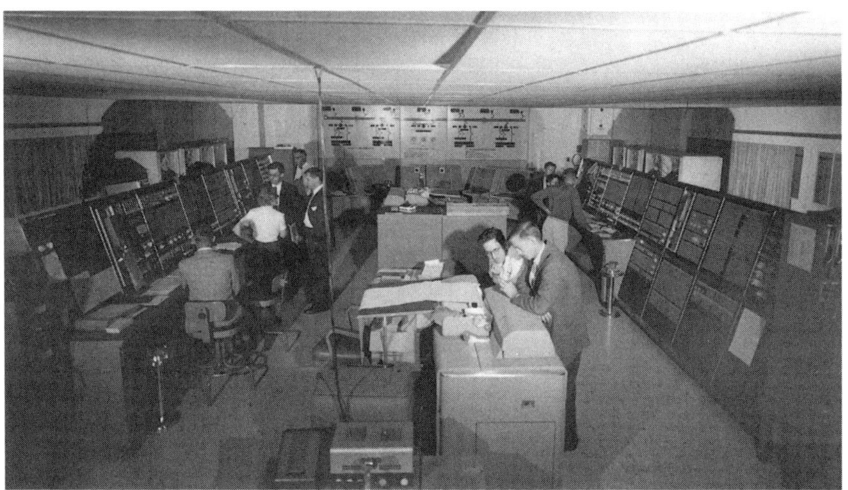

Figure 32.2
Control room for FSQ7 computer.

Figure 32.3
SAGE under construction.

guarded the oceanic and polar approaches to the North American continent.

In 1970, 12 years after he took command of the first operational SAGE site, retired Lieutenant General Arthur C. Agan recalled that the integration of SAGE into the existing manual system had been "the equivalent of laying a man out on the table and trying to keep him alive while we used the scalpel and took out his nervous system and put another one in."[6] In a more extensive 1976 interview, Agan elaborated on his earlier remarks, describing the impact of SAGE's installation as "tremendous, tremendous":

> It gave you an ability to see the entire situation dynamically. The radars were tied in so that from a single scope you could really see it all. You could select the size of area that you wanted to see; and you could see all of the radar information that you wanted and select the kinds of things that you wanted to see. If you wanted to see all tracks of certain classification, you could see that, like all friendlies or all hostiles. And each of these was tabbed to indicate their identification, so this is a vast improvement over any capability we had ever had before. It was actually an assist in working the intercept problem . . . instead of having to sit there with a pencil and draw vector lines and calculate it, we had already put these computations into the computer.[7]

Some other appraisals were less enthusiastic than LeMay's and Agan's. SAGE was sufficiently ambitious and innovative in scope and detail, from its conception through its birth and its development, that it inevitably drew criticism along the way, it unavoidably came under administrative and policy-level attack from time to time, and (as is so often the case where high technology is concerned) it was judged already obsolescent when installed. At the beginning of the 1950s, when Forrester and Valley had joined forces, the primary threat had been the manned intercontinental bomber, and the system had been designed and its computer programmed to meet that threat. But by the end of the 1950s, when the traditional, manually operated military system was being converted to semiautomatic operation, the threat of attack was being compounded by the advent of the intercontinental ballistic missile.

As these technical developments were taking place, the task of the system's computer remained specific. At the start it was to process the radar data supplied by geographically remote sensors and to provide tactical information essential to the selection of appropriate counteraction

Figure 32.4
The completed SAGE machine.

against hostile manned aircraft. But the defense concept developed with that specific mission in mind deliberately included the potential for limitless extension and applications. Consequently, the contemporary development of novel and superior offensive weapons did not mean that SAGE's approach had reached a dead end. Its electronic functions could be employed, as the Bull committee had foreseen, in a system that might be devised to respond to an ICBM attack.[8]

SAGE was a research and development effort that was definitely "high tech" for its day, even though it lacked the high-profile, cataclysmic glamour that atomic energy possessed after the A-bomb. As such, SAGE had to contend with the tensions that were unavoidably generated between the practices and policies of the older national defense tradition and the practices and policies of the newer R&D tradition. It was the business of R&D to bring tomorrow sooner, and in the military realm "tomorrow" included nuclear warheads, ICBMs, and digital computers. In contrast, and at a slower pace, it was—and always had been—the business of the military national defense tradition to make present-day technology work while preparing to use future technology, even though no one could be sure just when that technology would arrive, what its precise configuration would be, or how well it would work. One of the reasons for the uncertainty lay in the very character of the R&D process, which was reflected in all the scheduling and "stretch-out" problems that beset the men of Division 6 and Lincoln Laboratory while they were converting SAGE from concept to practice and from theoretical ideal to engineering reality.

Whether or not it would have been effective against enemy bombers (only combat could have provided the answer), SAGE was a historic accomplishment. It provided positive evidence that a radar system employing the electronic digital computer could, in the words of the historian Howard Murphy, "intelligently and almost instantaneously control complex operations throughout a wide area." A "real-time general-purpose computer," it could effectively operate "as the central and coordinating element in the conduct of various kinds of large and involved operations demanding both promptness and precision."[9] The predictions of Jay Forrester, Bob Everett, Perry Crawford, and other computer pioneers had been realized.

As a RAND study pointed out in the late 1960s, developing the SAGE system "may have had the effect of *deterring* the acquisition and deployment of a Soviet bomber force."[10]

The computer at the heart of the SAGE System (the AN/FSQ-7), and Whirlwind before it, contributed significantly to the advancement of the state of the art in computer hardware, in software, in applications, and in the creation of trained, experienced, knowledgeable, increasingly sophisticated computer engineers, programmers, and scientists. In military applications alone, the high-speed, real-time digital computer was the indispensable component in a whole family of command-and-control systems subsequently developed for national defense. The MITRE Corporation subsequently found both SAGE's underlying philosophy and its technology transferable to the tasks of planning for the NATO Air Defense Ground Environment System in Europe and the Base Air Defense Ground Environment System in Japan.[11] When Iraq invaded Kuwait, three decades after SAGE went into operation, Saudi Arabia was in possession of computer-equipped AWACS aircraft it had purchased from the United States.

Nor was the principle confined to military applications. The computer's ability to assimilate massive quantities of data and to process them instantly had many applications that would become commonplace,

Figure 32.5
SAGE operators in Surveillance Room.

including airline and railroad ticket reservations, stock inventories, magazine subscriptions, and air traffic control (the very application that MIT's Digital Computer Laboratory had been studying in 1949, when George Valley was urging Theodore von Karman of the Air Force Scientific Advisory Board to establish ADSEC). Designed for a specific mission, the AN/FSQ-7 was a more sophisticated and advanced version of Whirlwind, and both were products of a naturally continuing R&D program that had begun under the Navy's financial auspices during World War II. Continuity of R&D activity proved to be an essential feature of these historical developments.

The challenges met, the experience gained, and the enthusiasm shown in their R&D enterprise may not have been unique to the experiences of the members of the Whirlwind-SAGE team, but their accomplishments and contributions were a major factor in the development of an instrument that was to influence all areas of human activity and human institutions, from the superficial to the substantial. As the computer pioneer Herman Goldstine has pointed out, "much of interest and one thing vital to the computer field" came from the projects headed by Jay Forrester and Bob Everett.[12] Among the fruits of those projects were procedures for ensuring an extraordinarily high degree of component reliability and the development of a reliable and efficiently accurate memory system. The latter was "one of the most important steps in making computers reliable and capable of having very large memories."[13]

When Remington Rand had challenged the selection of IBM as a manufacturing source, more was at stake than commercial rivalry, profits, and prestige. Both companies swiftly recognized that the cooperative enterprise with Lincoln Lab's computer engineers promised access to significantly expanding roles on the frontier of a revolutionary technology. Tom Watson Jr. of IBM responded promptly and eagerly. IBM's participation in the program, Watson wrote in his memoirs, provided the "giant boost" essential to moving the corporation from punched-card machines to computers. The experience, Watson said, "enabled us to build highly automated factories ahead of anybody else, and to train thousands of new workers in electronics."[14]

IBM wasted no time in applying to its commercial line what it had learned from its collaboration with Division 6. In 1954 (the year in which it received its first contract for the production model AN/FSQ-7), IBM, facing stiff competition in the new industry, especially from Remington Rand, redesigned its commercial line to include technical advances derived from its collaboration with Division 6. In October of 1954, a year

Figure 32.6
SAGE building at McGuire Air Force Base.

after the insertion of the magnetic-core memory into Whirlwind, IBM established a core-manufacturing facility at Poughkeepsie. The following year, well ahead of its commercial rivals, it began delivery of the 704, a new model that incorporated the magnetic-core memory.[15] Also in 1954, IBM began to develop SABRE, "a real-time computer seat-reservation system," which 10 years later became operational as "the largest commercial real-time data-processing network in the world."[16]

Reviewing these developments, one participant concluded that "the decision to join MIT on Project SAGE was probably the most important decision management made during this period. It gave IBM an inside track to the most advanced computer technologies in the world and propelled the company into a leading position in ferrite core memories."[17] The sophisticated technology gained from the cooperative effort, subsequently modified and refined, helped push IBM to the leading position it occupied for a while in the new industry.

IBM was not alone in enjoying the benefits derived from working alongside the engineers of Division 6. Other major industrial and commercial firms benefited from Lincoln Lab's technical guidance until The MITRE Corporation took over the task that Lincoln had begun. Participating in building the air defense system, they improved their own technical proficiency during these cooperative endeavors. As was pointed out in a contemporary issue of the *Saturday Evening Post,* the SAGE System was

> the product of many brains and hands, many of them fiercely competitive in their civilian pursuits. Western Electric took over the job as overall coordinator. RCA, Bendix, and G. E. built the radars. Bell Telephone Laboratories worked out the complicated web of communications that would link together SAGE sectors across the country. Burroughs devised a miraculous machine which would translate raw radar data into mathematical language the computer could understand. System Development Corporation wrote the 7000 pages of plain English instructions which, reduced to a thousand pages of mathematical formulas and transferred to 3,000,000 punch cards, were fed into the computer so that the machine would know what it had to do.[18]

SAGE's programming demands brought the System Development Corporation into being. Initially organized as a division within the RAND Corporation, it was spun off from RAND, as MITRE was from Lincoln Laboratory, when its work began to conflict with the research responsibilities of its parent organization. SDC's work for SAGE, originating under the supervision of Lincoln Lab computer engineers, established the basis for much of SDC's later work. Thus, the utility programs prepared for the AN/FSQ-7 were "transplanted . . . into SDC's ADEPT timesharing system, and ten years later . . . the . . . COBOL data dictionary concept was invented in the SAGE compool."[19] "Everything was first done in SAGE" seemed for a while to be a theme that swept through the industry. "Many of the lessons learned as far back as SAGE," a May 1973 article in *Datamation* pointed out, "are often ignored in today's software developments although they were published over ten years ago . . . on the value of milestones, test plans, precise interface specifications, integrated measurement capabilities, formatted debugging aids, early prototypes and concurrent system development and performance analysis."[20]

SDC's costs for eight versions of SAGE's programs amounted to $150 million, approximately 2 percent of the generally accepted estimate of $8 billion for the total system (including site construction, computers, telephone communication lines, and sundry equipment and services).[21] Such figures indicate the overall impact on the national and regional

Figure 32.7
Ten of the principal architects of the SAGE Air Defense System in 1982. Front row, left to right: John F. Jacobs, Jay W. Forrester, Robert R. Everett, Robert Bright. Back row, left to right: Herbert D. Benington, John V. Harrington, Norman H. Taylor, Robert P. Crago, C. Robert Wieser, Albert R. Shiely Jr.

economies, and they provide incidental historical instance of the stimulus that military requirements gave the U.S. economy during the Cold War decades.

In a 1986 *Boston Globe* article titled "Reaping the Whirlwind," David Warsh attributed much of the prosperity of Massachusetts to MIT's work in computers under the leadership of Jay Forrester.[22] The economic ramifications were extensive and impressive. The technical experience and knowledge Kenneth Olsen gained during his years with Project Whirlwind and the SAGE program, when blended with his entrepreneurial talents, led to the founding of the Digital Equipment Corporation. The establishment of Lincoln Laboratory was, in part, a consequence of Forrester's and Valley's efforts to form the R&D enterprise that produced SAGE. The MITRE Corporation was another offspring of the SAGE R&D enterprise. All played major roles in the flourishing Massachusetts economy of the Cold War era. Forrester continued his work as a member of the MIT faculty, applying his understanding of the computer to developing his program of System

Dynamics, "his own approach to the modeling of societal systems using feedback loops similar to those used by engineers in the design of servomechanisms."[23]

Without the computer there would have been no SAGE air defense system, and without the magnetic-core memory there would have been no computer capable (as soon) of performing the required task. Both were the products of a team of young engineers, the nucleus of which first came together during World War II in Gordon Brown's Servomechanisms Laboratory on the MIT campus. The efforts of Jay Forrester and Bob Everett during the waning days of World War II generated Project Whirlwind before they knew it would require a machine that most people had not yet heard of: the digital computer. As early as 1947, their efforts to design, develop, and build Whirlwind for use with the projected (but never built) Aircraft Stability and Control Analyzer unwittingly began to lay the R&D groundwork for the continental air defense computer. They restored MIT, as Herman Goldstine put it, to the "leading role in the computing field," first established before World War II by the work of Vannevar Bush and Harold Hazen.[24] And in view of how the young Perry Crawford drew the attention of his fellow graduate student Jay Forrester to the prospects for the design of digital computers in the summer of 1945, the name of Crawford and that of one of his teachers, Professor Samuel Caldwell, should be added to the list.[25]

National recognition for Forrester's and Everett's accomplishments came in 1989 when President George Bush awarded them the National Medal of Technology for their "pioneering work in the development of the large-scale digital computer for real-time systems."[26]

In retrospect it is difficult to separate the R&D program that produced Whirlwind from the one that led to the SAGE air defense computer, even as it is difficult to differentiate between the AN/FSQ-7 and the SAGE System, for as the one became integral to the other, so the system approach to design became integral to both. Project Whirlwind and SAGE were historically distinct programs, warranting analysis as individual case studies in the history of the R&D tradition and in the advancement of computer technology. At the same time, they were closely interwoven. The development of the AN/FSQ-7 was really an extension of the development of Whirlwind, and the name first given it by its designers, "Whirlwind II," reflected their appreciation of what was happening in the laboratory at the design level.

The two programs overlapped and melded. As Whirlwind (completed during the planning phase of SAGE) was incorporated into the Cape

Cod System and provided information essential to the design and development of the AN/FSQ-7, so the Cape Cod System (with Whirlwind as its engine) provided information essential to the design and development of the SAGE computer and the SAGE air defense system.

In retrospect, Project Whirlwind and MIT's Digital Computer Laboratory made possible Division 6 of Lincoln Laboratory, out of which sprang The MITRE Corporation. Individually and collectively, the young engineers who came out of MIT's Servomechanisms Lab played major parts in advancing the state of the art in computer technology. They contributed heavily to the rise of command-and-control systems, setting the pattern for the military and space systems that followed. They contributed to laying the R&D foundation for the Massachusetts minicomputer industry. And they revolutionized the information industry by "spanning, in one inspired leap, the prehistoric computer era of serial batch processing and the modern world of interactive systems."[27]

As Bob Everett has noted, the SAGE program alone "trained hundreds of digital-system design engineers, thousands of computer programmers, and thousands of digital-computer field engineers who gave great impetus to the new field of digital computers." In the areas of hardware and of computer operations, SAGE's firsts included "computer-driven displays, on-line terminals, time-sharing, high-reliability computation, digital signal processing, digital transmission over telephone lines, digital track-while-scan, digital simulation, core memories, computer networking, duplex computers."[28]

It remained to be seen whether the Whirlwind-SAGE experience was too revolutionary and too unorthodox to serve as a model for pursuing R&D in the future, or whether it contained the seeds of a new orthodoxy of excellence in joining scientific research with engineering development and engineering development with industrial production in careful, ingenious ways that could be taught and which would replace the traditional, homely, poorly understood models of such romantic American heroes as Thomas Edison, the Wright Brothers, Henry Ford, and George Eastman.

Perhaps the best summary of the curious, elusive, diffuse impact of SAGE's basic contributions to the history of American industry and American scientific technology in the middle of the twentieth century comes from a technical observer who was not himself involved in the SAGE effort: "SAGE's technological significance doesn't derive from any particular technical innovation or set of innovations. Rather, it stems from a lesson. Above all, SAGE taught the American computer industry how to design and build large, interconnected, real-time data-processing systems. Through SAGE, Whirlwind's fabulous technology was transferred to the world at large, and computer systems as we know them today came into existence."[29]

Whether the techniques learned from that lesson would serve as a basis for further successful efforts in the United States and the rest of the world to place modern science and technology more efficiently and securely in the service of humanity remained to be seen.

Notes

Chapter 1

1. C. Robert Wieser to Jay W. Forrester, "Experimental Interceptions with Bedford MEW [Microwave Early Warning] Radar," Memorandum M-2092, April 23, 1951. (Except where other locations are noted, the technical "M-Notes" generated by the MIT Digital Computer Laboratory are in the archives of the MITRE Corporation in Bedford, Massachusetts.)

2. Interview with C. Robert Wieser by the authors, January 3, 1984.

3. C. Robert Wieser, "From World War II Radar Systems to SAGE," *Computer Museum Report* 22 (spring 1988), p. 16.

4. Hoyt S. Vandenberg to George E. Valley Jr., May 28, 1951. (Except where other locations are noted, cited correspondence is in the MITRE archives.)

5. Project Whirlwind was one of several projects funded by the Navy's Special Devices division to explore the design potential of the novel calculating machines called digital computers. Whirlwind turned out to be by far the most fruitful of these projects. The name may have been chosen by the Navy to symbolize the promise of speed in calculation.

Chapter 2

1. Report of War Department Equipment Board, February 23, 1946 (Box K2698, RG 156 WNRC, National Archives), cited on p. 25 of Koppes 1982.

2. Report of the Joint Army Navy Board, April 29, 1898 (BOF 3896, National Archives). See also Turnbull and Lord 1949, pp. 1–3; Goldberg 1957, p. 2.

3. Holly 1953, p. 172. See also William Mitchell, cited on p. 40 of volume I of Craven and Cate 1948. On the military use of air power during World War I, see Morrow 1993.

4. Final Report of War Department Special Committee on Army Air Corps, July 18, 1934, as quoted on p. 3 of Smith 1964. See also Futrell 1971, pp. 24–43; Goldberg 1957, pp. 36–43.

5. Futrell 1971, pp. 43, 53–55.

6. Smith 1964, p. 7. See also Volan (unpublished), pp. 18–23.

7. E. Kathleen Williams, "Deployment of the AAF on the Eve of Hostilities," in Craven and Cate 1948, p. 153.

8. Ibid., p. 155.

9. Volan (unpublished), pp. 18–23.

10. Ibid.; Futrell 1971, p. 58.

11. "Black Thursday" was reminiscent of another August day toward the end of World War I, in which Goering had served as a junior air officer. August 8, 1918, had been the day British forces pierced the Hindenburg (Siegfried) Line. Though not a major tactical victory, it had been a moral one, for it had convinced the German High Command that defeat was inevitable. See Liddell Hart 1930, pp. 429–438; Ropp 1959, p. 251.

12. Watson-Watt 1959, pp. 240–247; Stokesbury 1980, pp. 105–114; Bush 1970, p. 30.

13. "Activities of British Technical Mission from September 9, 1940," AVIA 10/2, Public Records Office (henceforth cited as PRO), Kew, Surrey, England. See also Memorandum from the British Technical Mission to the Canadian Defence Council, October 24 [1940], AVIA 10/1, PRO. See also these relevant letters (stored under AVIA 10/62, PRO): V. Bush to Sir Frank Smith, Ministry of Aircraft Production, March 13, 1942; R. A. Watson-Watt to V. Bush, May 2, 1942; V. Bush to Sir Frank Smith, June 6, 1942. For documents reporting meetings and conferences, see "Coordination of British and American Radio and Research Programmes," notes of meeting held June 30, 1942, and notes of meeting held July 30, 1942; "Standardization of British and American Radio Equipment," notes of meeting held 13th July 1942; these are stored under AVIA 10/62, PRO. See also British Radar Mission and U.S. Radar Working Committee, [1] "Minutes of Special Subject Conferences . . . 23 November 1943 . . . ," [2] "Minutes of Conference on Air Defense . . . 8 December 1943 . . . ," [3] "Minutes of Final Conference . . . 11 December 1943 . . . " (AVIA 10/65, PRO)

14. "Activities of British Technical Mission from September 9, 1940" (AVIA 10/2, PRO); Hartcup 1970, pp. 26–29; Bush 1949, pp. 20, 39; Kevles 1977, chapter XX.

15. Bush 1949, p. 39.

16. "British Technical Mission, Information Given to U.S. Authorities," October 28, 1940 (AVIA 10/2, PRO).

17. Memorandum, British Technical Mission to Canadian Defense Council, October 24 [1940] (AVIA 10/1, PRO).

18. Bowen to Tizard, "Recommendation from Microwave Section on Initial Development Contracts and NDC Microwave Laboratory," November 5, 1940 (AVIA 10/2, PRO). See also Smith 1964, pp. 8–9; Kevles 1977, chapter XX.

19. Memorandum from British Technical Mission to Canadian Defense Council, October 24 [1940] (AVIA 10/1, PRO); "Memorandum to Sir Henry Tizard on the Future Work Arising out of the British Technical Mission," November 25, 1940 (AVIA 10/2, PRO); Kevles 1977, pp. 302–323;. Bush 1970, p. 42, n. 318, and passim; Gray 1943, p. 65. W. A. Noyes Jr., "Relationship of NDRC to the Armed Services and to Allied Countries in Chemical Warfare," in Noyes 1948, pp. 152–154. (Noyes 1948 was originally issued by the Office of Scientific Research and Development as part of *Science in World War II,* a multi-volume study recording the work of the OSRD during that war.)

20. Millikan 1950, pp. 152–154; Dupree 1957, pp. 312–313.; Kevles 1977, pp. 120, 126–131, 142–145; Kent C. Redmond, "World War II, a Watershed in the Role of the National Government in the Advancement of Science and Technology," in Angoff 1968.

21. Major Gordon P. Saville, "Air Defense Doctrine," October 17, 1941 (National Archives, Records of the Army Air Forces, AAG 381, Air Defense Doctrines (S) "Bulky."). See also Smith 1964, pp. 8–9; Futrell 1971, p. 43.

22. Smith 1964, p. 12.

23. William A. Goss, "Air Defense of the Western Hemisphere," in Craven and Cate 1948, pp. 271–272.

24. Morison 1961, pp. 218–219. See also Smith 1964, p. 27, n. 55.

25. Watson-Watt report, 1959, quoted on pp. 291–292 of Goss, "Air Defense of the Western Hemisphere.

26. Ibid., p. 292. See also Watson-Watt 1959, pp. 308–318; Smith 1964, pp. 13–15.

27. Smith 1964, pp. 15–18.

28. Ibid., pp. 18–20.

29. "Research in Air Defense at the Institute of Science and Technology," issued by Institute of Science and Technology, University of Michigan (no date) (copy on file in Historical Office of Office of Information Services, Electronic Systems Division, Air Force Systems Command, Laurence G. Hanscom Field, Bedford, Massachusetts—hereafter abbreviated ESDHO.)

30. For a perceptive discussion of the RDB, its committees and panels, and the virtues and defects of this novel peacetime administrative arrangement, see Price 1954, preface and chapter 5. The episodes involving the two RDB Ad Hoc Panels relevant to this history support Price's evaluation.

31. Vandenberg to Bush, December 9 1947; Charles S. Fowler, "History of Air Force Cambridge Research Center" (prepared at ESDHO and hereafter cited in short form as Fowler, "History"), volume XV, part I, appendix 9. For a useful unpublished study completed in December 1964 at ESDHO, see Smith 1964, pp. 49–52 and p. 63, n. 90. Many of the points made in the pages cited here are developed more fully elsewhere in Smith's study. (See also "Project Lincoln: Case

History," volume I, document 2. This unpublished three-volume history, compiled under the auspices of the Historical Office of the Air Force Cambridge Research Center, contains copies of more than 100 letters, memoranda, reports and other documents spanning the period November 1946–August 1953. It opens with a 13-page chronicle which is identifiable by the endorsement at the top of the first page from the Vice Commander of the Cambridge Research Center: "C[ommanding] G[eneral], AFCRC, [Col.] H. W. Serig, 10 December 1952." Volume I contains items 1–27 (November 6, 1946–December 12, 1951); volume II contains items 28–52 (December 21, 1951–October 19, 1952); and volume III contains items 53–73 (October 30, 1952–August 12, 1953). Hereafter it will be cited in short form as "Project Lincoln.")

32. Vandenberg to Bush, December 9, 1947.

33. W. L. Barrow, Chairman, Panel on Radar, Committee on Electronics, RDB, to Normal L. Winter, Executive Director, Committee on Electronics, January 5, 1948, pp. 1–2. See also Fowler, "History," volume XV, part I, appendix 9; "Project Lincoln," volume I, document 2.

34. Barrow to Winter, January 5, 1948, pp. 2–4.

35. J. A. Stratton to Research and Development Board, "Early Warning System for Defense of the United States," December 15, 1947. Enclosed was an attachment: V. Bush to Secretary of Defense, "Research and Development Effort for an Early Warning System for Defense of the United States," December 15, 1947. See "Project Lincoln," volume I, document 2.

36. Smith 1964, pp. 50–51 and p. 63, n. 90.

37. E. M. Powers to Cmdg. Gen., AMC, "Development Program for Air Defense Ground Electronic Systems and Equipment," February 1949. The six items were actually described in these words: (1) "IFF"; (2) "modernization and improvement of postwar radars not in production (i.e., higher powered 'S' and 'L' band components, etc.)"; (3) "Anti-Jamming Techniques"; (4) "New ground control radar with significantly improved performance (Volir)"; (5) "Automatic Interceptor Director System"; (6) "Automatic Controller Computer System." See "Project Lincoln," volume I, document 3; Smith 1964, pp. 50–51.

38. Redmond and Smith 1980, chapters 1–3 and p. 112.

39. Ibid., p. 26.

40. R. A. Nelson to Jay W. Forrester, "Meeting of Ad Hoc Panel on Air Defense," March 18, 1949, pp. 1–2. (A copy of this memo appears on p. 44 of Forrester's MIT Computation Book No. 49.)

41. Ibid., p. 2. Goode joined the University of Michigan as Director of its Willow Run Research Center and played a major role in the discussions with the Air Force regarding the university's participation in the air defense program. See Harlan Hatcher Papers, Michigan Historical Collections, Bentley Historical Library, University of Michigan, Ann Arbor, especially the following letters:

Goode to Cmdg. Gen., Rome Air Development Center, attn.: Mr. Warren Dunn, February 2, 1953; Goode to Hatcher, February 26, 1953; Goode to Hatcher, March 2, 1953 (with proposed draft of letter from Hatcher to Lt. Gen. E. E. Partridge setting forth the conditions under which Michigan would enter into competition with MIT for development of an air defense system).

42. For a reference to the meeting held on May 26 and 27, 1949, at the MIT Servomechanisms Laboratory, see Forrester's dictated note on p. 52 of his MIT Computation Book No. 49. A measure of the secondary importance of the air defense problem at this time, at least in Forrester's mind, is the prominent identification he gave the group as "Representatives of the RDB's Committee on Scientific and Synthetic Analysis" (the Air Defense Panel's parent group to which the panel reported).

43. Forrester, "Notebook Supplement 50JWF2," dictated May 4, 1949, MIT Computation Book No. 50.

44. Forrester, Notebook Memorandum 49JWF67, December 27, 1949, MIT Computation Book No. 49, p. 67. Memorandum L-15, "A Visit to: Engineering Research Associates . . . , Aircraft Radiation Laboratories, Wright Field . . . , General Precision Laboratories . . . ," January 6, 1950.

45. Valley to von Karman, November 8, 1949 (MITRE archives), p. 1. See also George E. Valley Jr., "How the SAGE Development Began," *Annals of the History of Computing*, 7 (1985), no. 3: 199–200.

46. Fowler, "History," volume XV, p. 9; Smith 1964, pp. 53–54; Sturm 1967, pp. 38–39; Scientific Advisory Board to the Chief of Staff, USAF, "Progress Report on the Air Defense System Engineering Committee," May 1, 1950, p. 1; Valley, "How the SAGE Development Began," p. 199.

Chapter 3

1. Source: interviews by the authors with Col. G. T. Gould (July 16, 1957), Norman Taylor (January 10, 1983), and Gen. A. R. Shiely Jr. (January 13, 1983). Additional source: remarks of James Killian and Jerrold Zacharias, Project Charles oral history videotape, session 1. (These four videotaped reminiscences regarding Project Charles were furnished to us through the courtesy of the sponsoring organization, the Alfred P. Sloan Foundation.)

2. Valley to von Karman, November 8, 1949, p. 1; Valley, "How the SAGE Development Began," *Annals of the History of Computing*, 7 (1985), no. 3, pp. 198–199.

3. Valley to von Karman, November 8, 1949, p. 1. The article, "Limitations of an Air Defense System," by Lt. Col. Harry M. Pike, had appeared as an "In My Opinion" piece in the latest issue of *Air University Quarterly* (3 (1949), no. 2: 46–48).

4. Pike, "Limitations."

5. Valley to von Karman, November 8, 1949, p. 2.

6. Ibid., pp. 2–3.

7. Futrell 1971, p. 143.

8. Fairchild to DCS/Comptroller, Personnel, Operations, and Material, "Air Defense Technical Committee of the Scientific Advisory Board," December 15, 1949. See also "Project Lincoln," volume 1, item 5; Sturm 1967, p. 39; Division 2 Quarterly Progress Report (henceforth cited as QPR), June 1952, p. v; Valley, "How the SAGE Development Began," p. 199.

9. Vandenberg, quoted on p. 82 of Sturm 1967.

10. Verbatim quotes, Sturm 1967, p. 34; see also pp. 4 and 32–35. As members of the Historical Office in the headquarters of the Air Research and Development Command during the latter half of the 1950s, the authors witnessed at first hand the extraordinary influence of the Ridenour Report on the Command and on USAF Headquarters in Washington. In addition, their discussions and interviews with those responsible for air defense R&D left no doubt about the concern over both the nation's vulnerability to air attack and the need for an effective defense.

11. Futrell 1971, p. 138. See also Getting 1989, pp. 348–352.

12. Fairchild to Commanding General, Air Materiel Command, Wright-Patterson Air Force Base, "Air Defense System Engineering Committee, Scientific Advisory Board to the Chief of Staff, U.S. Air Force," January 27, 1950, pp. 1–2; "Project Lincoln," volume 1, item 5.

13. Division 2 QPR, June 1, 1952, "Introduction," pp. v–vii. This QPR includes a "Memorandum on Activities of the Air Defense Systems Engineering Committee," written by George Valley and dated April 7, 1950, as pp. vi–xv. With some changes in wording and organization, Valley's memorandum of April 7, 1950 and the ADSEC Report of May 1, 1950 are the same. See Valley, "How the SAGE Development Began," pp. 203 and 211.

14. "Scientific Advisory Board to the Chief of Staff, USAF, Air Defense System Engineering Committee," in "Project Lincoln," volume I, item 5. See also "Memorandum on Activities of the Air Defense System Engineering Committee," Division 2 QPR, June 1, 1952, p. vi; Sturm 1967, p. 39; Smith 1964, p. 67.

15. Vice Chief of Staff, USAF, to Cmdg. Gen., Air Materiel Command, "Air Defense Engineering Committee, Scientific Advisory Board to the Chief of Staff, U.S. Air Force," January 27, 1950; 1st Indorsement (to letter of January 27, 1950 from Vice Chief of Staff to the Cmdg. Gen., AMC), Major General St. Clair Streett, Cmdg. Gen. AMC to VCS, USAF, February 7, 1950. (In military parlance, 'indorsement' means that the writer endorses or concurs with the letter to which it is attached.) See also Hayden to Cmdg. Officer, 3160th Electronics Group, AFCRL, "Cooperation with the Air Defense System Engineering Committee, S.A.B.," February 2, 1950. These documents are included in "Project Lincoln" as items 5 and 6. See also Smith 1964, p. 68; Division 2 QPR, June 1, 1952, pp. vi–vii.

16. Fowler, "History," volume XV, p. 8; Smith 1964, p. 68, p. 92 n. 2; Valley, "How the SAGE Development Began," pp. 198–199.

17. Valley, "How the SAGE Development Began," pp. 198–199.

18. Division 2 QPR, June 1, 1952, p. v. See also "Scientific Advisory Board to the Chief of Staff, USAF, Progress Report on the Air Defense Systems Engineering Committee, 1 May 1950," pp. 4–6; "Project Lincoln," item 9; Valley, "How the SAGE Development Began," pp. 203–206.

19. ADSEC, Report of May 1, 1950, pp. 4–5. "Memorandum on Activities of the Air Defense System Engineering Committee," Division 2 QPR, June 1, 1952, pp. vi–xv; Smith 1964, pp. 68–69; R. R. Everett, C. A. Zraket, and H. D. Benington, "SAGE—A Data-Processing System for Air Defense," in Proceedings of the Eastern Joint Computer Conference, Washington, December 1957, p. 148.

20. "Memorandum on Activities of the Air Defense System Engineering Committee," pp. v–vi, viii; Smith 1964, pp. 70–71.

21. ADSEC, "Air Defense Systems," Report of October 24, 1950, pp. 9–10; "Project Lincoln," volume I, item 12. See also Smith 1964, pp. 77–78.

22. "Memorandum on Activities of the Air Defense System Engineering Committee," pp. xii–xiii.

23. Ibid. Also see Valley, "How the SAGE Development Began," pp. 206–207.

24. R. M. M. Serrel, Astrahan, G. W. Patterson, and I. B. Pyne, "The Evolution of Computing Machines and Systems," *Proceedings of the IRE*, May 1962, p. 1056.

25. "Requirement for an Electronic Computer for the Army Air Forces." (Though no date or author, is given, internal evidence indicates that this was prepared before June 1947.)

26. "A General Survey of Computer Development in the United States," talk delivered to Symposium on Modern Computing Machinery and Numerical Methods, University of California, Los Angeles, July 29, 1948 (draft copy dated August 12, 1948), p. 33.

27. E. M. Powers to Cmdg. Gen., AMC, "Development Program for Air Defense Ground Electronic Systems and Equipment," February 8, 1949. "Project Lincoln," volume I, item 3.

28. Lavington 1980, pp. 54–56.

29. "Report by British Telecommunications Research Limited," Ministry of Supply contract No. 1 6/WT/4683/CB14B (n.d.), p. 27 and appendix F, "Semi-Automatic Air Raid Reporting System" (Air 20/7334, PRO). The contract was issued January 11, 1949. The report was to be completed no later than the following September 30. See Lavington 1980, pp. 53–56. For a brief description of British defense computers, see J. W. Forrester et al., "Forecast for Military Systems

using Electronic Digital Computers," Report L-3, MIT Servomechanisms Laboratory, September 17, 1948 (MITRE archives.)

30. "Memorandum on Activities of the Air Defense System Engineering Committee," QPR, June 1, 1952, pp. vi, x–xii.

31. J. W. Forrester, "Notebook Memorandum 49JWF77a," January 20, 1950, MIT Computation Book No. 49, p. 77.

Chapter 4

1. J. W. Wiesner, Project Charles videotape. See also Redmond and Smith 1980, p. 174; Valley, "How the SAGE Development Began," *Annals of the History of Computing*, 7 (1985), no. 3, pp. 207–208.

2. Forrester, "49JWF83a," dictated January 30, 1950, MIT Computation Book No. 49, p. 83.

3. Interview with J. W. Forrester and R. R. Everett by the authors, July 23, 1963; Redmond and Smith 1980, pp. 174–175.

4. The Barta Building, at 2ll Massachusetts Avenue in Cambridge, built in 1904 for an industrial cleaning company, had been home to Project Whirlwind since 1948. It is now occupied by Information Systems, MIT's main computing and networking infrastructure support organization.

5. Forrester, "49JWF84," dictated February 1, 1950, MIT Computation Book No. 49, p. 84; Valley, "How the SAGE Development Began,"pp. 208–210.

6. Forrester "49JWF86," dictated February 15, 1950, MIT Computation Book No. 49, p. 86.

7. Ibid.

8. Forrester, "49 JWF88a," dictated February 15, 1950, MIT Computation Book No. 49, p. 88. ADSEC met on February 17 at the Cambridge Research Laboratories.

9. Major General D. L. Putt to Cmdg. Gen., AMC, "Information on the Research Program of the Air Defense System Engineering Committee," March 28, 1950. This letter can be found in Fowler, "History" (volume XV, parts I and II, appendix 12). See also Putt to Cmdg. Gen., AMC, "Air Defense Systems Engineering Committee Project," April 7, 1950, cited in Smith 1964, p. 68; Fowler, "History," p. 10.

10. Hqs, AMC, to Directorates, Research and Development, Supply and Maintenance, Procurement and Industrial Planning, "Air Defense System Engineering Committee Project," Technical Instruction TI2205-10A, June 2, 1950. See Fowler, "History," appendix 8.

11. These recommendations were later outlined in ADSEC's first formal report: Scientific Advisory Board to the Chief of Staff, USAF, "Progress Report on the Air Defense System Engineering Committee," May 1, 1950. See "Project Lincoln," volume I, document 9; Smith 1964, pp. 73–74.

12. Putt to Cmdg. Gen., AMC, "Information on the Research Program of the Air Defense System Engineering Committee" (Hqs, AMC, TI2205-10A), March 28, 1950; Col. Gilbert Hayden to CO, AFCRL, "Research Program of the Air Defense System Engineering Committee (Valley Committee)," March 31, 1950; Fowler, "History," appendix 12.

13. TI2205-10A.

14. Robert E. Rader, "New E. O. Number assigned to Air Defense Systems Engineering Committee," Routing and Record Sheet, AMC, July 26, 1950;R. E. Rader, "Memorandum for: Dr. George E. Valley," October 31, 1950, with enclosure re telephone call from Archie Brown to Capt. Collier, October 27, 1950. (The original documents, or copies of them, are on file in ESDHO.)

15. Directorate of Historical Services, Hqs, Continental Air Defense Command, Hqs, ADC, "A Decade of Continental Air Defense, 1946–1956," July 1956 (USAF, Historical Division, Archives Branch), pp. 61–66.

16. Killian, Project Charles videotape, session 1.

17. John V. Harrington, "Radar Data Transmission," *Annals of the History of Computing* 5 (1983), no. 4: 370–374; Malcolm H. Hubbard, Charles Videotape, session 1; Valley, "How the SAGE Development Began," p. 206.

18. ADSEC Report, October 24, 1950, pp. 29–30; "Project Lincoln," volume I, item 12.

19. "Conclusions—Scientific Advisory Committee Meeting," January 1, 1951, in "Project Lincoln," volume I, item 14. See also Smith 1964, pp. 77–78.

20. Von Karman 1967, p. 303.

Chapter 5

1. Before the war, Bush had left MIT to become president of the Carnegie Institution in Washington. For more on OSRD and Bush, see Baxter 1946, p. 14 and passim.

2. See *Historical Statistics of the United States, Colonial Times to 1957* (U.S. Dept. of Commerce, Bureau of the Census, 1960), Chapter W, "Research and Development," pp. 609–614. When McGuire AFB's SAGE center was activated, military R&D had already topped $1.8 billion per year. The only fiscal year between 1947 and 1957 in which the R&D expenditures of the Department of Defense declined was 1950, the same year in which ONR, compelled to accept a

smaller budget, was determined to reduce its dollar support of the Digital Computer Laboratory's Project Whirlwind.

3. For details, see chapters 2 and 3 of Sapolsky 1990. See also "The Bird Dogs: The Evolution of the Office of Naval Research," *Physics Today* 14 (August 1961): 66–73; Redmond and Smith 1980, pp. 103–106; *Research in the Service of National Purpose: Proceedings of the Office of Naval Research Bicentennial Convocation* (Washington, 1966).

4. Sapolsky 1990, pp. 48–49.

5. Ibid., pp. 46, 102–107.

6. For detailed discussion of these issues, see Redmond and Smith 1980, esp. chapters 5 and 8–10.

7. Stratton's reflections took the form of an 18-page (double-spaced) memorandum addressed to President James Killian Jr. and to Thomas K. Sherwood, dean of the School of Engineering. The memorandum was dated February 3, 1950.

8. C. V. L. Smith to J. W. Forrester, August 26, 1949.

9. "Summary Report No. 20, Third Quarter, 1949," quoted in Redmond and Smith 1980, pp. 196–197.

10. For a variant perspective on these events, see Redmond and Smith 1980, chapter 10, esp. p. 151.

11. Redmond and Smith 1980, pp. 151–154. The report, issued by the Committee on Basic Physical Sciences of the Research and Development Board, wsa titled Report of the Ad Hoc Panel on Electronic Digital Computers (PS 13/5, December 1, 1949). Compare the later, more favorable Report on Electronic Digital Computers by the Consultants to the Chairman of the Research and Development Board (PS 13-8, June 15, 1950).

12. See Redmond and Smith 1980.

13. Redmond and Smith 1980, pp. 32–33. The "distinctive élan" that characterized the Whirlwind team was carried over into Lincoln when the team became Division 6 and in part contributed to the differences that arose between it and the other divisions within Lincoln Lab.

14. Late in 1949, Forrester discussed the Whirlwind computer's situation in two letters, written a week apart: a 600-word letter (dated November 25) addressed to Captain J. B. Pearson of ONR, with copies directed to Alan Waterman and C. V. L. Smith of ONR, and a 22-page single-spaced letter (dated December 2) to Pearson. In the earlier letter he provided a thoughtful analysis of the engineering prospects, the engineering problems, and the opportunities that lay ahead where input-output equipment and electrostatic internal storage were concerned. In the longer letter he went into extreme detail.

15. J. W. Forrester to Captain Pearson, 25 November 1949, p. 1.

16. See the following Project Whirlwind technical M-Notes, filed in the MITRE archives: M-951, "Bi-Weekly Report, December 9, 1949," p. 20; M-960, "Bi-weekly Report, December 23, 1949," pp. 1, 7, 19, 20.

Chapter 6

1. Project Whirlwind had set to work designing an electrostatic storage tube in 1946. For details, see Redmond and Smith 1980, esp. chapters 4 and 12. For a contemporary description of the Williams tube, see F. C. Williams and T. Kilburn, "A Storage System for Use with Binary-Digital Computing Machines," *Journal of the Institution of Electrical Engineers (London)* 96 (1949), part 2, no. 50: 183–202.

2. J. W. Forrester, "Data Storage in Three Dimensions," MIT Servomechanisms Laboratory Memorandum M-70, April 1947. Among the Whirlwind engineers to whom Forrester put the idea were Everett, Fahnestock, David Brown, Steve Dodd, Pat Youtz, Bill Nolan, and J. Ross MacDonald. For further discussion of the data-storage problem, see also J. W. Forrester, "Coincident-Current Magnetic Computer Memory Developments at MIT" (paper given at Argonne National Laboratory Computer Symposium, August 4, 1953) and Redmond and Smith 1980, p. 181ff.

3. W. N. Papian, "Rough Resume of Magnetic Core Memory History," November 24, 1953 (p. 1 of two-page memo to R. R. Everett, written on Digital Computer Laboratory Inter-office Correspondence stationery and summarizing laboratory research efforts up to that time).

4. J. W. Forrester, "Chronology of the MIT Magnetic Storage Work and Assistance to the IBM Corporation," May 24, 1960 (p. 1 of ten-page memorandum).

5. An Wang and Way Dong Woo, "Static Magnetic Storage and Delay Line," *Journal of Applied Physics* 21 (1950), January: 49–54. See also Wang 1986, pp. 55–60.

6. For more on the work of Booth and Rajchman, see Goldstine 1972, pp. 309–311.

7. Goldstine 1972, p. 212.

8. Forrester, "Chronology," pp. 1–2.

9. Papian ("Rough Resume," p. 1) recalled he went to work "during the summer of 1949" on his seminar paper. Forrester ("Chronology," pp. 1–2) recalled having selected Papian in the autumn of 1949 "to continue the development work on magnetic core storage." Forrester was favorably enough impressed by Papian's seminar work to engage him in the master's thesis research that followed.

10. W. N. Papian, master's thesis proposal, MIT Electrical Engineering Department, January 24, 1950 (also issued as Servomechanisms Laboratory technical memorandum M-960).

11. "Engineering Note E-362, Storage Devices in Computing Systems, Text of an MIT E.E. Department, Colloquium Talk," R. R. Everett to J. W. Forrester, May 3, 1950, 9 pp. with 7 full-page drawings. This E-Note was issued formally as a Project Whirlwind, Servomechanisms Laboratory, MIT report, to be circulated internally within the laboratory, as was their custom.

12. Ibid, pp. 6–9 and drawings at end.

13. Ibid., pp. 3–4.

14. Ibid., p. 9.

15. Ibid., p. 5.

16. See S. H. Dodd, H. Klemperer, and P. Youtz, "Electrostatic Storage Tube," presented to North Eastern District Meeting of American Institute of Electrical Engineers, Providence, R.I, April 28, 1950, p. 1.

Chapter 7

1. Sage to Compton, September 17, 1948. Two versions of the report were prepared, one 10 pages and one 22 pages in length; each was designated "Report L-3," but they bore different titles. These two reports used the same four graphs and two charts to present the heart of the analysis. The longer version, with Sage's cover letter, went to Compton under the title "Forecast for Military Systems Using Electronic Digital Computers" (Report L-3, Servomechanisms Laboratory, MIT, September 17, 1948). The shorter version, dated three days earlier, was titled "Report L-3, A Plan for Digital Information Handling Equipment in the Military Establishment, Prepared for Dr. Karl T. Compton" (Servomechanisms Laboratory, MIT, September 14, 1948). The authors of both reports were, as listed, Jay W. Forrester, Hugh R. Boyd, Robert R. Everett, Harris Fahnestock, and Robert A. Nelson. For more on Compton's request, see Redmond and Smith 1980, pp. 111–114.

2. Forecast for Military Systems, pp. 5–6.

3. Ibid., pp. 7–8.

4. N. M. Sage, Memorandum of Conference between Dr. Compton and Messrs. Killian, Forrester, Foster and Sage on Project Whirlwind, September 8, 1948.

5. Forecast for Military Systems, p. 8. See also figure 1, "Comparison between Radar and Computer Development," p. 24.

6. Ibid., p. 12. See also figure 6, "Digital Information Handling Program."

7. Forecast for Military Systems, pp. 12 and 22 and figures 2–6 at end.

8. Ibid., pp. 13, 18.

9. W. Gordon Welchman, Conference Note C-96, Explanation of the Applications Study Group, April 20, 1949, p. 1. For a description of Welchman's work during the war, see Welchman 1982.

10. F. L. Foster, MIT, to 3151st Electronics Station, Watson Laboratories, AMC, attn.: WLBCF-1, Mr. Donald W. Callender, December 31, 1948. See also memorandum, J. W. Forrester to Dr. F. L. Foster, D. I. C., subj.: "Research and Investigation of Electronic Computers in Air Traffic Control for Watson Laboratories," December 28, 1948. The proposal (a formal response to Watson Laboratories Purchase Request No. 64565, dated December 15) was the culmination of many weeks of planning, of informal discussions by Forrester with Watson Laboratories personnel, and of preparation of an earlier "Tentative Proposal—Air Traffic Control," dated November 6, 1948, which had been drawn up by Forrester and Welchman in response to Paper 27-48/DO-12, "Air Traffic Control," prepared by Special Committee 31 of the Radio Technical Commission for Aeronautics (RTCA), dated May 12, 1948. The RTCA, a "cooperative association of all United States Government–Industry aeronautical telecommunication agencies," officially accepted the paper on February 17, 1948. For evidence of Navy support for the MIT proposal, see Notebook Supplement 49JWF10 (December 29, 1948) on p. 10 of Forrester's MIT Computation Book No. 49.

11. Forrester to F. L. Foster, Research and Investigation of Electronic Computers in Air Traffic Control for Watson Laboratories, December 28, 1948.

12. Ibid., pp. 2–3.

13. Ibid., p. 3.

14. Ibid., pp. 5–6.

15. Forrester, Notebook Supplement 49JWF15, dictated January 17, 1949, MIT Computation Book No. 49, p. 15. For further details of the Crawford-Forrester relationship, see pp. 26–28 of Redmond and Smith 1980.

16. The instructions were provided in two appendixes to the quarterly Air Traffic Control reports: Summary Report 1, March 1–April 25, 1949, pp. 9–33, and Summary Report 3, July 25–October 25, 1949, pp. 18–33. Ten of these quarterly reports were submitted to Watson Laboratories during the period of the contract, which ended June 20, 1951. All of the first five were devoted solely and explicitly to air traffic control and bore the title Interim Engineering Report on Electronic Flight Calculator for Air Traffic Control.

17. Air Traffic Control Summary Report 1, p. 9.

18. Bi-Weekly Report M-2032, January 20, 1950, p. 1.

19. Forrester, Notebook Memorandum 49JWF75, January 19, 1950, MIT Computation Book No. 49, p. 75. A duplicate was also logged into Computation Book 50 on p. 8, following Welchman's "Common System" memo.

20. Summary Report 6, April 25, 1950 –July 25, 1950, foreword, p. 3.

Chapter 8

1. Alan T. Waterman, diary note, "with copies to Code 400, 409, 430; Code 101, 100, and 280," March 6, 1950 (located by J. P. Hastings in Box 2 of Federal Records Center, Alexandria, Virginia; Catalog No. 16372, folder labeled "Project Whirlwind Correspondence, 1950," from Computer Branch), p. 1.

2. Ibid.

3. Ibid., pp. 1–2; Summary of Conference on Trip taken by Dr. C. L. Smith, 6–7 March 1950, Distribution Codes 409, 402, 400 (located by J. P. Hastings in Box 4, Federal Records Center, Alexandria, Virginia, Catalog No. 7408, folder labeled "Daily Pinks, July 1950 thru 31 Dec. 1950"), , p. 3.

4. Summary of Conference on Trip taken by Dr. C. L. Smith, p. 1. Smith had spent the morning with Whirlwind engineers Norm Taylor and Gus O'Brien discussing an ONR digital computer and its magnetic-drum, internal-storage unit projected to be installed on the Berkeley campus of the University of California. By implication this was the sort of level of effort that ONR would be willing to support for Whirlwind; it was a less sophisticated and less expensive level than Whirlwind's electrostatic tubes incurred, and it avoided the "bugs" that electrostatic R&D was encountering.

5. Ibid., pp. 1–2.

6. Waterman, diary note, p. 2.

7. Smith, Summary of Conference, p. 2. The "Fink report" had been issued by the RDB Ad Hoc Panel the preceding December. Smith's report appears to have been dictated, and the dictating machine may have failed to catch "whole-heartedly."

8. Ibid., pp. 3–4.

9. Waterman, diary note, p. 2.

10. J. W. Forrester, entry 49JWF94, Computation Book No. 49, p. 94.

11. [J. W. Forrester], Reliability in Digital Computers. On the typescript draft copy that we found in Robert Everett's files, Forrester's initials and the date "3-10-50" are penned in the upper right corner. This "appendix A" version, 7½ pages long, is a part of Report L-21, Analysis of Digital Computer Laboratory Proposed 1950–51 Budget, by Jay W. Forrester, Robert R. Everett, Hugh R. Boyd, and Harris Fahnestock. Copy 4, which we consulted, comprises 74 unnumbered pages and three foldout sheets. All but the first five pages of the text are devoted to four appendixes, the first of which is the essay on reliability.

12. Ibid., p. 1.

13. Ibid.

14. Ibid., pp. 2–3.

15. Ibid., pp. 3–4. In 1950 the machine was moved to the Navy's Bureau of Ordnance Proving Ground at Dahlgren, Virginia.

16. Ibid., p. 4.

17. Ibid., p. 5.

18. Ibid.

19. Ibid., p. 6. For the circumstances of the invention of marginal checking, see Redmond and Smith 1980, pp. 80–81, 85–88.

20. Ibid., p. 7.

Chapter 9

1. Marchetti to Cole, March 21, 1950 (ASC-23-25, MITRE archives).

2. Compare the foreword of "Summary Report 5, January 25–April 25, 1950" with that of "Summary Report 6, April 25, 1950–July 25, 1950," submitted by Project DIC 6673, Air Traffic Control Project, Servomechanisms Laboratory, MIT, under Contract AF 28(99)-45. The work was officially redirected and continued under the same Air Force contract until April 1951. The quarterly Summary Reports continued to be submitted to the Watson Lab until July 1950, when they were diverted to the Air Force Cambridge Research Laboratories. On April 2, 1951, administrative control of these Laboratories was transferred from the Air Material Command to the recently created Air Research and Development Command, and they were redesignated the Air Force Cambridge Research Center. For more on "track-while-scan," see chapter 24.

3. "Summary Report 6," p. 7.

4. Division 6 QPR, June 1, 1952, pp. 6-6, 6-29, 6-30.

5. Division 2 QPR, June 1, 1952, p. 2-67.

6. "Summary Report 6," p. 7.

7. Ibid., pp. 4, 8. "The sub-program adds (or subtracts) an increment to the spot coordinates at a fixed rate. The effect is movement of the spot in either or both axes at a constant speed proportional to the increment. The size of the increment is controlled by means of a joy stick. The joy stick consists of a mechanical coordinate resolver which actuates switches. There are three switches for each of the four directions of motion (up, down, right, left), and this provides three spot speeds in each direction or any combination. The increment switches operate relays which control crystal gates in one of the flip-flop storage registers of WWI. In this way the joy stick sets the speed increment and inserts it directly into WWI." (ibid., pp. 8–9)

8. Summary Report 6, pp. 5–6.

9. "Summary Report 7, July 25–October 25, 1950," p. 1; see also the abstract preceding p. 1. This was the second of the four Air Defense quarterly reports submitted by the erstwhile Air Traffic Control Project of the MIT Servomechanisms Laboratory under redirected ATC Contract AF 29(099)-45. (The laconic titles of these four reports and of the "Final Report" covering the entire contract period from March 1, 1949 to June 30, 1950, give only the contract number and never identify explicitly what technical project they are reporting on.)

10. "Biweekly Report, Project 6673, August 4, 1950," M-2065, Electronic Computer Division, Servomechanisms Laboratory, MIT, pp. 2, 5.

11. M-2073, "Bi-Weekly Report, Project 6673, October 13, 1950," p. 1.

12. "Summary Report 7," pp. 3–7.

13. Ibid., pp. 1, 7. MTI: Moving Target Indicator.

14. Ibid., pp. 1, 2.

15. "A Perspective on SAGE: Discussion," *Annals of the History of Computing*, 5 (1983), no. 4: 391.

16. "Summary Report 8, October 25, 1950–January 25, 1951," p. 1 and abstract preceding p. 1.

17. M-2084, "Bi-Weekly Report, Project 6673, February 2, 1951," p. 8.

18. M-2087, "Bi-Weekly Report, February 16, 1951," p. 13.

19. "Summary Report 8," pp. 1, 3.

20. Ibid., p. 2.

21. Ibid., pp. 4–5.

22. Ibid., pp. 2–4, 6.

23. Ibid., p. 8.

24. Ibid, pp. 5, 7. "There is now in existence," according to Summary Report 8, "a collection of programs for sine, cosine, arc sine, arc cosine, arc tangent and square root, analyzed according to storage requirements, time requirements, and accuracy." The special difficulty they encountered in arriving at a method of approximation sufficiently accurate over the range of all values of the variable for the arc sine and arc cosine was resolved when they found a helpful program among those used on the EDSAC computer at Cambridge University.

25. Ibid., pp. 6, 9.

26. Ibid., pp. 9–10. Indeed, according to Summary Report 8, "a member of the analysis group is writing a doctor's thesis entitled 'Treatment of Digital Control Systems and Numerical Processes in the Frequency Domain,'" and "the method of

analysis that is being developed in this thesis has already thrown light on current problems."

27. Ibid., p. 10.

28. "Summary Report 9, January 25, 1951–April 25, 1951," p. 1 and passim.

29. Wieser and Welchman to 6673 Group, "Temporary Allocation of Responsibilities in Project 6673," M-2086, February 15, 1951, p. 1.

30. Ibid.

31. Ibid., pp. 2–4.

32. Ibid., p. 2.

33. "Summary Report 9," p. 2.

34. Memorandum 2090, "Bi-Weekly Report, Project 6673, March 30, 1951," p. 1. The words are those of Bob Wieser, who on March 1 had replaced Gordon Welchman as Group Leader upon the latter's departure to take a job with Engineering Research Associates in St. Paul. Forrester had announced Welchman's approaching departure to the laboratory in January (Administrative Memorandum A-113, January 26, 1951).

35. C. R. Wieser to J. W. Forrester, "Bedford Experiments, Bulletin #1, Week of April 2," M-2094, May 2, 1951, pp. 1–2.

36. Wieser to Forrester, "Bedford Experiments, Bulletin #2, Week of April 9," M-2095, May 2, 1951, pp. 1–2.

37. Wieser to Forrester, "Bedford Experiments, Bulletin #3, Week of April 16," M-2096, May 2, 1951, p. 1.

38. Ibid., pp. 1–2.

39. Wieser to Forrester, Experimental Interceptions with Bedford MEW Radar, M-2092, April 23, 1951, p. 1.

40. Ibid., pp. 1–2. See also "Summary Report 9," p. 3.

41. Servomechanisms Laboratory, MIT, "Summary Report 6, April 25, 1950–July 25, 1950," for Project DIC 6673 to Watson Laboratories, AMC, under AF Contract AF28(099-45), p. 2.

42. M-2098, "Bi-Weekly Report, May 11, 1951," p. 1; M-2183, "Division 6 Bi-Weekly Report, May 22, 1953," p. 6.

Chapter 10

1. Valley, "How the SAGE Development Began," *Annals of the History of Computing*, 7 (1985), no. 3, p. 212. For another personal account of the background to the SAGE program, see Getting 1989, pp. 223–244.

2. Vandenberg to James R. Killian Jr., December 15, 1950. See also Fowler, "History," volume XV, part 1, p. 10 and appendix 13; Smith 1964, pp. 84–85; Killian, Project Charles videotape, session 1.

3. Vandenberg to Killian, December 15, 1950.

4. For background information, see Maj. Gen. Donald L. Putt to McCone, "MIT Air Defense Laboratory," March 17, 1950. See also "Project Lincoln," volume I, item 19; Smith 1964, pp. 84, 88; Fowler, "History," volume XV, part 1, p. 10.

5. J. A. Stratton, Lecture Notes, Air War College, Montgomery, Alabama, May 22, 1953 (AC 4, MIT archives).

6. Stratton to Killian, October 20, 1950 (AC 4, MIT archives).

7. Killian, Project Charles videotape, session 1; Leslie 1993, pp. 32–43.

8. Killian, "To Members of the Staff," August 22, 1950.

9. Hubbard and Killian, Project Charles videotape, session 1; Killian to F. Wheeler Loomis, February 16, 1951; "Project Lincoln," volume I, items 14 and 16; Smith 1964, pp. 85–86; Valley, "How the SAGE Development Began," p. 213. For a brief discussion of the RLE's origins, activities, and impact, see Leslie 1993, pp. 25–32.

10. Basic Agreement AF18(600)-11, August 6, 1951, corrected May 5, 1952; Contract AF19(122)-458, January 30, 1951; N. M. Sage to Lt. Gen. E. E. Partridge, January 10, 1952; "Project Lincoln," volume 2, item 37; Howard R. Murphy, The Early History of The MITRE Corporation (unpublished draft, June 30, 1972; in MITRE archives), volume I, pp. 116–122

11. The center had been known first as the Cambridge Field Station, then as the Air Force Cambridge Research Laboratories.

12. Hubbard, Project Charles videotape, session 3.

13. Killian, Project Charles videotape, session 1; Problems of Air Defense, Final Report of Project Charles, August 1, 1951, volume I, pp. xvi–xx.

14. Killian et al., Project Charles videotape, sessions 1 and 2. See also Valley, "How the SAGE Development Began," pp. 213–214.

15. This session took place on February 23, 1951. See Problems of Air Defense, volume I, pp. xvi–xx; volume II, appendixes P-1 and P-2.

16. Problems of Air Defense, volume I, p. xviii; volume II, appendix P-2.

17. R. F. McMullen, Air Defense and National Policy, 1951–1957, Historical Study No. 24, Air Defense Command, USAF, p. 10. See also Problems of Air Defense, volume I, p. xvii.

18. Problems of Air Defense, volume I, pp. 1–10.

19. J. B. Wiesner, Project Charles videotape, session 3; Problems of Air Defense, volume I, pp. xxii–xxiv, 9–29.

20. Problems of Air Defense, volume I, pp. xi, xix.

21. Ibid., pp. xix–xx, 31. The "Quick Fix" program was discontinued early in 1953 when the Air Force rejected its recommendations and requested that its funds be applied to "more urgent" work. For further details, see "Conclusions—Scientific Advisory Committee Meeting," January 19, 1951; Division 2 QPR, July 15, 1953, p. 3; C. L. Grant, The Development of Continental Air Defense to 1 September 1954, USAF Historical Study No. 126, USAF Historical Division, Research Studies Institute, Air University, pp. 73–74; Valley, "How the SAGE Development Began," pp. 213–214, 222.

22. See Problems of Air Defense, volume I, p. xx. Volume I was a general discussion of air defense; divided into eight sections, it presented the deliberations on the dangers of attack, both immediate and future, and reviewed and summarized the various working committees' proposals as to how such threats could be thwarted. Volumes II and III contained more detailed administrative information regarding Project Charles's composition and procedures, and technical information regarding its findings and recommendations.

23. Problems of Air Defense, volume _, pp. 203–206.

24. Ibid., pp. xxvii–xxxii, 125–198, 209–212.

25. Ibid., pp. vi, 85.

26. Ibid., pp. 29–82, esp. p. 32.

27. Ibid., pp. xxiv–xxvi, 29–82, 85–122.

28. Ibid., pp. xxv, 88–93.

29. Problems of Air Defense, volume I, pp. 85–88; volume II, appendix IV-1, "Communications."

30. Problems of Air Defense, volume II, appendixes III-5 and IV-4. This GOC report was the result of a study made at the request of Project Charles by the Control Systems Laboratory at the University of Illinois. See also B. A. Harrison, R. Hobert, and L. R. Bruner, "Analysis of Ground Observer Corps—Participation in Air Defense Exercises—December 1951," Willow Run Research Center Report No. UMM-91, Project MIRO, Air Force Contract No. AF306029 (University of Michigan Engineering Library).

31. Problems of Air Defense, volume I, pp. 96–98; volume II, appendix IV-4.

32. Problems of Air Defense, volume I, pp. 98–100.

33. Ibid,, volume I, pp. 108–116; volume II, appendixes IV-6 and IV-7. See also Norman H. Taylor, "Marginal Checking as an Aid to Computer Reliability," Project Whirlwind Report R-178, March 28, 1950.

34. Problems of Air Defense, volume I, p. 117; volume II, appendix IV-5.

35. Problems of Air Defense, volume I, pp. x, 118–120.

36. Daniel E. Dustin and Jay W. Forrester, Project Charles videotape, sessions 4 and 3, respectively.

Chapter 11

1. Lincoln Steering Committee, minutes of meetings of April 7, 14, and 21, 1952; Murphy 1972, volume I, p. 23. For details of the name change, the Steering Committee's dissatisfaction with the impermanent connotation of the word 'Project', and the Air Force's suggestion that "Lincoln Laboratory" without "MIT" would be preferable, see the minutes of the Lincoln Steering Committee meeting of April 14, 1952, and the attached letter from Loomis to Killian. See also p. 2 of the minutes of the meeting of April 21, 1952, at which, upon hearing objections from Hill, Wiesner, and Zacharias, the committee concluded that MIT's name should appear on the nearly finished building in Lexington.

2. Lt. Gen. K. B. Wolf to Cmdg. Gens. Air Material Command and Air Research and Development Command, "Emergency Mobilization of Scientists and Engineers for Air Force Research and Development," February 19, 1951.

3. Ibid.

4. AFDRD-EL to Cmdg. Gens. AMC and ARDC, "MIT Air Defense Laboratory Policy," March 29, 1951, enclosure 2, "Proposed General Policy for the Establishment of the Air Defense Laboratory (MIT Contractor)." For this and other relevant documents, see "Project Lincoln," volume I, item 20.

5. Vandenberg to VCS, "Organization for Research and Development in the USAF," October 12, 1950; Ridenour to von Karman, "Interim Report on Activities of the Special Working Group, SAB," May 7, 1951; Ridenour, memorandum for the record, August 7, 1951. See also "Minutes of the 6th July 1951 Meeting of the Special Working Group of the SAB," cited on p. 43 of Sturm 1967.

6. Fowler, "History," volume XV, part I, p. 11.

7. "Charter for the Operation of Project Lincoln," July 26, 1951. See also Fowler, "History," appendix 18; "Project Lincoln," volume I, item 20.

8. Brig. Gen. D. N. Yates to Cmdg. Gen., ARDC, "Project Lincoln." See also Fowler, "History," appendix 21; interview with Col. G. T. Gould Jr. by the authors, July 16, 1957. It is quite possible that the failure was not an oversight.

9. Schedule to Contract No. AF19(122)-458. See also "Project Lincoln," volume II, item 37. The reference to "(A)" in item 2 concerns the statement to be found in the contract schedule (not copied here) that immediately precedes this list.

10. "Charter for the Operation of Project Lincoln," July 26, 1951. See also Fowler, "History," appendix 18; "Project Lincoln," volume I, item 20.

11. "Charter for the Operation of Project Lincoln," July 26, 1951. See also Fowler, "History," appendix 18; "Project Lincoln," volume I, item 20. See also F. W. Loomis, "Proposed Statement of Task for the Bedford Air Defense Laboratory," June 4, 1951; "Project Lincoln," volume I, item 20.

12. For "Minutes for the Project Lincoln Advisory Committee Meeting of 24 September 1951," see Fowler, "History," appendix 21. Committee members included representatives from the Air Force (Brig. Gen. D. N. Yates, Principal and Chairman; Col. B. K. Holloway, Alternate and Recorder), from the Army (Col E. R. Petzing, Principal; Col. G. F. Moynahan, Alternate), and from the Navy (Rear Adm. C. N. Bolster, Principal; Capt. E. O. Wagner, Alternate).

13. See Fowler, "History," appendix 20, "Proposed Program for Project Lincoln," August 6, 1951, and appendix 21, "Minutes for the Project Lincoln Advisory Committee Meeting of 24 September 1951." See also Loomis to Forrester, October 19, 1951, and minutes of Lincoln Steering Committee meeting of October 29, 1951. The major divisions of Lincoln Laboratory by the end of the first year were as follows: Division 1, "Business Management and Services" under Malcolm M. Hubbard. Division 2, "Aircraft Control and Warning" under George E. Valley Jr. Division 3, "Communications and Components" under Albert G. Hill. (Hill was replaced in March 1952 by W. H. Radford, when Hill became Deputy Director of Lincoln pending Loomis's departure in July.) Division 4, "Weapons" under William M. Pease. Division 5, "Special Projects" under Carl F. J. Overhage. (Overhage was replaced by W. H. Radford in July 1952, when Overhage returned to Eastman Kodak.) Division 6, "Digital Computer" under J. W. Forrester and R. R. Everett. Division 7, "Engineering Design and Technical Services" under Joseph A. Vitale. In July 1952, problems associated with the growth of the Lab (the total number of Lincoln personnel had risen from 85 in October 1951 to 1175 in July 1952) caused Division 7 to split off from Division 1. See A. G. Hill to Lincoln Laboratory personnel, "Administrative Bulletin No. 33—Administrative Changes in the Lincoln Laboratory," July 9, 1952. Loomis realized that it made administrative sense for Forrester's group in the MIT Digital Computer Laboratory in late October 1951 to formally become Division 6, even though they remained housed in the Barta Building until space became available at Lincoln Laboratory.

14. Lincoln Steering Committee meeting, March 3, 1952; LSC meeting, May 5, 1952. Several major personnel changes occurred when Hill succeeded Loomis as Director of Lincoln Lab in July: Radford, who had been a Group Leader in MIT's Research Laboratory of Electronics, replaced Hill as Chief of Division 3. Zacharias became head of Division 5, replacing Overhage, and Smullin became head of Division 4, replacing Pease. Hubbard was relieved of his divisional administrative duties to assume broader responsibilities on "matters of policy concerned with all phases of the Laboratory's operations," and P. Cusick, Fiscal Officer of the Division of Industrial Cooperation, replaced him as Chief of Division 1, while Division 1 was split and Division 7 was added under J. A. Vitale. Rollefson left later that summer, to return to the University of Wisconsin. For further details, see A. G. Hill, "Administrative Bulletin No. 33," July 9, 1952; LSC meeting, July 7.

15. Killian to Members of the Staff, January 2, 1953 (AC 4, MIT archives).

16. LSC meeting, May 21, 1956. It is not clear what happened to the Division of Defense Laboratories; presumably its functions were taken over by Cochrane's office.

17. See minutes of LSC meetings, March 17, 1952, July 7, 1952, October 31, 1955, September 16, 1957, November 8, 1957, and April 12, 1958.

18. Lt. Gen. K. B. Wolfe to Cmdg. Gens. AMC and ARDC, "Emergency Mobilization of Scientists and Engineers for Air Force Research and Development," February 19, 1951; Brig. Gen. D. N. Yates to Cmdg. Gen. ARDC, "Project Lincoln," August 27, 1951.

19. Yates, "Project Lincoln," August 27, 1951.

20. 1st Ind. (Yates to Cmdg. Gen ARDC, August 27, 1951), Brig. Gen. J. W. Sessums Jr. to Cmdr., AFCRC, September 7, 1951. (Sessums was "indorsing" the action proposed in Yates's letter and ordering its implementation.)

21. LSC meeting, September 12, 1951, with attachments: "Project Lincoln Budget" and "Supplementary Information Concerning Programs Now Planned for Project Lincoln," September 10, 1951.

22. LSC meeting, October 8, 1951.

23. LSC meetings, October 8 and December 10, 1951.

24. LSC meeting, December 10, 1951.

25. M. M. Hubbard, "Meeting in Dr. Killian's Office, December 11, 1951," January 7, 1952, attached to a note from Hubbard to the President's Office, January 7, 1952 (AC-4, MIT archives).

26. Ibid.

27. See Horace S. Ford to Maj. Gen. D. L. Putt, March 4, 1953, in "Project Lincoln," volume I, item 62. For a historical parallel involving MIT, Project Whirlwind, and the Navy in 1948, see Redmond and Smith 1980, chapters 7 and 8, esp. pp. 126–127.

28. LSC meeting, December 17, 1951.

29. Loomis to Killian, December 21, 1951. For letters from Loomis to Killian, Killian to Secretary Finletter, and Finletter to Killian, see minutes of LSC meetings of December 17and December 31, 1951.

30. Killian to Thomas K. Finletter, December 21, 1951.

31. Yates to Loomis, December 26, 1951 (AC-4, MIT archives). In view of the speed with which these events were moving, it is reasonable to assume there was informal telephone contact.

32. Loomis to Yates, January 3, 1952 (AC 4, MIT archives).

33. Finletter to Killian, February 5, 1952. See also "Project Lincoln," volume II, item 33. For detailed comments on Lincoln's role in the air defense program, see "Minutes of the Advisory Committee of Project Lincoln, 11 February 1952" ("Project Lincoln," volume II, item 34).

34. Early in 1953 the progress of technical and budgetary events caused the Air Force to cancel its funding of the "quick fix" modifications that Lincoln proposed, identify them as "operationally undesirable," and request that both manpower and funds be diverted to "more urgent work." For details, see "Conclusions—Scientific Advisory Committee Meeting," January 19, 1951; Division 2 QPR, July 15, 1953, p. 3; C. L. Grant, The Development of Continental Air Defense to 1 September 1954, USAF Historical Study No. 126, USAF Historical Division, Research Studies Institute, Air University, pp. 73–74.

35. "Supplementary Information Concerning Programs Now Planned for Project Lincoln," attachment to minutes of LSC meeting of September 12, 1951.

36. Finletter to Killian, February 5, 1952.

37. LSC meeting, March 3, 1952.

38. LSC meeting, March 31, 1952.

39. LSC meeting, April 7, 1952.

40. LSC meeting, July 21, 1952.

41. J. A. Giordano, interview with the authors, January 15, 1982.

42. Jay W. Forrester and Robert R. Everett, interview with the authors, January 11, 1983. See Redmond and Smith 1980, pp. 108–128, 144–167.

43. LSC meeting, January 11, 1952.

44. LSC meetings, October 15, 1951, March 17, 1952, November 18, 1952, January 27, 1953, and May 2, 1955.

45. LSC meeting, November 19, 1956. The source documents cited in this brief discussion of funding are only a part of the materials available. Readers who wish to pursue the subject in greater depth should consult the minutes of the meetings of the Lincoln Steering Committee *seriatim*, since these are replete with discussions of contemporary and future budgets.

46. Hq ARDC to all Centers, TWX RDCB-6-E, June 29, 1957. For further information and documentation, see Study of Air Power, Hearings before the Subcommittee on the Air Force of the Committee on Armed Services, U.S. Senate, 84th Congress, 2d session (Government Printing Office, 1956). See also Research and Development (Office of the Secretary of Defense), 32d Report by the Committee on Government Operations, 85th Congress, 2d session, House Report Nr. 2552, Union Calendar Nr. 1070 (Government Printing Office, 1958).

47. LSC meetings, May 13 and July 15, 1957.

48. LSC meeting, September 27, 1957.

49. LSC meeting, October 7, 1957.

50. LSC meetings, April 18, April 25, and May 2, 1958.

Chapter 12

1. Jay W. Forrester, MIT Computation Book No. 49, entry of September 28, 1951, p. 139. See also "Change in Laboratory Name," Digital Computer Laboratory Administrative Memorandum A-124, September 20, 1951; The Digital Computer Laboratory of the Massachusetts Institute of Technology, DCL Report R-199, September 20, 1951.

2. Lincoln Steering Committee, minutes of meeting of September 12, 1951, pp. 3–4. Attending this first meeting were F. Loomis (presiding), G. Brown, A. Hill, M. Hubbard, C. Overhage, R. Rollefson, and G. Valley. An indication of the LSC's management philosophy is to be found in the LSC's decision that "one copy of the CHARLES report is to be made available to each LINCOLN group leader. . . . All staff members will be urged to acquaint themselves with this report." On the ONR-Whirlwind funding story, see Redmond and Smith 1980, esp. chapters 5, 7, 8, and 10.

3. Forrester, MIT Computation Book No. 49, entry of September 28, 1951, p. 139.

4. Ibid. See also LSC meeting, September 21, 1951, p. 2.

5. LSC meeting, October 2, 1951, p. 2.

6. Loomis earlier in the meeting had reported that "Dr. Getting had recently suggested a half-day visit to Raytheon should be made by a number of the people from Project Lincoln," and the members had agreed to send "Loomis, Forrester, Hubbard, Overhage, Rollefson, Pease, Valley, Van Voorhis." In the meantime, the LSC should realize that the proposal to build ten computers "urged by Griggs and Ridenour . . . will be vigorously pushed by a committee under Getting."

7. LSC meeting, October 2, 1951, p. 3. "Forrester is satisfied that the key personnel has adequate competence for the new effort," the minutes conclude.

8. Jay W. Forrester and Robert R. Everett, Digital Computers for Air Defense System, DCL Report L-30, October 5, 1951, pp. 1–2.

9. Ibid., p. 2.

10. Ibid., figure 1: "Times when Indicated Activities Predominate. Air Defense Computer."

11. Ibid., figure 2: "Personnel for Air Defense Computer Design—Not including WWI or Cape Cod System."

12. LSC meeting, October 8, 1951, pp. 2–3.

13. LSC meeting, October 29, 1951, with Loomis's "Draft Letter" to Forrester attached. See also "6889 Air Defense Bi-Weekly, October 26, 1951," DCL Memorandum M-1314.

14. By January 1952, three engineering groups were listed—Group 61: Cape Cod. Group 62: Whirlwind II. Group 63: Magnetic Materials. The first Division 6 Quarterly Report listed three more groups in June 1952—Group 60: Administration and Services. Group 64: Whirlwind I Computer. Group 65: Storage Tube Development. Between early 1955 and mid 1957, three more groups were added. These reflected the changing needs of the division as its program became larger and more complex, advancing from computer design and development through manufacturing and programming to testing and installation—Group 66: Production Coordination Office. Group 67: Program Production. Group 68: System Office. See Jay W. Forrester to Digital Computer Staff, "Division VI Organization List," DCL Memorandum M-1366, January 10, 1952 (AC-11, MITRE archives); M-1314; "6889 Air Defense Bi-Weekly, October 26, 1951"; Division 6 Report L-34, "Group Leaders' Meeting, March 26, 1952"; Division 6 QPR, June 1, 1952. QPR, March 15, 1955, p. 63. QPR, March 15, 1956, p. 73. QPR, June 15, 1957, p. 87.

15. Division 6 QPR, June 1, 1952, p. 6-139.

16. Valley (1985, p. 210) recalls that, after a March 1950 meeting at which, speaking for ADSEC, he had "agreed that the Air force would support Whirlwind's budget," he had been "snubbed in the halls of MIT by a personage very high in its administration." For Stratton's temperate reflections, see pp. 11–12 of his memorandum to President Killian and Dean Sherwood, "Project Whirlwind," dated February 3, 1950 (AC-12, MIT archives).

17. Division 6, "Group Leaders' Meeting, August 30 1954," Report L-164, August 31, 1954; John C. Proctor, interview with the authors, July 13, 1964; Valley, "How the SAGE Development Began," *Annals of the History of Computing*, 7 (1985), no. 3, pp. 214–217.

18. Jay W. Forrester, Division VI Program, July 1952–June 1953, Report L-32, January 7, 1952; Brief Summary of Activity Planned by Project Lincoln during F/53 (PLA-126), January 9, 1951 (AC-15, MITRE archives).

19. Forrester, "Staff Organization," Administrative Memorandum A-111 to all personnel of Electronic Computer Division, Servomechanisms Laboratory, December 11, 1950.

20. J. W. Forrester, R. R. Everett, and N. H. Taylor, Visit to University of Illinois and Engineering Research Associates, M-1152, January 22, 1951, p. 3. (The actual author of this six-page technical note with five graphs attached was apparently Taylor; only his signature appears at the end.)

21. Forrester to Valley, January 29, 1951, pp. 1–3.

22. Ibid., pp. 1–2.

23. Ibid., pp. 2–3.

24. Ibid., pp. 3–4.

25. John von Neumann to Valley, "Memorandum on the Aircraft Sorting and Tracking Problem," April 27, 1951.

26. R. R. Everett, entry dated "5-10-51," MIT Computation Book No. 32, p. 10.

27. R. R. Everett, "Air Intercept Using A Whirlwind Type Computer," Engineering Note E-2022, May 25, 1951. Everett's introductory abstract includes the following sentences: "It appears that by the addition of auxiliary drums and special operations to facilitate sorting and coordinate conversion, a Whirlwind type computer should be able to handle the problem, including intercept calculation, within a 15 second scan time. The following somewhat rambling report has been taken without polishing from a notebook (32 RRE 10-23) analysis made about 2 weeks ago. The problem considered is that described by Von Neumann in his report of April 28, 1951."

28. Ibid., pp. 1–2.

29. Ibid., p. 2.

30. Ibid.

31. Ibid., pp. 2–3.

32. Ibid., pp. 4–9.

33. Ibid., pp. 10–11.

34. Norman H. Taylor, "Visit to the 1101 Computer," M-1243, July 11, 1951, pp. 1–2; Forrester to Valley (copies to Everett and Taylor), August 23, 1951. See also N. H. Taylor, "Possible Uses of Magnetic Drums," M-1245, July 11, 1951, p. 1. Because M-1245 presented significant and useful new information, Taylor addressed it to eleven other project members: W. N. Papian, E. S. Rich, J. A. O'Brien, C. W. Watt, C. R. Wieser, R. R. Everett, H. Fahnestock, C. W. Adams, D. Israel, J. W. Forrester, R. P. Mayer. Presumably a copy was also filed in the laboratory library.

Chapter 13

1. Forrester, "Staff Organization," Administrative Memorandum A-111 to all personnel, Electronic Computer Division, Servomechanisms Laboratory, December 11, 1950. Both Steve Dodd and Norm Taylor, as Group Heads, would be reporting directly to the "Project Supervisor" (explicitly Forrester); implicitly, by long informal practice, they would be reporting to Everett.

2. For a record of the November 1951 meetings, see the following M-Notes: B. E. Morriss, Meeting on Air Defense Computer, M-1318, November 5; B. E. Morriss, Second Meeting on Air Defense Computer, M-1319, November 6; B. E. Morriss,

Third Meeting on Air Defense Computer, M-1320, Nov. 7; B. E. Morriss, Fourth Meeting on Air Defense Computer, M-1321, November 8; D. R. Brown and B. E. Morriss, Summary of the First Four Meetings on Air-Defense Computer, M-1327, November 9; B. E. Morriss, Fifth Meeting on Air Defense Computer, M-1322, November 9.

3. For more on Morriss, see the following M-Notes: M-1206, p. 14, M-1215, p. 19; M-1219, p. 17; M-1233, p. 15.

4. For more on Walquist, see M-2064, pp. 1–2.

5. On the Cape Cod Muldar (multiple radar), see E-2023, June 22, 1951.

6. See M-2032, p. 2; M-2063, July 21, 1950.

7. B. E. Morriss, Meeting on Air Defense Computer, M-1318, November 5, 1951, p. 1; D. R. Brown and B. E. Morriss, Summary of the First Four Meetings on Air-Defense Computer, M-1327, November 9, 1951, pp. 1–2.

8. Morriss, M-1318, p. 1; Brown and Morriss, M-1327, pp. 1–2.

9. M-1318, pp. 1–6.

10. M-1319, pp. 1–2.

11. M-1319, pp. 3–7. A simple schematic block diagram was appended to this M-Note.

12. Morriss, Fourth Meeting on Air Defense Computer, M-1321, November 8, 1951, pp. 1–3. ("Mr. Israel said that with 1000 aircraft, 5000 conversions to *x,y* coordinates which now requires 40 orders, 1000 inverse tangent solutions which now requires 40 orders, and 1000 square root solutions which now require 20 orders would be necessary in the program for each scan of the radar. This would be 260,000 orders of which the *x,y* conversion is by far the largest part.")

13. M-1320, p. 4.

14. M-1321, p. 3.

15. Ibid., pp. 3–4.

16. Ibid., p. 4.

17. Ibid.

18. Brown and Morriss, Summary of First Four Meetings on Air-Defense Computer, M-1327, November 9, 1951, pp. 1–2.

19. M-1322, p. 2-3.

20. N. H. Taylor, Whirlwind II Meeting of February 1, 1952, M-1397, February 6, 1952, p. 1. The number 6889, assigned by MIT's Division of Industrial Cooperation to work done for the air defense computer, appeared at the top of each of the technical "M-Notes" generated by the air defense computer project.

21. Taylor, Whirlwind Meeting of February 8, 1952, M-1402, February 18, 1952, pp. 1–2. The nine engineers were Everett, Fahnestock, H. Grosch, W. Hosier, J. Jacobs, R. Jeffrey, W. Linvill, W. Papian, and Taylor.

22. The name was truthful, as the engineers viewed it, but its implications to newcomers and outsiders were nevertheless pejorative—how could one speak sensibly about half a machine?—and a month later the name was changed to WW IA. See N. H. Taylor and R. P. Mayer, Whirlwind II Meeting of February 15, 1952, M-1407, issued February 26, 1952, p. 1; N. H. Taylor and R. P. Mayer, Whirlwind II Meeting of March 21, 1952, M-1432, issued March 25, 1952, p. 1.

23. M-1407, pp. 1–2.

24. M-1397, p. 1; M-1407, pp. 2–4.

25. M-1407, p. 4.

26. Taylor and Mayer, Whirlwind II Meeting of March 14, 1952, M-1428, issued March 18, 1952, p. 1. Eighteen project members were present. For the intervening meetings of February 29 and March 7, see M-1417 and M-1420.

27. M-1428, pp. 1–2.

28. Ibid., p. 2.

29. Ibid., p. 4.

30. Ibid.

31. Taylor and Mayer, Whirlwind II Meeting of March 21, 1952, M-1432, issued March 25, p. 1.

32. N. H. Taylor and R. P. Mayer, Whirlwind II Meeting of April 4, 1952, M-1435, issued April 9, 1952, p. 3. Twenty-two project members were in attendance.

33. N. H. Taylor to J. W. Forrester, Time Schedule for WWII Computer, M-1444, April 3, 1952, p. 1.

34. M-1444, pp. 2–3.

Chapter 14

1. Whirlwind II Block Diagrams Meeting of April 8, 1952, M-1449, p. 1. The sixteen meetings held between April 8 and June 3, 1952, were reported in the following M-Notes: M-1449 (first meeting, April 8), M-1457 (second meeting, April 10), M-1466 (third and fourth meetings, April 16 and 17), M-1469 (fifth meeting, April 22), M-1475 (sixth meeting, April 24), M-1481 (seventh meeting, May 1), M-1486 (eighth and ninth meetings, May 6 and 8), M-1504 (tenth and eleventh meetings, May 13 and 15), M- 1506 (twelfth and thirteenth meetings, May 20 and 22), and M-1511 (fourteenth through sixteenth meetings, May 27, May 29, and June 3). W. A. Hosier wrote up all the minutes except those of the fifth meeting,

which were reported by R. P. Mayer. All these reports were formally directed to "WWII Planning Group."

2. M-1449, p. 3.

3. M-1457, pp. 1–2. For technical details on Olsen's matrix switch, see M-1282.

4. M-1466, p. 1.

5. M-1469, p. 3.

6. N. H. Taylor, R. P Mayer, and W. Papian, "Whirlwind II Meeting of April 25, 1952," M-1495, issued May 21, 1952, p. 1. Twenty-three project members were present.

7. Digital Computer Laboratory Bi-Weekly Reports M-1092 (September 1, 1950), p. 6; M-1112 (October 13, 1950), p. 4; M-1119 (October 27, 1950) p. 5; M-1125 (November 10, 1950), p. 4; M-1136 (December 8, 1950), p. 6.

8. Bi-Weekly Reports M-1136 (December 8, 1950), p. 6; M-1202 (April 13, 1951), p. 4.

9. J. W. Forrester, "Chronology of the MIT Magnetic Storage Work and Assistance to the IBM Corporation," May 24 1960, pp. 1–3. This ten-page document, prepared for reference during patent litigation, is headed simply "Massachusetts Institute of Technology, Memorandum."

10. Forrester to Valley, October 2, 1950, pp. 1–2; Forrester to Everett and Papian, October 2, 1950; Forrester to Leslie K. Gulton, Glenco Corporation, October 2, 1950.

11. Papian to Everett, "Rough Resume of Magnetic Core Memory History," interoffice correspondence, November 24, 1953, p. 1. A marginal handwritten note in the author's copy reads: "I believe Hazen first suggested ferroelectrics. RRE." See also Forrester, "Chronology," p. 3; Bi-Weekly Reports M-1219 (May 25, 1951), p. 8; M-1228 (June 8, 1951), p. 6; M-1233 (June 22, 1951), p. 5.

12. Bi-Weekly Reports M-1215 (May 11, 1951), pp. 6–7; M-1219 (May 25, 1951), pp. 7–8.

13. Page 5 of Bi-Weekly Report M-1233 (June 22, 1951) cites Engineering Note E-406, "Preliminary Tests on the Four-Core Magnetic-Memory Array." See also Papian, "Rough Resume," p. 2; Everett, "Selection Systems for Magnetic Core Storage," Engineering Note E-413, August 7, 1951, p. 1.

14. Bi-Weekly Reports M-1241 (July 6, 1951), p. 5; M-1269 (August 31, 1951), p. 9.

15. Bi-Weekly Reports M-1255 (August 17, 1951), p. 8; M-1269 (August 31, 1951), p. 9.

16. An electric-circuit diagram showing the principle of operation of the crystal-matrix switch (signed "D.R.B" and dated 5/23/47) was included as Figure 1 in Olsen's master's thesis proposal, which also appeared as M-1282, "A Multi-position

Magnetic Switch and Its Incorporation into a Magnetic Memory" (September 21, 1951) (MC 140, Box 5, MIT archives). Olsen's thesis, completed the following May, was supervised by Everett (M-1282, p. 7). See also Divison 6 QPR, June 1, 1952, pp. 6-81–6-87; Redmond and Smith 1980, p. 216.

17. Bi-Weekly Report M-1278, September 14, 1951, pp. 6, 8.

18. M-1282 (Olsen's thesis proposal), p. 5.

19. Bi-Weekly Reports M-1278 (September 14, 1951), p. 5; M-1290 (September 28, 1951), pp. 7–8.

20. Bi-Weekly Reports M-1336 (November 23, 1951), p. 7; M-1348 (December 7, 1951), p. 7; M-1364 (January 4, 1952), p. 6.

21. M-1364, p. 6.

22. Bi-Weekly Report M-1378 (January 18, 1952), pp. 7–8.

23. Bi-Weekly Report M-1386 (February 1, 1952), p. 8.

24. Bi-Weekly Report M-1426 (March 14, 1952), p. 8.

25. M-1495, p. 1.

26. Forrester to Whirlwind II Logical Planning Group, "Boundary Conditions for Whirlwind II Design," M-1468, April 29, 1952, p. 1; M-1495, p. 1.

27. M-1468, pp. 1–2.

28. Ibid., p. 1.

29. M-1475, p. 2.

30. Ibid., pp. 3–4.

31. M-1481, p. 1.

32. Ibid., pp. 2–3 (May 13 meeting) and pp. 4–5 (May 15 meeting).

33. N. H. Taylor and R. P. Mayer, "Whirlwind II Meeting of May 9, 1952," M-1498, issued May 23, 1952, p. 2. Twenty-seven project members were in attendance.

34. N. H. Taylor and W. Mosier, "Whirlwind II Meeting of May 16, 1952," M-1500, issued May 23, 1952, p. 1. Twenty-six project members attended.

35. Ibid., p. 2.

36. Ibid.

37. Ibid., p. 3.

38. Ibid., p. 5.

39. Both the attendance and the membership of these planning groups were flexible and informal. Men were expected to attend in order to exchange knowl-

edge and views, not to constitute a quorum. There was no casting of votes. Everyone knew who was in charge. During the formative weeks between July 25 and August 22, Olsen and Taylor attended all of the six meetings on record, Everett missed only one, and Forrester attended only one. Attendance ranged from six at the MTC Power Supply meeting (M-1592) to thirteen at two of the meetings. "Regulars" attending at least four of the meetings included (in addition to Everett, Taylor, and Olsen) W. A. Hosier, R. Hughes, A. W. Ogden, H. K. Smead, Papian, and R. von Buelow. The earliest WWIA Planning Group "Distribution List" on record is that of July 7, 1952 (M-1547); it included Everett, Taylor, Olsen, R. L. Best, Ogden, Papian, Smead, and von Buelow. Six attended the August 12 meeting and thirteen the August 4 and August 22 meetings. The appropriate "MTC Meeting" notes are, in sequence: M-1566 (July 25meeting), M-1605 (August 4), M-1592 (August 11power supply meeting), M-1601 (August 15) M-1602 (August 18), M-1611 (August 22).

40. W. A. Hosier to WWIA Planning Group, "Initial Decisions on WWIA Block Diagrams," M-1547, July 7, 1952, p. 1. See also the following reports of these later Whirlwind II meetings (all but one of which—see M-1562—Olsen attended): M-1518 (May 23, 1952 meeting), M-1585 (June 13, 1952 meeting), M-1562 (July 18, 1952 meeting), M-1570 (July 25, 1952 meeting), M-1640 (September 12, 1952 meeting).

41. For a detailed account of Mayer's and Papian's work as presented at the WWII Planning Group meeting of July 18, 1952, see pp. 2–8 of M-1562.

42. M-1495, pp. 1–4.

43. Ibid., pp. 4–5.

44. Ibid., p. 8. According to M-1585 (p. 4): "Briefly, the z axis refers to the operation of a digit plane driver for inhibiting the writing of a '1' whenever a '0' is desired."

45. M-1585, pp. 4, 6. Papian's detailed analysis of the "signal-to-noise ratio" problem can be found on these pages.

46. M-1562, pp. 7–8.

47. M-1547, pp. 1–3; Taylor to Forrester, "Memory Test Computer," Inter-Office Correspondence, July 21, 1952 (AC 23-1, MITRE archives; AC 140, MIT archives). For a detailed description tied to Mayer's preliminary block diagram of the computer, see pp. 2–7 of M-1562.

48. M-1562, p. 2.

49. M-1518, p. 2.

50. Everett, "Logical Design," Division 6 QPR, June 1, 1952, section 3.2, pp. 6-73–6-75.

51. Ibid.

52. For further details and a complete roster of names, see QPR, *Division 6*, June 1, 1952, pp. 6-6, 6-7, and passim.

Chapter 15

1. "Group Leaders' Meeting, June 9, 1952," Report L-46, issued June 11, 1952, p. 2. On March 26, 1952, Forrester had begun to call the DCL's Group Leaders to weekly meetings in order to discuss common administrative problems and share significant technical information. According to page 1 of "Group Leader's Meeting, March 26, 1952" (Report L-34): "Those attending were Forrester, Everett, Fahnestock, Wieser, Taylor, Dodd, Youtz, and Adams. . . . Since Brown was absent from the first meeting he was unanimously chosen for Secretary . . . and will prepare minutes each week."

2. "Group Leaders' Meeting, June 23, 1952," Report L-49, issued June 24, 1952, p. 2. See also "Group Leaders' Meeting, July 14, 1952" (Report L-53, issued July 15, 1952), p. 2; J. W. Forrester to James Birkenstock, July 24, 1952; Birkenstock's reply, July 30, 1952. Describing IBM's involvement as seen from an IBM perspective, Pugh (1984, p. 93) cites his interviews with J. C. McPherson (June 11, 1982) and Birkenstock (May 25, 1982).

3. L-46, p. 2.

4. Lincoln Steering Committee minutes, July 21, 1952, p. 1. As Forrester explained later that year in a letter to P. V. Cusick (November 5), it had become "desirable to bring an industrial manufacturer into the program to assist . . . with the completion of the development and to prepare manufacturing and maintenance information" needed for production.

5. LSC minutes, July 21; Forrester to Cusick. See also Forrester to Cusick, "Background related to selection of IBM to work with MIT on Whirlwind II Digital Computer Development," memorandum, November 5, 1952; Forrester to A. G. Hill, "Selection of a Company to Work with the Lincoln Laboratory on the Transition System," Report L-95, May 12, 1953, p. 1.

6. L-95, p. 1; Forrester to Birkenstock, July 24, 1952.

7. L-95, p. 1; E. E. Minett to D. L. Gunter, "Lincoln Computer," RCA Company Memorandum, August 4, 1952 (MC 140, Box 7, MIT archives). Compare Minett to L. F. Jones, "Competitive Information on International Business Machines," March 23, 1951 (MC 140, Box 4, MIT archives).

8. L-95, pp. 1–2; "Group Leaders' Meeting, August 18, 1952," L-57, issued August 18, 1952, p. 2; "Group Leaders' Meeting, August 25, 1952," L-58, issued August 27, 1952, pp. 2–3.

9. Birkenstock to Forrester, July 30, 1952.

10. L-95, pp. 1–6; Forrester to Cusick, "Background to selection of IBM to work with MIT on Whirlwind II Digital Computer Development," November 5, 1952; "Total of Ratings by Forrester, Everett, Taylor, and Wieser made Aug. 22, 1952."

11. On May 8, 1953, Taylor provided Forrester with "Notes Regarding Trip to Raytheon to Evaluate Its Possible Participation in the MIT Program," "Notes Regarding Trip to Remington Rand to Evaluate Its Possible Participation in the MIT Program," and "Notes on the Choice of IBM as Participant in the MIT Program" (RRE files, Box ASC-23, MITRE archives).

12. Lincoln Steering Committee minutes, August 25, 1952, p. 3.

13. Forrester to Cusick, November 5, 1952, pp. 1–2.

14. Birkenstock, quoted on p. 94 of Pugh 1984.

15. For other accounts, see Stern 1981, pp. 146–148; Erwin Tomash, "The Start of an ERA: Engineering Research Associates, Inc., 1946–1955," in Metropolis et al. 1980, pp. 492–495; Augarten 1984, pp. 162–164, 182–184.

16. Forrester, "Technical and Administrative Notes," entries for April 14, May 5, and May 12, 1953, MIT Computation Book No. 53 (MC 140, Box 81, MIT archives).

17. Forrester to Cusick, November 5, 1952; L-95, May 12, 1953.

18. Forrester to G. E. Valley, "Investigation of Lincoln Subcontract to IBM by the House of Representatives—Committee on Appropriations—October 26, 1955," in Forrester, MIT Computation Book No. 54, p. 26. See also "Re: The Semi-Automatic Ground Environment System (SAGE) [and] The Distant Early Warning System (DEW)," Flatley Report, January 1956, Memorandum for the Chairman, House Appropriations Committee, from Charles G. Haynes, Director, Surveys and Investigations, January 14, 1956, p. 42.

19. Watson and Petre 1990, pp. 203–207, 232, 243.

20. A. Kromer to N. H. Taylor, "Contract with IBM re WWII," Inter-Office Correspondence, Servomechanisms Laboratory, September 19, 1952, p. 1. See also "Research and Development Services in Connection with a Digital Computer and Related Terminal Equipment," an undated, anonymous, seven page, typewritten document (ASC 23-1, MITRE archives), which begins: "As a part of a new improved digital air defense system being developed by MIT, IBM has been requested to submit a proposal for research, development, and construction of prototype models of the central digital information processing system and to make plans for production and field maintenance of air defense equipment. To insure the proper integration of this equipment in the air defense system being developed by the MIT Lincoln Laboratory under Contract AF 19(122)-458, it is understood that the Lincoln Laboratory will exercise technical direction over this program until prototype models are approved for production. This technical direction from MIT shall include the establishment of design parameters, determination of functional performance and operating characteristics, as well as determining any necessary design changes and modifications in experimental or prototype equipment." The document then describes a program of eleven "phases," presents a proposal "to perform research and development services under technical direction of MIT," identifies six such services, and offers detailed

specifications for a "cost-plus-a-fixed-fee form of contract, with appropriate priority rating." The document closes with a request for "more specific engineering data regarding specifications . . . in order to project estimated cost beyond the scope of the above proposal" and indicates the need for prior agreement regarding disposition of patents. See also Kromer to Forrester, "Discussion of contract status with IBM," Report L-66, October 17, 1952, which details a discussion Forrester held with IBM regarding the "Michigan ACDS system" and "the present status of negotiations regarding a contract for IBM to work with MIT."

21. Kromer to Taylor, "Contract with IBM re WWII," September 19, 1952, p. 2.

22. Youtz to Everett, "Williams System of Electrostatic Storage," Inter-Office Correspondence, September 4, 1952.

23. Everett to Forrester, "IBM Defense Calculator," Inter-Office Correspondence, Servomechanisms Laboratory, March 13, 1951 (MC 140, Box 4, MIT archives).

24. Pugh 1984, pp. 28–33; Augarten 1984, pp. 191–193; Watson and Petre 1990, pp. 203–207; Cuthbert C. Hurd, "Computer Development at IBM," in Metropolis et al. 1980; James W. Birkenstock, "Preliminary Planning for the 701," *Annals of the History of Computing* 5 (1983), no. 2: 112–114; Nathaniel Rochester, "The 701 Project as Seen by Its Chief Architect," *Annals of the History of Computing* 5 (1983), no. 2: 115–117; "Review—Type 701 Electronic Data Processing Machine Programs," undated IBM publication (Box ASC-23, MITRE archives).

25. For a personal account of the perceived importance to IBM of the contract for manufacturing the SAGE air defense computer, see pp. 230–233 of Watson and Petre 1990.

26. "Discussion of Contract Status with IBM," DCL Memorandum L-66, October 17, 1952, pp. 1–2; Lincoln Steering Committee, minutes, October 20, 1952, item 4; Forrester, "Technical and Administrative Notes," MIT Computation Book No. 53, pp. 1–2. (Page 1 of the last item is actually a two-page entry by Forrester concerning the possible use of the 701 and contractual negotiations with IBM. Page 2 is actually a five-page "Memorandum for Record" by Arthur P. Kromer, dated October 17, 1952, on contract discussions held with IBM the preceding day, during which the use of the 701 was brought up by Forrester. The memorandum was essentially the basis for L-66.)

27. Ibid. A number of other documents fill in the picture: R. P. Mayer to N. H. Taylor and M. M. Astrahan, "Preliminary Notes on the Logical Design of the IBM 701 Computer," DCL Inter-Office Correspondence, September 26, 1952 (MC 140, Box 8, MIT archives); R. P. Mayer to M. M. Astrahan and N. H. Taylor, "Notes on the Logical Design of the IBM 701 Computer," DCL Engineering Note E-487, October 8, 1952; J. W. Sheetz, "Rough Comparison of the Whirlwind II versus the IBM 701 Systems," October 8, 1952 (MC 140, Box 8, MIT archives); C. Corderman, N. Daggett, S. Dodd, R. Everett, J. O'Brien, N. Taylor, and R. von Buelow to J. W. Forrester, "Comments on the IBM 701 Computer," L-68, November 4, 1952. These reports bore the caveat "Commercially Confidential," and their circulation at MIT was restricted. See Harris Fahnestock to All Staff and

Secretaries in Group 62, Librarian, A. Falcione, J. Ulman, and R. Rathbone, "IBM Industrially Classified Information," DCL Memorandum M-1663, October 3, 1952; Fahnestock to N. H. Taylor, S. H. Dodd, K. Olsen, R. von Buelow, N. Daggett, J. Arnow, and J. A. O'Brien, "Dissemination of Information on IBM 701," M-1677, October 16, 1952 (MC 140, Box 8, MIT archives). A visit to IBM late in November to discuss crystal diodes, plug-in circuits, and production problems associated with the 701 provides an example of the quality of the technical information being exchanged, even though this trip was made after the verdict was in: see C. W. Watt and B. B. Paine to N. H. Taylor, "Visit to IBM, Poughkeepsie, N. Y.," M-1740, December 3, 1952 (AC-18, MITRE archives).

28. L-68, pp. 5, 8, 11–13, and passim.

29. "Group Leaders' Meeting, October 20, 1952," L-67, October 23, 1952, p. 3. The 14-page report referred to in this document is L-68.

30. Pugh 1984, pp. 94–95.

31. Kromer to Forrester, Everett, and all Group Leaders, "Engineering Relationships with IBM," M-1653, September 30, 1952, pp. 1–2. The assignments, including those not yet filled, were as follows. Design Engineer: N. H. Taylor (MIT), J. M. Coombs (IBM). Liaison Engineer: A. P. Kromer (MIT), T. A. Burke (IBM). Logical Design and Block Diagrams: R. P. Mayer (MIT), M. M. Astrahan and E. H. Goldman (IBM). Programming: C. R. Wieser (MIT), M. M. Astrahan (IBM). Arithmetic Element and Transistors: J. F. Jacobs (MIT), D. J. Crawford (IBM). Vacuum Tube Circuitry: R. L. Best (MIT). Magnetic Materials Development: D. R. Brown (MIT), L. P. Hunter (part time, IBM). Terminal Equipment, including Drums, Tapes, Cards: W. E. Triest and J. E. Heywood (IBM). Control Equipment: K. H. Olsen (MIT), H. D. Ross (IBM). Power Supplies: K. H. Olsen (MIT). Standards: A. P. Kromer (MIT), J. Goetz (part time, IBM). Mechanical Design and Packaging: A. P. Kromer (MIT). For later MIT and IBM personnel assignments, see the following M-Notes: A. P. Kromer to J. W. Forrester, H. Fahnestock, All Group Leaders, and All Group 62 Staff Personnel, "Field of Work," M-1892, March 9, 1953; A. P. Kromer to All Group Leaders and Group 62 Staff Members, "IBM Project High Engineering Organization," M-2347, August 10, 1953.

32. Kromer to Forrester, Brown, Everett, Fahnestock, Wieser, and Section Heads of Group 62, "Schedule of Design Objectives, WWII," M-1661, October 2, 1952, p. 1. Five copies were sent to IBM.

33. Report L-30, October 5, 1951.

34. Taylor to Forrester, "Time Schedule for WWII Computer," M-1444, April 3, 1952, pp. 1–3.

35. M-1661, p. 1.

36. M-1444, pp. 1–3.

37. For an IBM view, see Pugh 1984, especially chapter 4.

38. DCL Memorandum M-1661 consisted of only one page of text and a two-page bar graph.

39. M-1661, pp. 1–3.

40. David R. Brown, "Group Leaders' Meeting, October 20, 1952," L-67, issued October 23, 1952, p. 3. Brown served as secretary at the meetings. Compare L-30 and M-1444.

41. Taylor to Forrester, Best, Brown, Everett, Frost, Kromer, and Youtz, "Plans for a trip to study reliable tube production," Inter-Office Correspondence, Digital Computer Laboratory, October 23, 1952, p. 1. See also "Summary of IBM-MIT Collaboration, October 27, 1952 to November 30, 1952 inclusive," M-1739, December 3, 1952, p. 1. The plants listed included General Electric's at Owensboro, Kentucky, Sylvania's at Emporium, Pennsylvania, and RCA's at Harrison, New Jersey. IBM participated in this trip to the extent of 23 man-days and MIT to the extent of 27 man-days. See also "Minutes of Joint MIT–IBM Conference Held at Hartford, Connecticut January 20, 1953," M-1810, January 25, 1953, p. 10.

Chapter 16

1. Fahnestock to all Staff members [of] Digital Computer Laboratory, "Liaison with IBM Corporation," M-1696, October 28, 1952.

2. Kromer to Forrester, Everett, Fahnestock, Wieser, Taylor, Brown, Dodd, and Youtz, "Summary of IBM-MIT Collaboration, October 27, 1952 to November 30, 1952 inclusive," M-1739, December 3, 1952, p. 2. Twenty-three of these monthly collaboration summaries, all written by Kromer to Forrester, Everett, and the Division 6 Group Leaders, spanned the period from November 1952 through December 1954. Then a new Coordination Committee of MIT and IBM representatives began to meet weekly, as reported in Kromer's final collaboration memorandum (6M-3270, January 7, 1955). Also, a Production Coordination Office was set up in February 1955 (see 6M-3371), and Group 66 was created in Division 6. For a complete list of the 23 collaboration summaries issued in chronological order as internal Laboratory M-Notes, see the bibliographic note at the end of this chapter, where each summary is identified and work loads are given in man-days.

3. Kromer to Taylor, "Conference between IBM and MIT engineering groups," DCL Inter-Office Correspondence, November 7, 1952.

4. Ibid.

5. Kromer, Mayer, and Taylor to Whirlwind II Planning Group, "Design and Construction Schedule, WWII Prototype," M-1753, December 11, 1952. The meeting, held on December 5, appears to have been the last meeting of the Whirlwind II Planning Group, which had flourished during 1952 but which now was dying a natural death as a consequence of the expanding operations with IBM.

6. Ibid, pp. 1–2.

7. Ibid., pp. 2–4. (Page 3 is a bar graph of the schedule.)

8. Ibid., pp. 4–5.

9. Kromer to all conferees, Fahnestock, Wieser, Brown, Dodd, and Youtz, "Minutes of Joint MIT–IBM Conference Held at Hartford Connecticut January 20, 1953," M-1810, January 26, 1953 (AC-18, MITRE archives), pp. 1–2. Those attending included the following. From MIT: J. W. Forrester, R. R. Everett, N. H. Taylor, A. P. Kromer, H. E. Anderson, R. L. Best, J. F. Jacobs, R. P. Mayer, D. J. McCann, J. H. McCusker, K. R. Olsen, W. N. Papian, R. von Buelow. From IBM (Poughkeepsie): J. M. Coombs, T. A. Burke, M. M. Astrahan, D. J. Crawford, N. P. Edwards, J. A. Goetz, E. H. Goldman, B. Housman, R. L. Palmer, H. D. Ross, B. L. Sarahan, W. H. Thomas, C. W. Watt. From IBM (New York): G. Solomon.

10. Ibid., p. 2. For other accounts of the meeting by conferees who were there, see Morton M. Astrahan and John F. Jacobs, "History of the Design of the SAGE Computer —The AN/FSQ-7," IBM Research Report No. RJ3117 (38413), April 10, 1981, subsequently published under the same title in *Annals of the History of Computing* (5 (1983), no. 4: 340–349); Jacobs 1986, chapter 13.

11. M-1810, p. 2.

12. Ibid., p. 3.

13. Ibid., p. 4: "Analysis of the types of operations encountered in the Air Defense problem indicates that the computer time is divided roughly as follows: Addition and similar operations 50%[;] Multiply and similar operations 10%[;] Divide and similar operations 1%[;] Shift operations 5%[;] Transfer operations 30%[;] Subprograms 20%." (Since some of these operations could be carried on simultaneously by different parts of the machine, the percentages totaled more than 100%.)

14. Ibid.

15. Ibid., pp. 4–5.

16. Ibid., pp. 5–6: "At the moment, the following are considered as the circuits that will have to be developed. (Those marked * have only little known about them. Those marked ** represent complete unknowns.) *Pulse Logic*[:] High-speed flip-flop (20 volts)[,] Cathode follower[,] Gate tube[,] Buffer amplifier[,] Pulse standardizer[,] Delay unit[.] *Diode Logic*[:] *Flip-flop (40 volts)[,] * Cathode follower[,] *Diode circuits[,] *Inverter[,] *Gate standardizer[.] *Memory*[:] *Current pulse generator[,] * Sensing amplifier[.] *In-Out Logic*[:] Flip-flop[,] Magnetic core circuits[,] Phone line coupling circuits[.] *In-Out Equipment*[:] *Decoder (digital to analog converter)[,] *Scope deflection amplifier[,] **Character generator[,] **Drum circuits[,] **Tape circuits[,] *Photoelectric pickup circuit[,] Contacts."

17. Ibid., pp. 6, 10. IBM Report IM-9 dealt with four phases of the standardized-circuits problem; IM-10 provided a preliminary list of preferred tube types.

18. M-1810, p. 7. Astrahan, in his comments, referred specifically to IBM Reports IM-6, IM-7, and IM-8, already issued.

19. M-1810, pp. 7–8.

20. Ibid., p. 8.

21. Ibid., pp. 8–9.

22. Ibid., p. 9.

23. Ibid., pp. 10–11. ("The schedule for this work is covered in more detail in Memorandum M-1753 issued by the Digital Computer Laboratory.") See also Jacobs 1986, pp. 31, 55–56: ". . . at that time, about 20 percent of that number were assigned."

24. M-1810, pp. 11–12.

Chapter 17

1. "Summary of IBM-MIT Collaboration, January 1, 1953 to January 31, 1953 inclusive," M-1817, February 3, 1953, pp. 1–2; Kromer to Forrester et al., "Summary of IBM-MIT Collaboration, February 1, 1953 to February 28, 1953 inclusive," M-1891, March 9, 1953, pp. 1–2; "Summary of IBM-MIT Collaboration, March 1, 1953 to March 31, 1953 inclusive," M-1955, April 3, 1953, pp. 1–2; "Summary of IBM-MIT Collaboration, April 1, 1953 to April 27, 1953 inclusive," M-2125, April 30, 1953, pp. 1–2; "Summary of MIT-IBM Collaboration, April 28 through May 30, 1953 inclusive," M-2221, June 8, 1953, pp. 1–3.

2. E. H. Goldman to all conferees, Zollinger, and Kromer, "Minutes of Joint MIT-IBM Conference Held at Hartford, Connecticut March 17, 1953," IBM Memorandum, March 20, 1953, pp. 1–5 (AC-18-l, MITRE archives). See also Jacobs 1986, p. 57.

3. Kromer to all conferees, Fahnestock, Wieser, Brown, Dodd, and Youtz, "Minutes of Joint MIT-IBM Conference Held at Hartford, Connecticut April 21, 1953," M-1998, April 23, 1953.

4. "MIT Lincoln Laboratory Subcontract No. 20, Prime Contract No. 19(122)458, Report of Completion Article I Task 1 (b)," April 27, 1953 by Project High, p. 2 (AC-18, MITRE archives).

5. "Minutes of Basic Circuits Meeting at Hartford April 30, 1953" (MC 140, Box 9, MIT archives).

6. M-2191, "Meeting on Packaging of Whirlwind II, Hartford, Connecticut, May 21, 1953," May 27, 1953, pp. 1–3; "Joint Meeting on Packaging of WWII," M-2246, June 17, 1953, pp. 1–2; Astrahan and Jacobs, "History of the Design of the SAGE Computer: The AN/FSQ-7," p. 345.

7. Astrahan and Jacobs, "History of the Design of the SAGE Computer," p. 345; Jacobs 1986, chapter 13.

8. M-1891, March 9, 1953, p. 2.

9. M-2125, pp. 1–2; Kromer to Forrester et al., "Summary of MIT–IBM Collaboration, June 1 through June 30, 1953 inclusive," M-2272, July 3, 1953, p. 1. For further details of the arithmetic element, see the following monthly collaboration summaries: M-1817 (January 1953), M-1891 (February), M-1955 (March), M-2221 (May).

10. M-1891, p. 2; M-1955, p. 2.

11. M-2125, p. 2; M-2221, p. 2; Astrahan and Jacobs, "History of the Design of the SAGE Computer," p. 345.

12. M-2221, p. 2.

13. Ibid. See also Forrester to Solomon, May 21, 1953 (ASC 23-15, MITRE archives), p. 1. (Forrester wrote more than one letter to Solomon on this date; the letter here cited begins: "We are in general agreement with the proposal, 'A Program for Research, Development, and Pilot Production of Magnetic Cores. . . ." Forrester's penned initials, "JWF," appear at the end.)

14. Forrester to Solomon, May 21, 1953, p. 1.

15. Ibid., pp. 1–2.

16. Ibid., p. 2.

17. Ibid.

Chapter 18

1. Kromer to all conferees, Fahnestock, Wieser, Brown, Dodd, and Youtz, "Minutes of Joint MIT-IBM Conference Held at Hartford Connecticut January 20, 1953," Digital Computer Laboratory Memorandum M-1810, January 26, 1953 (AC-18, MITRE archives), p. 10.

2. Kromer to Forrester, "Suggested items for consideration by Mr. Forrester during visit to IBM on Thursday, June 11, 1953," Inter-Office Correspondence, June 9, 1953, p. 1. In a covering memo on Taylor's file copy (MC 140, Box 10, MIT archives), Kromer wrote: "At Mr. Forrester's request attached was prepared for his consideration in connection with his trip to IBM. . . ."

3. Ibid., p. 1.

4. Ibid., p. 2.

5. Ibid.

6. "Division 6 Biweekly Report, June 19, 1953," M-2257, p. 36.

7. J. A. O'Brien and K. E. McVicar to Forrester et al., "Visit to IBM to Observe Work on Magnetic Drums, April 16 and 17, 1953," M-2223, June 8, 1953, pp. 1, 5. "There was very little discussion of electronic circuits to be used in the drum

system, and our impression was that they have done relatively little work in this direction. Somewhere along the line, mention was made of the fact that they have been using a condenser discharge writing circuit having a maximum writing speed of 1 KC and using 5687 tubes." (p. 5.)

8. Astrahan and Jacobs, "History of the Design of the SAGE Computer—The AN/FSQ-7," p. 345.

9. Ibid., pp. 345–346.

10. J. W. Forrester and R. R. Everett, interview with the authors, January 11, 1983.

11. N. H. Taylor to I. Aronson et al., Inter-Office Correspondence, June 22, 1953, pp. 1–2. Taylor envisioned covering the following topics on Wednesday, Thursday, and Friday: slowed-down video, video mappers, power supplies, marginal checking, magnetic drums (all phases of basic circuits as well as logic). On Wednesday, Thursday, and Friday of the next week, at Poughkeepsie, they would cover input registers, marginal checking again, and magnetic memory, and wind up with a further discussion of magnetic drums. The list "is only about 50% complete," Taylor added.

12. The findings of the seven meetings were reported in the following M-Notes: M-2266 (1st Day), June 24 (14 from MIT, 8 from IBM); M-2267 (2nd Day), June 25 (20 from MIT, 12 from IBM); M-2268 (3rd Day), June 26 (11 from MIT, 8 from IBM); M-2283 (4th Day), June 30 (8 from MIT, 10 from IBM); M-2284 (5th Day), July 1 (11 each from MIT and IBM); M-2285 (6th Day), July 2 (11 from MIT, 12 from IBM); M-2359 (7th Day), July 15 (10 from MIT, 18 from IBM). According to the records, additional "Project Grind" meetings were held on August 30 and October 6, 1953. The minutes of the August 30 meeting were identified by IBM Control Number CD # 904-88-I, those of the October 6 meeting by CD # 904-149 I. (See MITRE Archives AC-131-1.) At the sixth session (July 2), the engineers agreed to meet biweekly, beginning July 15 at Cambridge; however, the only records available of meetings held after July 15 are the August and October minutes cited above. If other meetings were held and minutes were recorded by IBM personnel, those minutes could have been filed in the IBM archives, which are not open to researchers. Even Emerson Pugh of IBM, in his book *Memories That Shaped an Industry*, cites only the Division 6 minutes for the first meeting (M-2266).

13. "Lincoln Steering Committee, Meeting of June 8, 1953, PLSCM-57," item 6, p. 2. It had become obvious that "Project Lincoln," which MIT originally had agreed to carry on for the Air Force, was in a position to take on additional technical air defense projects; accordingly, "the phrase 'Project Lincoln' has been replaced by 'Lincoln Laboratory'" (item 4, p. 2).

14. Kromer and Mayer to AN/FSQ-7 Planning Group, "Project Grind Meeting of June 24, 1953 (First Day)," M-2266, June 29, 1953, p. 1; "Project Grind Meeting of July 15, 1953 (Seventh Day)," M-2359, August 13, 1953, p. 1.

15. Taylor, Inter-Office Correspondence, June 22, 1953, p. 2.

16. The report of each meeting listed the attendees by name and indicated if any member was present for only part of the meeting. See "A. P. Kromer, R. P. Mayer to AN/FSQ-7 Planning Group, Project Grind Meeting of June 24, 1953 (First Day)," M-2266, June 29, 1953, p. 1; Division 6 Biweekly Report, M-2280, July 3, 1953, pp. 37–40; Division 6 Quarterly Progress Report, September 15, 1953, p. 78. Listed among "members present" at meetings of Project Grind were (according to the M-Notes reporting the meetings) R. R. Everett, N. H. Taylor, J. F. Jacobs, R. P. Mayer, and K. H. Olsen from MIT, and J. M. Coombs, M. M. Astrahan, D. J. Crawford, and H. D. Ross from IBM. Coombs attended the last four meetings part time. Crawford attended meetings 4, 5, and 7 part time. Ross attended the seventh meeting part time. R. P. Crago of IBM missed only the third meeting. A. P. Kromer attended only the first two.

17. "Project Grind Meeting of July 15, 1953 (Seventh Day)," M-2359, August 13, 1953, p. 1.

18. The exercise was well named. The thirteen major areas that the engineers examined in detail broke down into 31 topics for separate discussion, a few of which consisted of further discussions on the same subject, e.g., drums, power supplies, crosstelling. The 31 topics were taken up in the following order at the seven meetings: *First Day* (M-2266): Slowed-down video inputs; Video mappers; Slowed-down video input registers. *Second Day* (M-2267): Marginal checking; Power supplies; Magnetic memory. *Third Day* (M-2268): Magnetic drums. *Fourth Day* (M-2283): Output display system; Manual-input buffer drum. *Fifth Day* (M-2284): Cross-telling; Output drums; Output links; Digital information display; Maintenance console; Mechanical design. *Sixth Day* (M-2285): Vacuum-tube types; High-speed flip-flop; High-speed gate-tube circuit; Diode circuits and cathode followers; Pulse lengths; Low-speed flip-flop; Cathode-ray tubes; Miscellaneous circuits; Standards Committee. *Seventh Day* (M-2359): Mapper Subcontract; Crosstelling; A Review on Drums; Card and Paper Tape Machines; Review of Input Counters; Manual Inputs; Power Supplies.

19. "Project Grind Meeting of July 15, 1953 (Seventh Day)," M-2359, August 13, 1953. See also "Division 6 Biweekly Report, July 17, 1953," M-2311, p. 37.

20. M-2266, pp. 1–2.

21. M-2267, p. 1. Ten of the MIT engineers attended part time, as their specialized expertise was required. Five of the IBM engineers attended part time.

22. "Project Grind Meeting of June 25, 1953 (Second Day)," M-2267, June 29, 1953, pp. 4–5; "Division 6 Biweekly Report, July 3, 1953," M-2280, pp. 38, 42.

23. Ibid.

24. There should be a "standby off" provision that would turn off all the DC voltages but leave the filaments burning. "Normal off" would shut down all the voltages, DC and AC, in sequence, while an "emergency off" button would cut all

power abruptly, except that "each piece of auxiliary equipment will have its own power supply." Voltage and temperature detectors would be provided to warn the human operators of any variations either above or below the norms, and if the operators took no action while the condition worsened, then automatic sequencing of "power off" would occur. The operators should also receive a warning signal from a "spike" detector on the power supply.

25. M-2359, pp. 4–6.

26. M-2280, p. 38.

27. Ibid. According to M-2267: "Memorandum M-2202 states the existence and function of the marginal checking group. Although further study need be made by this group, it is desirable at the moment to obtain some idea of how to break down marginal checking lines so that electronic design of the logical progress can proceed and so the power supplies can be worked out."

28. M-2268, p. 1; M-2283, p. 1. Though all eight of the MIT engineers were present throughout the fourth day, four of the ten IBM engineers were present part time.

29. M-2268, p. 1: "It was agreed that there should be 6 fields per physical drum, with 32 data tracks per field plus 2 status tracks per field (except in some cases), with probably 2 heads per track except for the auxiliary memory fields. There will be 62 spots per circumferential inch, 20 tracks per axial inch (0.050 inch spacing), and the tracks will be 0.020 inches wide. There will be 204 tracks per physical drum (including timing), and space for 408 heads. There will be 2,048 spots per revolution, using an 'open' timing track (not quite filling the circumference). A study of the use of drum fields must be made before deciding on the number of fields and spares required. It has been suggested that 5 physical drums may be desirable, providing eight spare fields." See also M-2283, pp. 3–4.

30. M-2283, pp. 1–3. Of the ten IBM engineers present, four were there only part time. The character of the approach taken to design and performance problems involving the output display system is reflected in the following excerpt from the report of the meeting, written by Art Kromer and Rollin Mayer: "It is estimated that it will take two seconds to display all tracks, flight plans, history points, uncorrelated data, and geography (assuming 50 microseconds per point, 25 microseconds per vector, and 100 microseconds per character). This does not include "DID" [Digital Information Display] (see Project Grind, Fifth Day). This figure of 2 seconds will be assumed in the rest of the discussion since there seems to be little advantage in displaying different categories at different rates. Thomas and Clark will prepare a program for MTC and WWI . . . for the purpose of studying phosphors, filters, psychological aspects of display rates, etc. . . . (The display test program must provide data randomly distributed over the scope face and over the various time intervals.) . . . Ed Rich will train an IBM man concerning light guns." (ibid., p. 2.)

31. Ibid., pp. 2–3.

32. Ibid., p. 2.

33. M-2284, pp. 2–3. Six of the engineers from MIT and seven from IBM attended part time.

34. Ibid., pp. 5–6.

35. M-2285. One engineer from MIT and seven from IBM attended part time. As this M-Note reminds us, the Division 6 engineers did not hesitate to use the minutes of these meetings to provide additional information and notice of things to be done: "The following tube types were accepted at the Project Grind meeting: 7 AK7, 5965, 5998, and 2D21." Any designer wishing to use a tube of a different type had to first get the approval of the Tube Standards Committee. "Not mentioned at the meeting was the fact that improved types which are electrically-equivalent tubes will probably be substituted as standards in place of 7 AK7, 5965, and 2D21, and that a low level amplifier (6136 or 5963) must be investigated." (p. 2)

36. M-2359, p. 2.

Chapter 19

1. M-2311, p. 37.

2. "Summary of MIT-IBM Collaboration, August 1 through August 31, 1953," M-2406, p. 2. "The continued engineering work on AN/FSQ-7 equipment during this period has covered considerable activity in the fields of standardization on components, materials, etc.; circuits and vacuum tubes; electrical design of arithmetic element has proceeded, as well as logical design for the drum system." (p. 1)

3. Taylor to Forrester, "Organization and Planning of the AN/FSQ-7 (XD-1) Combat Information Central Program," Engineering Note E-562, June 30, 1953 (ASC-24, MITRE archives), p. 1. The heart of this Engineering Note was an attachment entitled "Summary of Program for AN/FSQ-7 (XD-1) Combat Information Central" and consisting of three pages of bar graphs charting the work to be done by IBM and Division 6, along with two pages scheduling the work to be done "exclusively by Divisions of the Lincoln Laboratory or other agencies." The five pages of bar graphs had received Forrester's signature of approval on July 3, 1953. For subsequent biweekly technical progress reports on the XD-1, see M2322, M-2633, M-2676, M-2698, M-2730, M-2767, and M-2831 (AC-353, MITRE archives).

4. These eight technical memoranda, all dated July 15, 1953, are as follows: M-2300, "Planning, Scheduling and Administering the AN/FSQ-7 Program (with Broad Principles of MIT-IBM Relationships and Some Detail on Division 6 Procedures)." M-2301, "Planning, Scheduling and Administering the AN/FSQ-7 Program: Scheduling Office (Supplement to M-2300)." M-2302, "Planning, Scheduling and Administering the AN/FSQ-7 Program: Approval of Schedule

Changes (Supplement to M-2300)." M-2303, "Planning, Scheduling and Administering the AN/FSQ-7 Program: Progress Reports to the Scheduling Office (Supplement to M-2300)." M-2304, "Planning, Scheduling and Administering the AN/FSQ-7 Program: Transfer of Responsibility from MIT to IBM (Supplement to M-2300)."M-2305, "Planning, Scheduling and Administering the AN/FSQ-7 Program: Approval of Basic Circuits for AN/FSQ-7 (Supplement to M-2300)." M-2306, "Planning, Scheduling and Administering the AN/FSQ-7 Program: Approval of Block Outlines and Block Schematics for AN/FSQ-7 (Supplement to M-2300)." M-2307, "Planning, Scheduling and Administering the AF/FSQ-7 Program: MIT-IBM Communication (Supplement to M-2300)."

5. M-2307, p. 1; M-2305, p. 1; M-2302, p. 1; M-2303, p. 1.

6. M-2306, p. 1. At the time, technical information was passed on to the Air Force through the Air Force Cambridge Research Center.

7. M-2304, p. 1: "In addition to giving the title of the particular job and references to reports, conferences, drawings, etc., it should also, under the heading 'General Remarks and Requirements,' summarize the basis for proceeding with the work. Included here should be information concerning accepted specifications for design, performance, etc.; the exact scope of the work to be done (e.g., to what extent the unit is to be tested before delivery); any elements of time schedules that must be met; and the basis for maintaining liaison with MIT during work (name of the MIT engineer assigned to follow the work, steps at which formal reports to MIT are desired, etc.). Copies of the letter should be sent to Forrester, Division 6 group leaders, Group 62 section leaders, Hill and Valley of MIT, to Palmer, Zollinger, Solomon, Coombs and Burke of IBM, and to Kromer for the Planning and Control Office to transmit to AFCRC."

8. "Division 6 Biweekly Report, May 22, 1953," M-2183, pp. 49, 38. See also Papian's report on p. 43 of this Biweekly Report.

9. M-2280, p. 58.

10. "Group Leaders' Meeting, May 25, 1953," Digital Computer Laboratory Report L-98, May 27, 1953, p. 2.

11. "Group Leaders' Meeting, June 15, 1953," L-101, June 17, 1953, pp. 1–2.

12. "Division 6 Biweekly Report, July 17, 1953," M-23ll, p. 47. "The most important of these [proposals] is to add an IBM drum, divided into three fields to simulate auxiliary, input buffers, and output buffers respectively. H. E. Anderson has been transferred from MTC to Dick Best's group to work on circuitry for this drum. Attention is now being given to the question of what sort of re-casting of MTC control will promote early and fruitful use of the drum and associated devices." However, before the end of July a Group Leaders' Meeting was discussing transferring the MTC memory bodily to the Whirlwind machine. (See "Group Leaders' Meeting, July 27, 1953," L-111, July 27, 1953, pp. 1–2. Only one item was on the agenda of this meeting: "Transfer of MTC memory to WWI.")

The implications of these developments for FSQ-7 design work were not lost on the MIT and IBM engineers. See also "Group Leaders' Meeting, August 3, 1953," L-112, pp. 1–2.

13. M-2267, p. 3: "It was definitely decided that everything should have sufficient capacity so that a complete second memory can be added later without redesigning of logic, air conditioning, power supply, room space, etc., although a second memory may not actually be built for sometime."

14. "Summary of MIT-IBM Collaboration, Sept. 1 thru Sept. 30, 1953," M-2455, pp. 1, 3. For a discussion of the role and the importance of the Systems Office, see Henry S. Tropp, moderator, "A Perspective on SAGE: Discussion," *Annals of the History of Computing* 5 (1983), no. 4: 392–393.

15. Astrahan and Jacobs, "History of the Design of the SAGE Computer—The AN/FSQ-7," p. 346. See also chapter 15 of Jacobs 1986. For an official description and discussion of the functions of the Systems Office a year later (November 1954), see Forrester to Lincoln Steering Committee, "Organization and Tasks of Division 6," Memorandum 6L-173, November 16, 1954, p. 22 and figures A and H. For a personal view of the political tensions between the two divisions, see chapter 20 of Jacobs 1986.

16. Astrahan and Jacobs, "History of the Design of the SAGE Computer—The AN/FSQ-7," p. 346.

17. Ibid., pp. 346–347. "The design practices and disciplines developed for the SAGE computer," these two participants noted, "later helped IBM to standardize the hardware of its commercial product line."

18. Ibid., p. 68; Division 6 Quarterly Progress Report, December 15, 1953, p. 24. The preceding QPR, dated 15 September 1953, had referred to Project Grind only in these general terms: ". . . a series of large-scale formal joint meetings has reviewed the status of development and design work. At these meetings continuation of development along existing lines in certain areas was confirmed, and programs for additional work were outlined in other areas." (A. P Kromer, p. 78) That QPR went to press too early to include any reference to the "engineering concurrence" procedures that were set up after the last meeting of Project Grind.

19. M-2455, p. 3.

20. Division 6 Staff to Forrester, "Biweekly Report, September 11, 1953," M-2407, p. 27; M-2455, p. 3.

Chapter 20

1. David R. Brown to Group Leaders, "Group Leaders' Meeting, February 1, 1954," Division 6 Memorandum L-139, p. 2.

2. Ibid.

3. Lincoln Steering Committee minutes, February 1, 1954, p. 2. (Everett, as recording secretary, issued the minutes.)

4. L-140, "Group Leaders' Meeting, February 8, 1954," issued February 9, 1954, item 1, Time Schedules for FSQ-7 (XD-1), pp. 1–3. (Obviously, "fulse" should be "false.")

5. Ibid., p. 2.

6. Ibid., pp. 2–3; "Group Leaders' Meeting, February 15, 1954," L-141, items 3 and 4, pp. 2–3.

7. LSC minutes, February 15, 1954, pp. 2–3: "Hill asked Forrester to make an outline of his discussion including the kind and amount of help needed." This outline became a memo: Forrester to Lincoln Steering Committee, "AN/FSQ-7 (XD-1) Prototype Schedule and Installation," L-142, 26 February 1954.

8. L-142, p. 3.

9. "Program for Getting Back on Schedule," signed by J. M. Coombs, n.d. [ca. February 5, 1954] (AC-59, MITRE archives); L-140, pp. 1–3; "Group Leaders Meeting, March 1, 1954," L-143, issued March 3, 1954, p. 3; Lincoln Steering Committee minutes March 1, 1954.

10. "Group Leaders' Meeting, March 22, 1954," L-145, p. 3.

11. "Group Leaders' Meeting, March 29, 1954," L-146, p. 2.

12. "Group Leaders' Meeting, April 20, 1954," L-147, p. 3. The views of Dodd and Taylor are reported in Brown's words.

13. "Group Leaders' Meeting, April 26, 1954," L-148, p. 1.

14. "Group Leaders' Meeting, May 10, 1954," L-149, p. 2. (Pages 2–4 of this L-Note are mislabeled "Memorandum M-149.")

15. "Group Leaders' Meeting, June 7, 1954," L-153, p. 2; "Group Leaders' Meeting, July 6, 1954," L-157, p. 1; "Group Leaders' Meeting, July 19, 1954," L-159, p. 3.

16. L-159, p. 3; "Group Leaders' Meeting, July 26, 1954," L-160, p. 160.

17. "Group Leaders' Meeting, August 9, 1954," L-161, p. 2; Watson and Petre 1990, p. 232.

18. "Group Leaders' Meeting, August 16, 1954," L-162, pp. 1–2; Robert P. Crago, IBM "File Memorandum," August 18, 1954, with attached organization charts; "Organization of Project High," September 1, 1954, approved by J. M. Coombs (ASC-23, MITRE archives).

19. "Group Leaders' Meeting, September 20, 1954," 6L-166, pp. 2–3; "Group Leaders' Meeting, September 27, 1954," 6L-167, pp. 1, 3. (In September 1954, in

response to moves to centralize authority and clarify internal information channels within Lincoln Laboratory, the prefix 6 was added to Division 6 L-Notes.)

20. "Group Leaders' Meeting, October 18, 1954," 6L-169, pp. 2–3.

21. "Group Leaders' Meeting, November 15, 1954," 6L-172, p. 1: "At a meeting held November 12, MIT agreed to accept delivery of AN/FSQ-7(XD-1) in January, 1955, without correction of the –15 volt mesh [to ground]. IBM estimated that from 1 to 3 months' delay would be necessitated if the difficulty were corrected in Poughkeepsie, but agreed to correct it after delivery if serious trouble appears. The magnetic drums are behind schedule and will be tied into the system after delivery in Lexington, probably sometime in February. The essential parts of the air conditioning system in Building F will be ready by January 1, 1955. IBM estimates that equipment will begin to arrive three weeks after the power is turned off [to the XD-1] in Poughkeepsie. Presumably power will be turned off on December 31, 1954." See also "Group Leaders' Meeting, November 29, 1954," 6L-174, pp. 1–2.

22. "Group Leaders' Meeting, November 15, 1954," 6L-172, pp. 1, 3: "J. W. Forrester visited Poughkeepsie on November 11 to talk to G. Cullen and H. Beattie. He told both that MIT still believes IBM is the right company to do the FSQ-7 job. In general the situation is good, but some flaws exist. The leadership of Project High is particularly weak. MIT knows of [some]one in the Project High organization who could be promoted to a position of leadership. The present difficulties are largely the fault of IBM management. . . . J. Zollinger and C. F. McElwain will see J. W. Forrester on November 16. J. W. Forrester also plans to discuss the present difficulties with R. Palmer and W. W. McDowell. If necessary, [MIT Vice-President] Adm. E. L. Cochrane might see T. J. Watson."

23. Ibid.; Forrester, MIT Computation Book No. 54, entry for November 22, 1954, p. 20 (MC 140, Box 81, MIT archives); Forrester to H. S. Beattie and G. A. Cullen, IBM Engineering Laboratory, Poughkeepsie, November 19, 1954 (MC 140, Box 13, MIT archives); Watson and Petre 1990, pp. 232–233.

24. Ibid.

25. "Group Leaders' Meeting, December 6, 1954," 6L-175, p. 1: "A meeting was held on December 1 in Admiral E. L. Cochrane's office to discuss the management of Project High with the following representatives of IBM: J. Zollinger, C. F. McElwain, J. M. Coombs, and M. M. Astrahan. . . . C. F. McElwain is now in a position to correct weaknesses in the management of Project High. A joint IBM-MIT guidance committee was set up to hold its first meeting on December 9. Membership of the guidance committee is as follows: C. F. McElwain (Chairman), J. Zollinger, G. R. Solomon, R. C. Sampson, G. A. Cullen, J. M. Coombs, and T. A. Burke (Secretary) from IBM; and J. W. Forrester, R. R. Everett, N. H. Taylor, and S. H. Dodd from Lincoln Laboratory." (R. P. Crago of IBM was added to the list, according to memorandum 6M-3270, cited directly below.) See also Lincoln Steering Committee minutes, December 7, 1954, item 13, p. 4; "Summary of MIT-IBM Collaboration—November 1 thru December 31, 1954," Division 6

Memorandum 6M-3270, issued January 7, 1955 (AC-72, MITRE archives); IBM File Memorandum, "Project High Coordination Meeting, December 9, 1954 (No. 1)," issued December 10, 1954; and subsequent memoranda in MITRE Archival Collection ASC-23.

26. Watson and Petre 1990, p. 232.

27. Augarten 1984, pp. 208–209.

28. Forrester to Lincoln Steering Committee, "Organization and Tasks of Division 6," 6L-173, November 16, 1954, p. 21; Division 6 QPR, March 15, 1955, p. 63.

29. 6L-173, p. 22; Division 6 QPR, March 1955, p. 75.

30. "Group Leaders' Meeting, December 27, 1954," 6L-178, p. 2.

31. "Group Leaders' Meeting, January 3, 1955," 6L-179, p. 4.

32. Ibid.; "Group Leaders' Meeting, January 3, 1955," 6L-179, p. 4; Division 6 QPR, March 15, 1955, pp. ix, 23–25.

33. Forrester to C. F. McElwain, January 18, 1955 (MC 140, Box 14, MIT archives).

34. Ibid., pp. 1–2.

35. Ibid., p. 3.

36. Ibid., p. 2.

37. Ibid.

38. "Group Leaders' Meeting, February 21, 1955," 6L-186, pp. 1–2.

39. Ibid.

40. "Group Leaders' Meeting, February 28, 1955," 6L-187, p. 2.

41. "Group Leaders' Meeting, January 3, 1955," 6L-179, p. 4: "Three specific instances were noted. While IBM's attitude might be reasonable in one or two of these cases, the general attitude is sufficiently prevalent to warrant more investigation and discussion of the problem."

42. "Group Leaders' Meeting, January 10, 1955," 6L-180, p. 1.

43. Division 6 QPR, 15 June 1955, pp. 27, 31–32.

44. "ADES-Lincoln Biweekly Meeting, 26 July 1955," 6M-3799, issued August 1, 1955 (AC-72, MITRE archives).

45. Division 6 QPR, June 15, 1955, pp. 27, 75.

46. For a preliminary overview, see Forrester, Everett, and Wieser to Lincoln Steering Committee, "A Proposal for Accomplishing the SAGE System Computer

Programming Tasks Outlined in Memorandum 6M-3416," 6L-189, March 11, 1955, and "Proposed Responsibilities for Tasks in Memorandum 6M-3416, prepared by Jay W. Forrester, Robert R. Everett, C. Robert Wieser for Conference at Western Electric, March 17, 1955," 6L-190, issued March 16.

Chapter 21

1. Problems of Air Defense, Final Report of Project Charles, August 1, 1951, volume I, pp. 118–120; Division 2 QPR, p. 2-5; Division 6 QPR, p. 6-9.

2. Problems of Air Defense, volume I, p. 118; Division 6 QPR, June 1, 1952, p. 6-11.

3. S. H. Dodd to R. E. Wetherbee, "Facilities of the Experimental SAGE Sector," Division 6 Memorandum 6M-4981, May 6, 1957, p. 1 (AC-72, MITRE archives); D. Sternight to All Interested Staff, "SAGE System and Air Defense Sector Boundaries," Memorandum 66D-HQ328-1, File: O-2, Group 66 Headquarters, April 24, 1957 (AC 38-5, MITRE archives).

4. For a personal account of the reasons for the change in name, see Valley, "How the SAGE Development Began," *Annals of the History of Computing*, 7 (1985), no. 3, pp. 224–225.

5. H. W. Boehmer and C. R. Wieser to Valley and Forrester, "Automatic Aircraft Guidance," Memorandum M-1730, January 14, 1952, pp. 1–2.

6. D. R. Israel to C. R. Wieser, "Overall Organization of the Cape Cod Information and Direction Center (Supplement to M-1815: 1953 Cape Cod System)," M-1839, February 11, 1953, p. 1.

7. Ibid,. pp. 1–5.

8. See "Decoding GOC Post Locations," M-1542, July 1, 1952; "Discussion of Group 61 GOC Test of July 15, 1952 and July 29, 1952," M-1597, issued August 13, 1952; "Discussion of Group 61 GOC Test of August 26, 1952 and September 2, 1952," M-1629, issued September 5, 1952; "Discussion of Group 61 GOC Tests of October 14, 1952 and October 21, 1952," M-1695, issued October 28, 1952.

9. E. S. Rich to J. V. Harrington and H. W. Boehmer, "Design of Receiving and Conversion Equipment for SDV Data," M-1630, September 8, 1952, p. 1.

10. MIT Lincoln Laboratory, "A Proposal for Air Defense System Evolution: The Transition Phase," Technical Memorandum 20, January 2, 1953 (hereafter cited as TM-20), p. 5. A first draft had been finished by early December, circulated informally for comments, revised, and superseded by this second draft.

11. Lincoln Steering Committee (hereafter cited as LSC), minutes, August 25, 1952, item 14; meeting of December 2, 1952, item 10.

12. LSC minutes, December 2, 1952, p. 3; December 15, 1952, p. 3.

13. Division 6 QPR, December 15, 1953, p. iii; TM-20, p. 5.

14. LSC minutes, December 2, 1952; TM-20, pp. 1, 3–4.

15. TM-20, p. 1.

16. Ibid., pp. 4–6.

17. Ibid.

18. This report was later revised and reissued as R-158.

19. TM-20, pp. 166–168; John V. Harrington, "Radar Transmission," *Annals of the History of Computing* 5 (1983), no. 4, p. 371.

20. TM-20, p. 166.

21. Ibid., pp. 7–17; Hqs, ADC, Report: Operational Plan, Semi-Automatic Ground Environment System for Air Defense (SAGE), March 7, 1955 (MITRE archives), pp. 3, 5. The latter report was a joint production of ADC and Lincoln Laboratory, assisted by Western Electric, Bell Telephone Laboratories, and International Business Machines Corp. It was commonly referred to as the "Red Book" or the "SAGE Operational Plan." See also Murphy, Early History of MITRE, volume 1, pp. 28–30; volume 2, p. 320, notes 1 and 2 for p. 29.

22. C. R. Wieser, "Cape Cod System," *Annals of the History of Computing* 5 (1983), no. 4, pp. 364–365.

Chapter 22

1. C. R. Wieser, "From World War II Radar Systems to SAGE," *Computer Museum Report* 22 (spring 1988), p. 16. For a personal discussion of the reservations held by many concerning the SAGE System, see Valley, "How the SAGE Development Began," *Annals of the History of Computing,* 7 (1985), no. 3, pp. 217–220.

2. Institute of Science and Technology, University of Michigan, "Research in Air Defense at the Institute of Science and Technology" (n.d.). We encountered a copy of this document in ESDHO. For a participant's view of the rivalry between ADIS and SAGE, see Valley, "How the SAGE Development Began," pp. 220–224. Other points of view within ARDC were expressed in our interviews with Col. G. T. Gould Jr. (July 16, 1957) and with A. G. Wimer Jr. (July 17, 1957).

3. "Research in Air Defense at the Institute of Science and Technology," Problems of Air Defense, volume 1, p. 91; E. E. Partridge to J. R. Killian Jr., January 28, 1953, in "Case History on Project Lincoln," in W. H. Wood, Case History on Project Lincoln (a collection of undated documents found in the Historical Branch of the AFCRC's Office of Information Services, hereafter cited as Wood, Case History), document 25; Partridge to Harlan Hatcher, January 28, 1953, in Harlan Hatcher Papers, Michigan Historical Collections, Bentley Historical Library, University of Michigan; interview with Gould, July 16, 1957.

4. Richard F. McMullen, "The Birth of SAGE, 1951–1958" (Archives, Air University, Maxwell Air Force Base, Alabama), pp. 10–11, 17–18.

5. Partridge to Killian, January 28, 1953; Partridge to Hatcher, January 28, 1953.

6. Forrester, MIT Computation Book No. 53, "Technical and Administrative Notes," October 17, 1952, entries for October 17, "IBM Contract Developments," p. 1, and Arthur P. Kromer, Memorandum for the Record, subj.: "Discussion of Contract Status with IBM," October 17, 1952, p. 2. (Forrester occasionally copied or pasted in his Computation Book material he considered important.)

7. Forrester, Project Charles videotape, session 3.

8. Forrester, MIT Computation Book No. 53, Forrester's and Kromer's entries for October 17, 1952, pp. 1–2; LSC minutes, October 20, 1952, item 4.

9. David R. Brown to Group Leaders, "Group Leaders' Meeting, November 24, 1952," Memorandum L-71, Digital Computer Laboratory, MIT, item 1: "Air Defense." See also reports of Group Leaders' meetings of January 1953 in L-79 (item 3) and L-80 (item 1); interview with Col. G. T. Gould Jr., Hq, ARDC, by the authors, July 16, 1957.

10. Killian to Finletter, January 9, 1953, in Wood, Case History, document 23; interview with A. G. Wimer Jr. by Redmond, July 17, 1953. See also Killian, Project Charles videotape, session 1.

11. Kent C. Redmond and Harry C. Jordan, Air Defense Management, 1950–1960: The Air Defense Systems Integration Division, unpublished study prepared by Historical Branch, Office of Information, AFCCDD, L. G. Hanscom Field, ARDC, February 1961 (ESDHO), p. 17.

12. Forrester, Project Charles videotape, session 3.

13. Lt. Col. P. J. Schenk to Lt. Gen. L. C. Craigie, "Recent developments on the Killian-Finletter correspondence concerning Project Lincoln," January 19, 1953 (cited in Redmond and Jordan, "Air Defense Management," p. 18). See also Finletter to Killian, January 15, 1953, in Wood, Case History (document 24) and in "Project Lincoln" (volume 3, item 59).

14. Finletter to Killian, January 15, 1953 (Wood, Case History, document 24; "Project Lincoln," volume 3, item 59).

15. Redmond and Jordan, "Air Defense Management," pp. 17–18.

16. Partridge to Killian, January 28, 1953 (Wood, Case History, document 26; "Project Lincoln," volume 3, item 60). See also Partridge to Hatcher (Hatcher Papers).

17. Partridge, "Revision of Command Policy Pertaining to Lincoln Transition System," "Reorientation of Policy for ADEE," and "Revision of Command Policy Pertaining to ADEE (all in "Project Lincoln," volume 3, item 66).

18. Harry H. Goode to Hatcher, February 26, 1953; Goode to Hatcher, March 2, 1953, with attached draft of Hatcher's reply to Partridge's letter of January 28, 1953; Hatcher to Partridge, March 12, 1953 (all in Hatcher Papers).

19. Partridge, "Memorandum for the Record," subj. "Visit at University of Michigan," March 13, 1953; Hatcher to Partridge, March 12, 1953 (cited in Redmond and Jordan, "Air Defense Management," pp. 18–19). See also interview with Wimer, July 17, 1957.

20. LSC minutes, April 27, 1953, item 6.

21. "Project Lincoln," volume 1, p. 12; interview with Wimer, July 17, 1957; Partridge to Hatcher, May 6, 1953; Putt to Cmdr, WADC, May 6, 1953, in "Project Lincoln," volume 3, item 66.

22. Interview with Wimer; interview with Albert R. Shiely Jr. by the authors, January 13, 1983.

23. TM-20, pp. 4–5.

24. Division 6 QPR, March 15, 1953, p. iii.

Chapter 23

1. Report of War Department Equipment Board, February 23, 1946 (Box K2698, RG 156 WNRC, National Archives), cited in Koppes 1982, pp. 25–29.

2. Peter J. Schenk to John W. Marchetti, February 10, 1947 (ESDHO).

3. Charles J. Smith, "History of the Electronics Systems Division, January–June 1964," volume 1, pp. 39–47.

4. Gimbel 1968, pp. 201–206. See also Hartmann 1965, pp. 30–52.

5. *New York Times*, October 13, 1949, p. 5.

6. Richard F. McMullen, "Air Defense and National Policy, 1940–1950" (unpublished study, Air University Archives), pp. 51–61. For the sense of urgency created by the Korean War, see also "A Report to the National Security Council by the Executive Secretary on United States Objectives and Programs for National Security," National Security Council document no. NSC 68/3, December 8, 1950, p. 1.

7. Goldberg 1957, pp. 129–137; interview with Col. G. T. Gould Jr. by the authors, July 15, 1957.

8. Huntington 1961, pp. 307–312. See pp. 298–312 for Huntington's discussion of "strategic deterrence."

9. Quoted in Huntington 1961, p. 339. See *New York Herald Tribune*, October 16, 1953, p. 10; *New York Times*, October 7, 1953, p. 6.

10. Huntington 1961, pp. 308–310.

11. U.S. Atomic Energy Commission, "In the Matter of J. Robert Oppenheimer" (transcript of hearings before Personnel Security Board, April 12, 1954–May 6,

1954). See especially the testimony of David Tressel Griggs (pp. 742–770) and Jerrold Zacharias (pp. 596–602, 922–933). See also Major 1971, pp. 147–176.

12. Huntington 1961, pp. 307–312, 336–357.

13. Glenn H. Snyder, "The 'New Look' of 1953," in Schilling et al. 1962, pp. 386–387.

14. "A Report to the National Security Council by the Executive Secretary on Basic National Security Policy," NSC 162/2, October 30, 1953. (The "/2" indicates that this was the second version of an earlier document, NSC 162.)

15. Gerhard Colm and Marilyn Young, *Can We Afford Additional Programs for National Security?* (Planning Pamphlet 84, National Planning Association, October 1953), pp. 6, 17 (cited in Schilling et al. 1962, pp. 212–213).

16. Warner R. Schilling, "The Politics of National Defense: Fiscal 1950," in Schilling et al. 1962, pp. 298–312.

17. "A Report to the National Security Council by the Executive Secretary on United States Objectives and Programs for National Security," NSC 68, April 14, 1950, p. 3; Paul Y. Hammond, "NSC 68: Prologue to Rearmament," in Schilling et al. 1962, p. 49.

18. NSC 68, p. 1.

19. NSC 68/2, introductory note by Executive Secretary.

20. Hammond, "NSC-68," p. 361.

21. NSC 68, p. 60.

22. Ibid. , p. 37.

23. "A Report to the National Security Council by the Secretaries of State and Defense and the Director for Mutual Security on Reexamination of United States Programs for National Security,"NSC 141, January 19, 1953, pp. i, 1. See also Hammond, "NSC 68," pp. 359–363.

24. NSC 141, p. i.

25. Ibid., pp. 76–77.

26. Ibid., pp. 78–80.

27. Ibid., p. 80.

28. Huntington 1961, pp. 329–331. See also "In the Matter of J. Robert Oppenheimer," pp. 598–600, 763–769, 922–933.

29. NSC 141, p. 80.

30. Ibid., pp. 78, 82.

31. Hammond, "NSC-68"; Snyder, "The 'New Look' of 1953," pp. 361–363, 389–390.

32. Stevens R. Rivkin, The Decision-Making Process for National Defense Policy, quoted in Huntington 1961, p. 333.

33. Huntington 1961, p. 333.

34. NSC 162/2, p. ii. See also Snyder, "The 'New Look' of 1953," p. 436.

35. Huntington 1961, pp. 331–333; Charles J. V. Murphy, "The U.S. as a Bombing Target," *Fortune*, November 1953, p. 118ff. See also "A Report to the National Security Council by the NSC Planning Board on Continental Defense," NSC 159/3, September 16, 1953, p. 20. The Bull Committee operated under the administration of the Office of Defense Mobilization, headed by Arthur Fleming.

36. Lincoln Steering Committee minutes, July 6, July 20, and October 26, 1953.

37. NSC 159, "A Report to the National Security Council by the Continental Defense Committee on Continental Defense," July 22, 1953, pp. 40, 44–46, 53.

38. Huntington 1961, pp. 331–332; Division 2 QPR, April 15, 1953, p. 3.

39. Gilpin 1962, pp. 125–126.

40. Huntington 1961, pp. 307, 357; LSC minutes, October 20, 1952, item 16.

41. "Defense and Strategy," *Fortune*, December 1953, pp. 77–84.

42. Charles J. V. Murphy, "A New Strategy for NATO," *Fortune*, January 1953, p. 80ff.; Huntington 1961 pp. 63–64; Snyder, "The 'New Look' of 1953," pp. 388–389.

43. *New York Herald Tribune*, March 16 and 20, 1953, quoted on p. 332 of Huntington 1961.

44. "A Report to the National Security Council by the Executive Secretary on Continental Defense," NSC 159/4, September 25, 1953, p. 4. A Statement of Policy contained in this report explained: "In recent years we have emphasized the elements of peripheral defense, offensive capabilities, and mobilization base more than we have emphasized the element of 'continental defense'. Yet this latter element is necessary for the protection of our vitals and for the survival of our population and our Government in the event of attack. 'Continental defense' is now clearly inadequate." For the corrective measures recommended, see NSC 162/2, issued October 30, 1953.

45. NSC 162/2, p. 2.

46. Ibid., pp. 7–8, 19.

47. Snyder, "'New Look' of 1953," pp. 462–463.

48. Murphy, "The U.S. as a Bombing Target," p. 118ff.; Huntington 1961, pp. 334–337; Major 1971, pp. 252–261; "In the Matter of J. Robert Oppenheimer", pp. 743–770, 922–933; Killian, Project Charles videotape, session 1; James R. Killian Jr. and A. G. Hill, "For a Continental Defense," *Atlantic Monthly*, November 1953, p. 37ff.

49. Huntington 1961, pp. 338–341.

Chapter 24

1. For further details of their early efforts, see Redmond and Smith 1980. The 1947 proposal appeared as Jay W. Forrester and Robert R. Everett, "Information Systems of Interconnected Digital Computers," Limited Distribution Memorandum L-2, Project Whirlwind, Servomechanisms Laboratory, October 15, 1947. In 1949 it was revised and reissued as "Report R-158, Information System of Interconnected Digital Computers for Anti-Submarine Naval Group" (April 12). See also Forrester and Everett, "Digital Computation for Anti-Submarine Problem," L-1, October 1, 1947 (also identified as "Memorandum M-108 changed to L-1"). Indicative of the authors' continuing interest in military applications of digital computers is Report L-3, which appeared a year later in a shorter and a longer form: "A Plan for Digital Information Handling Equipment in the Military Establishment, Prepared for Dr. Karl T. Compton," September 14, 1948, and "Forecast for Military Systems Using Electronic Digital Computers," September 17, 1948. The co-authors of both versions were Forrester, Everett, Hugh R. Boyd, Harris Fahnestock, and Robert A. Nelson.

2. TM-20, Second Draft, 2 January 1953, p. 113.

3. Ibid., p. 3; Operational Plan, Semi-Automatic Ground Environment System for Air Defense, pp. 81–86. The latter (issued by Headquarters, Air Defense Command, Ent Air Force Base, Colorado Springs, on March 7, 1955) was said to have been "Prepared jointly by: Air Defense Command, United States Air Force; Lincoln Laboratory, Massachusetts Institute of Technology; with the assistance of Western Electric Company, Bell Telephone Laboratories, International Business Machines Corporation."

4. TM-20, p. 5.

5. Ibid., pp. 1, 3. See also ibid., chapter VI, esp. pp. 141–142, 145–146.

6. Ibid., p. 10.

7. Ibid., pp. 1, 10–11.

8. See, e.g., "Transition System for Air Defense, Estimated Cost and Delivery for the Initial Sector," Memorandum VI–L-84, February 18, 1953, prepared by Division 2 and Division 6; Kromer to Forrester, "Discussion with IBM Regarding Estimate of Cost for Initial Sector, Lincoln Transition Air Defense System," L-88, March 20, 1953; Hill to Putt, May 13, 1953; Putt to Hill, July 6, 1953; Hill to Putt, July 21, 1953.

9. Kromer, "Joint investigation of telephone-line transmission characteristics is under way by IBM, MIT, and the Bell System (Project ADES—Air Defense Engineering Service)," Division 6 QPR, December 15, 1953, p. 24.

10. Forrester and Everett compared the relative merits of TM-20's original scheme and the scheme accepted by the Air Force in their introduction to the Division 6 QPR of December 15, 1953.

11. Nelson to Forrester, "Meeting on Preparation of Material for Colorado Springs 17 August Meeting," M-2315, July 21, 1953. See also "Preliminary Plans for Installing the Transition System in the 26th Air Division Sector (1955 Boundaries)," memorandum VI–L-113, August 6, 1953, prepared by Division 2 and Division 6 for the Colorado Springs meeting "to list the actions which must be budgeted by the Air Force in installing the Transition Air Defense System" (p. 1). See also Kromer to all Group Leaders, "Summary of Conference at ADC—August 17 to August 21 Inclusive," M-2375, August 26, 1953, p. 2; LSC minutes, August 14, 1953, item 3; LSC minutes, August 31, 1953, item 8; Division 6 QPR, September 15, 1953, p. 63.

12. Division 6 QPR, 15 December 1953, chapters I and II, esp. pp. 17, 23, 28, 30, 34; Division 6 QPR, 15 March 1954, chapter I, pp. 21–25, chapter II, esp. pp. 29, 35. 37. The details of the physical installation of the XD-1 prototype are set forth in the Division 6 QPR, of June 15, 1954 (part I, pp. A1–A17).

13. LSC minutes, May 11, 1953, item 4. See also Fagen 1978, pp. 571–581.

14. LSC minutes, January 12, 1953, item 4, p. 2; January 27, 1953, item 6; March 2, 1953, item 9; March 23, 1953, item 7.

15. LSC minutes, April 27, 1953, item 12.

16. Proposed memorandum for Secretary of the Air Force from the Chief of Staff, "'Package' Contract for Implementation of Semi-Automatic Direction Centers," n.d., attached to Air Staff Summary Sheet prepared by Electronics Systems Division, Director of Communications, DCS/O, Hq, USDAF, n.d. (Internal evidence places these documents about December 1953.) An extensive account and analysis of these events can be found in Redmond and Jordan, Air Defense Management (ESDHO). For another account of these events, see pp. 320–321 of Murphy 1972.

17. Division 6 Memorandum VI-L-135-1, "Comments on BTL Report #3 to ARDC, 'The Ground Environment Problem in Air Defense: An Appraisal of the Lincoln Transition System,'" prepared by the staff of the Lincoln Laboratory, January 26, 1954, p. 1. (We were not able to obtain a copy of the BTL report itself, which was issued December 17, 1953 under Contract AF-18(600)-652.)

18. LSC minutes, January 18, 1954, item 10.

19. Memorandum VI-L-135-1, "Comments on BTL Report #3 to ARDC, 'The Ground Environment Problem in Air Defense: An Appraisal of the Lincoln Transition System'," prepared by staff of Lincoln Laboratory, January 26, 1954,

pp. 6–7. "SDV" refers to "slowed-down video" equipment which was designed to modify the rapidly pulsed information from the gap-filler radars in order to send it over telephone lines to the computer at the control center.

20. "Group Leaders' Meeting, February 8, 1954," L-140, issued February 9, 1954, p. 4.

21. These events could be taken as lending support to the apprehensions that Eisenhower would voice at the end of his presidency: there appeared to be emerging on the national scene a new "military-industrial complex," which might be less concerned with acting in the national interest than with influencing military spending to further its own ends. Provocative though such implications might appear, the record shows that this was not the situation which involved BTL, MIT, and the Air Force over management of the SAGE R&D program. The circumstances and the issues at stake in this instance were more immediate and complicated; they rested on genuine differences over the proper conduct of R&D, and all parties were agreed that protecting the nation from air attack was paramount.

22. Crabb to Chief of Staff, Hq, USAF, "ADC Transition System Program," January 27, 1954 (ESDHO). To expedite the program, the ADC commander urged that funding be provided "as a complete 'package.'"

23. Proposed memo for Secretary of the Air Force from the Chief of Staff, "'Package' Contract for Implementation of Semi-Automatic Direction Centers," cited in Redmond and Jordan, Air Defense Management, p. 24.

24. Letter Contract AF33(600)-26451, Project ADES (Air Defense Engineering Service), January 18, 1954, pp. 28–30, cited in Redmond and Jordan, Air Defense Management, p. 24.

25. Letter Contract AF33(600)-26451, Project ADES, "Exhibit 'A'," January 8, 1984; Maj. Gen. James McCormack to Maj. Gen. D. L. Putt, February 11, 1954. See also Lt. Col. Lee V. Gossick to Col. Ernest N. Ljunggren, "Conference on Air Force–Lincoln Laboratory" (n.d., ca. November 1954); interview with Arthur G. Wimer by Redmond, June 1, 1959, cited in Redmond and Jordan, Air Defense Management, p. 24.

26. LSC minutes, February 1, 1954, item 4.

27. Gen. E. W. Rawlings to Gen. O. R. Cook, "Statement of Work for the Air Defense Engineering Service (ADES) Group," February 12, 1954. Cited in Redmond and Jordan, Air Defense Management, p. 24.

28. DF, RDTSD-ADES Office (Att'n. L. C. Watts), /s/ Lt. Col. Albert Shiely Jr., Ass't Chief for R&D, ADES Project Office, "Statement of Work for Western Electric Services Contract," December 13, 1954, with enclosure: "Statement of Work for Western Electric Management Services Contract." Cited in Redmond and Jordan, Air Defense Management, p. 25.

29. LSC minutes, April 26, 1954, items 7 and 9.

30. LSC minutes, June 7, 1954, item 4; meeting of June 21, 1954, item 1; Brigadier General Floyd B. Wood, "Meeting of 5 June 1954 on Semi-Automatic Direction Center Program," n.d. A copy of the memorandum was forwarded by A. G. Hill to Admiral Cochrane, G. Valley, and J. W. Forrester with the accompanying comment: "Herewith is Wood's version of the minutes of the meeting of 5 June (black Saturday)." General Wood and his staff were strong and persistent instigators of efforts to establish a central point of managerial authority for the SAGE Air Defense System and to bring Lincoln Laboratory into conformity with traditional contractual relationships.

31. Kromer to Everett and Forrester, "Conference Regarding Proposed Modification to Delivery Schedule for AN/FSQ-7 Duplex Centrals—June 9 and 10, 1954," L-154, June 11, 1954, p. 1. Copies went to Taylor and Valley. The report states: "This meeting was called by the AMC Joint Progress Office headed by Colonel R. Osgood." Both "Progress" and "Project" are terms associated with ADES; see, for example, "Division VI Group Leaders' Meeting—15 August 1955," 6L-216, p. 1. The JPO did not officially become the full ADES-ADC-AMC-ARDC Joint Project Office until September of 1954.

32. Ibid.

33. L-154, p. 1.

34. Ibid., p. 2.

35. Ibid., p. 5.

36. Ibid., p. 6.

37. Ibid, p. 7.

38. Ibid.

39. Ibid, p. 8. No IBM personnel were included in the announcement.

40. Interview with A. G. Wimer Jr. by Redmond, July 17. 1957; interview with Col. D. A. Detwiler by Redmond, August 20, 1956 (ESDHO); Murphy, Early History of MITRE, volume 1, p. 31.

41. "Memorandum for File: Meeting on June 16, 1954 in Room 4E337, Pentagon, to discuss scheduling of introduction of Transition System," undated anonymous draft (ASC 23-32, MITRE archives), pp. 1–2. (The upper right corner of the first page is stamped "July 1, 1954.")

42. LSC minutes, July 6, 1954, item 5; meeting of July 19, 1954, item 1.

43. Preliminary draft of minutes of the meeting held at MIT on July 2, 1954, p. 1. Undated, this draft is accompanied by a covering memorandum from M. M. Hubbard to J. W. Forrester, dated July 9, 1954

44. LSC minutes, July 6, 1954, item. 5.

45. Operational Plan, Semi-Automatic Ground Environment System for Air Defense (Formerly Designated the Transition System), Short Title: SAGE, Headquarters, Air Defense Command, Ent Air Force Base, Colorado Springs, March 7, 1955.

46. DF, RDTDR to RDTD, /s/ Col. Gordon T. Gould Jr., Chief, Radar and Communications Division, DC/TO, Hq, ARDC, "Staff Study on Lincoln Laboratory and the Air Defense Electronic Environment System," January 14, 1955, with enclosure: "Staff Study on the Management of the Air Defense Ground Electronic Program." Cited in Redmond and Jordan, Air Defense Management, p. 25.

47. Interview with Col. D. A. Detwiler by Redmond, August 20, 1957.

48. Fagen 1978, p. 574; Jacobs 1986, pp. 84–85; Gen. E. W. Rawlings to Gen. O. R. Cook, "Statement of Work for the Air Defense Engineering Service (ADES) Group," February 12, 1954; cited in Redmond and Jordan, Air Defense Management, p. 24. See also Maj. Gen. Floyd B. Wood to Cmdr, AFCRC, "Implementation of the Lincoln Transition System," September 10, 1954, in Wood, Case History, document 47.

49. Murphy, Early History of MITRE, volume I, p. 31; Jacobs 1986, pp. 108–109.

50. Murphy, Early History of MITRE, volume I, p. 31; interview with Col. G. T. Gould Jr. by the authors, July 16, 1957; interview with Col. D. A. Detwiler by Redmond, August 20, 1957; interview with A. G. Wimer by Redmond, July 17, 1957; Lt. Col. L. V. Gossick to Col. E. M. Ljunggren, "Conference on Air Force–Lincoln Laboratory" (n.d.); DF, RDTR from RDTD, /s/ Col Gould, "Staff Study on Lincoln Laboratory and Air Defense Electronic Environment System," January 14, 1955, with enclosure, "Staff Study on the Management of the Air Defense Ground Electronic Environment Program." Carbon copy of remarks by J. Keto, Technical Director, WADC (identified as "J. Keto's Comments," by James V. Burke, RDRW and formerly of WADC, and given as an accurate representation of the views of Keto and many others at WADC at the time). Cited in Redmond and Jordan, Air Defense Management, p. 26.

51. Wood, Case History, p. 14.

52. Ibid., pp. 14–15. For the give and take of the negotiations, see "Proposed Charter for the Operation of the Lincoln Laboratory" (n.d.); "Proposed Charter for the Joint Services Advisory Committee–Lincoln Laboratory" (n.d.); "Proposed Charter for Joint Services Advisory Working Group" (n.d.) (Wood, Case History, documents 60, 61, and 62, respectively); LSC minutes for January 13, 1955, April 25, 1955, August 1, 22, and 29, 1955, September 19 and 26, 1955, November 28, 1955; "Proposed (Lincoln Laboratory 26 Aug 55) Charter for the Operation of Lincoln Laboratory," Enclosure "A" to M. G. Holloway, Director Lincoln Laboratory, to Lt. Gen. D. L. Putt, Hq, USAF, September 1, 1955 (Wood, Case History, documents 57 and 58); and "Charter for the Operation of the Lincoln Laboratory," January 11, 1956 and "Charter for Joint Services Advisory

Committee–Lincoln Laboratory," January 11, 1956 and "Charter for Joint Services Advisory Committee Working Group," January 11, 1956 (Wood, Case History, documents 78 and 79, respectively).

53. Lt. Gen. Thomas S. Power to Cmdr, AFCRC, att'n.: CRK-1, December 20, 1955, 1st Indorsement to William J. Irwin to Cmdr, ARDC, "Administration of Contract No. AF19(122)-458, Massachusetts Institute of Technology," November 2, 1955 (Wood, Case History, documents 70 and 71, respectively). Interview with Col. G. T. Gould Jr. by Redmond, July 16, 1957.

Chapter 25

1. TM-20, p. 155.

2. "Cape Cod System and Demonstration," talk by C. R. Wieser, Memorandum VI (L-86), March 13, 1953 (AC-15, MITRE archives), p. 1. This talk was given to visitors attending demonstrations of the system.

3. Ibid., pp. 1–2; TM-20, pp. 151–160; C. R. Wieser, "Cape Cod System," *Annals of the History of Computing*, volume 5, no. 4 (1983): 366–367.

4. Wieser, "Cape Cod System and Demonstration," p. 2.

5. Ibid.

6. Ibid., pp. 2–3.

7. TM-20, p. 157.

8. Ibid., pp. 155–159; Forrester, "Project Lincoln, Division VI Program, July 1952–June 1953," January 7, 1952 (AC-15, MITRE archives).

9. TM-20, p. 159.

10. "Air Defense Biweekly Report, Memorandum M-1805, January 16, 1953," p. 1.

11. B. Morriss and G. Young to C. R. Wieser, "The Introduction of Data from the North Truro CPS-6B and Small Gap-Filler Radar Directly into the Computer . . . ," M-1963, April 8, 1953, pp. 1–2.

12. "Air Defense Biweekly, February 29, 1953," M-1890, p. 1; K. E. McVicar to W. W. Butler, "Auxiliary Drum Testing—Summary # 3," M-2119, April 28, 1953, p. 1.

13. R. L. Walquist, "Automatic Monitoring of Data Screening for the Cape Cod System," M-1704, October 30, 1952, pp. 1–2.

14. "Air Defense Biweekly, March 13, 1953," M-1921, p. 1; "Air Defense Biweekly, April 24, 1953," M-2113, p. 1.

15. Wieser, "Cape Cod System and Demonstration," pp. 1–2.

16. O. T. Conant to C. R. Wieser, "Telephone Intercommunication System . . . ," M-2187, June 3, 1953, pp. 1–2.

17. W. W. Attridge Jr. to C. R. Wieser, "Smoothing and Prediction Functions in the 1953 Cape Cod System . . . ," M-2175, May 25, 1953, p. 1. See also a technical M-Note prepared for instructing Air Force personnel: J. Levenson to C. R. Wieser, "Introduction to Track-While-Scan Functions in the 1953 Cape Cod System (Supplement to 1815)," M-2916, August 9, 1954.

18. Narrative reports of the activities of Group 61 during 1953 are provided in the Division 6 QPRs issued 15 March 1953, 15 June 1953, 15 September 1953, and 15 December 1953; in each issue of the QPR the relevant account begins on p. 3.

19. Division 2 QPR, June 1, 1952, pp. 2-25–2-27; "Project Lincoln," volume I, att. document 24A: G. W. Fox, L. W. Brown, C. R. Whelan, W. Janvrin, "Survey of Sites for Cape Cod Air Defense System," issued October 11, 1951.

20. Ibid., pp. 2-25–2-26, 2-29–2-31.

21. Ibid., p. 2-31; Division 6 QPR, June 1, 1952, p. 6-11.

22. Division 2 QPR, July 15, 1953, p. 9.

23. Division 2 QPR, June 1, 1952, p. 2-67. See also John V. Harrington, "Radar Data Transmission,"*Annals of the History of Computing*," volume 5, no. 4 (1983): 370–374.

24. See editor's note to Harrington, "Radar Data Transmission."

25. Harrington, "Radar Data Transmission," pp. 370–373.

26. "Air Defense Biweekly Report, May 8, 1953," M-2150, p. 4; "Group Leaders' Meeting, May 18, 1953," Memorandum L-97, issued May 21, 1953, p. 2.

27. Division 6 QPR, 15 March 1953, pp. 101–103; 15 June 1953, pp. iii, vii, 83–84, 89–90. L-111, "Group Leaders' Meeting, July 27, 1953," issued July 27, 1953. Both pages of this L-Note were devoted to a single topic: "Agenda: Transfer of MTC Memory to WWI."

28. Ibid.; "Group Leaders' Meeting, June 8, 1953," Memorandum L-100, issued June 9, 1953, p. 3; "Group Leaders' Meeting, June 29, 1953," L-106, issued June 30, 1953, p. 3.

29. W. N. Papian to R. R. Everett, "Rough Resume of Magnetic Core Memory History," November 24, 1953, pp. 1–2. This two-page memorandum was typed on the standard carbon triplicate "Digital Computer Laboratory, Inter-Office Correspondence" form.

30. "Division 6 Biweekly Report, May 22, 1953," M-2183, p. 43; "Group Leaders' Meeting, July 27, 1953," L-111, issued July 27, 1953, pp. 1–2; "Group Leaders' Meeting, August 3, 1953," L-112, issued August 3, 1953, pp. 1–2; interview with C. R. Wieser by the authors, January 3, 1984.

Chapter 26

1. "Group Leaders' Meeting, August 2, 1954," L-161, issued August 9, 1954, p. 2. See also L-159 (meeting of July 19, 1954), L-160 (meeting of July 26), and L-162 (meeting of August 16, 1964). Biweekly Reports M-2955, "Biweekly Report for July 30, 1954," and M-2981, "Biweekly Report for August 13, 1954," chronicled how widely the tasks were spread. See also chapter 19 of Jacobs 1986.

2. LSC minutes, July 19, 1954, item 10. Hill said he would suggest to JSAC Chairman D. L. Putt that he "invite operational people to the meeting, i.e. Generals Smith, Blake. . . ." See also Jacobs 1986, chapter 19; Valley, "How the SAGE Development Began," *Annals of the History of Computing,* 7 (1985), no. 3, pp. 224–225.

3. Division 6 QPR, 15 December 1954, p. 6.

4. LSC, minutes, meeting of August 16, 1954, item 7.

5. Operational Plan, Semi-Automatic Ground Environment System for Air Defense (Formerly Designated the Transition System), Short Title: SAGE, 7 March 1955, prepared jointly by Air Defense Command, United States Air Force, and Lincoln Laboratory, Massachusetts Institute of Technology, with the assistance of: Western Electric Company, Bell Telephone Laboratories, International Business Machines Corporation.

6. SAGE Operational Plan, p. 5: "This represents a tremendous gain in capacity over the present system, both in the area of air surveillance and weapons guidance."

7. "1954 Cape Cod System Flight Crew Manual of Experimental Standard Operating Procedures," Division 6 Memorandum 6M-2936; P. O. Cioffi to C. R. Wieser and Distribution List, April 5, 1955 (AC-16, MITRE archives), p. 6.

8. "D. R. Israel to C. R. Wieser, Group 61, All Persons on M-1815 Distribution List, "Memos Concerning the 1954 Cape Cod System," Division 6 Memorandum M-2706, March 3, 1954. The memo read, in part: "Procedures are described for issuing memos relating to plans, programs, equipment, or schedules for the 1954 Cape Cod System. . . . In light of current plans for preparation of the 1954 Cape Cod System, it is desirable to start a new series [of M-Notes]."

9. E. O. Martin, ed., "Summary Report of the 1953 Cape Cod System Tests," 6M-5066, p. 29. Martin's report is dated 1 November 1955 and appears to have circulated widely. It is described in its "Abstract" on p. 1 as providing not "any new analysis" but instead "a summary of the many reports written about the System." In general, these reports were written in 1953 and 1954. The latest sources Martin cites in his appendix are dated November 1954. The dating he gives of two of these, however, is suspect (6M-5059 and 6M-5070 on p. 31), since they are dated in the MITRE archives master list of M-Notes as 9/7/55 and 11/3/55 respectively. To complicate further the dating of Martin's sources, the 6M-5000 block of numbers was "used as a Block by Groups 61 and 62 for their joint interests," as noted

on "Page 263" of the MITRE archives master list. As we have already remarked, that which was designated "Experimental SAGE Subsector" at the start was subsequently renamed "Experimental SAGE Sector."

10. Operational Plan, p. 83; Division 6 QPR, June 15, 1954, part II, p. 8.

11. Division 6 QPR, December 15, 1954, pp. 3–4.

12. Israel to Wieser, "Action on the Recommendations of ADC Visitors," M-2711, March 4, 1954, p. 1. Israel wrote, summarizing: "A number of the recommendations . . . have been incorporated into the plans for the 1954 Cape Cod System, while others have been deferred for future action and possible inclusion in the XD-1 and succeeding systems. . . ." See also M-2619; C. C. Grandy and D. R. Israel, "Schedule for Nine-Day Familiarization with the Cape Cod and Transition Systems," M-2619-1, January 4, 1954 (AC-15, MITRE archives).

13. C. H. Gaudette and C. C. Grandy, "Air Defense Familiarization Program," M-2726, March 12, 1954, pp. 1–2; "Syllabus for the Cape Cod Familiarization Program," M-2726 (Supplement #1), March 22, 1954, pp. 1–21.

14. Gaudette and Grandy, "Cape Cod Familiarization Program: Revision 1," M-2726, Supplement #2, April 24, 1954, pp. 1–11.

15. Division 6 QPR, June 15, 1954, part II, pp. 3, 9; Division 6 QPR, September 15, 1954, part II, pp. 27, 59.

16. Division 6 QPR, December 15, 1954, p. 3.

17. Gaudette and Knapp to Wieser, "Proposal for the 1954 Cape Cod Simulation Programs," M-2775, April 1, 1954, p. 1.

18. 1. Israel to Wieser, "Equipment Changes," M-2746, March 25, 1954, p. 1. Since Forrester and Everett long had made a practice of identifying responsibility with authority, this order too had to receive formal approval before it could be issued.

19. The equipment was obtained from the Raydist Navigation Corporation of Hampton, Virginia.

20. Stahl to Group 61, "The Raydist Tracking System for Use in Cape Cod Calibration and Tracking Accuracy Measurement," M-2979, July 15, 1954, p. 1.

21. Ibid., pp. 1–4.

22. Division 6 QPR, June 15, 1954, part II, pp. 3, 9; Division 6 QPR, September 15, 1954, part II, p. 27; LSC minutes, August 30, 1954.

23. Attridge and Zraket to Wieser, "Implementation of Training Program for Air Force Personnel, 1954 Cape Cod System," M-2947, July 30, 1954, pp. 1–2.

24. L. J. Murray to C. R. Wieser, "Fighter Data Storage," M-2965, August 2, 1954, pp. 1–13. For further information on the array of programs into which this program fitted, see H. D. Benington, D. R. Israel, and C. A. Zraket to C. R. Wieser, "Programs and Program Responsibilities for the Weapons Direction Section of

the 1954 Cape Cod System," M-2958, August 9, 1954; S. Knapp and C. Gaudette to C. R. Wieser, "Master Control Programs for the 1954 Cape Cod System," M-2933, August 9, 1954 (AC-16, MITRE archives).

25. W. Z. Lemnios et al. to C. R. Wieser, "Program Specifications for the Weapons Assignment (WA) and Intercept Direction (IND) Program Group: 1954 Cape Cod System," M-2959, August 17, 1954, pp. 1, 3.

26. M-2933, pp. 1–8.

27. H. Peterson et al. to C. R. Wieser, "Monitoring Functions in the 1954 Cape Cod System," 6M-2952, September 24, 1954, p. 1. The proposal received its required preliminary approval from R. L. Walquist before the customary 105 copies were run off.

28. Benington and Zraket to Wieser, "Operational Specifications for the Weapons Direction Section of the 1954 Cape Cod System," 6M-2957, September 24, 1954, p. 4.

29. "Biweekly Report for 19 November 1954," 6M-3175, p. 2. On the "live testing," see also BiWeekly Reports 6M-3207 (for 3 December 1954), pp. 2, 3; 6M-3233 (for 17 December 1954), pp. 2, 3.

30. LSC minutes, December 7, 1954; Robert A. Nelson to Steering Committee, "Cape Cod Demonstration, 16 December 1954," December 10, 195; R. W. Nelson to A. G. Hill, G. E. Valley, J. W. Forrester, R. R. Everett, C. R. Wieser, "Tentative Attendance at Scheduled Cape Cod Demonstrations," January 13, 1955 (ASC-23, MITRE archives). See also the following BiWeekly Reports: 6M-3175 (for 19 November 1954, pp. 1, 2; 6M-3207 (for 3 December 1954), pp. 2, 3; 6M-3233 (for 17 December 1954), pp. 2, 3.

31. LSC minutes, March 7, 1955.

32. P. O. Cioffi to C. R. Wieser and Distribution List, "1954 Cape Cod System Flight Crew Manual of Experimental Standard Operating Procedures," 6M-2936, April 5, 1955, pp. 1, 2, 5. Cioffi gives an illuminating and detailed overall description of the system in this document.

33. Montauk is at the eastern tip of Long Island's South Fork.

34. Ibid., pp. 5–6.

35. Ibid., pp. 6–7.

36. W. Z. Lemnios, "Test Specification: Interceptor Tracking Accuracy Tests, Non-maneuvering Courses," 6M-5013, April 8, 1955, pp. 1–4.

Chapter 27

1. Forrester to Lincoln Steering Committee, "Organization and Tasks of Division 6," 6L-173, November 16, 1954, pp. 1–28.

2. LSC minutes, December 7, 1954, pp. 1–3 (esp. items 3 and 8).

3. "Group Leaders' Meeting, December 13, 1954," 6L-176, issued December 17, 1954, pp. 3–4 (item 10).

4. "Biweekly Report for 17 December 1954," 6M-3233, p. 1; "Biweekly Report for 31 December 1954," 6M-3257, pp. 2, 10: "In brief," reported Israel, "results from newspaper ads were somewhat higher in quantity than had been expected but somewhat lower in quality. Results of interviews with professors and placement personnel were extremely encouraging and point towards successful interviews with the students during February and March."

5. "Biweekly Report for 14 January 1955," 6M-3287, p. 7. See also "Biweekly Report for 28 January 1955," 6M-3345, pp. 1–2, 9. Israel reported: "A goodly portion of time has been devoted to getting the college recruiting campaign under way. This appears to have been accomplished."

6. "Biweekly Report for 11 February 1955," 6M-3375, pp. 1–4.

7. For detailed listings of the recruiting trips, see the following Biweekly Reports: 6M-3375, pp. 2–4; 6M-3405, pp. 1–2; 6M-3457, pp. 1–2; 6M-3482, pp. 1–2.

8. "Biweekly Report for 31 December 1954," M-3257, p. 1; Division 6 QPR, December 15, 1954, pp. ix, 3; Division 6 QPR, March 15, 1955, pp. 3, 10, 17.

9. "Biweekly Report for 25 March 1955," 6M-3482, p. 2; "Group Leaders' Meeting, March 28, 1955," 6L-193, p. 2.

10. 6M-3482, p. 10. Further technical details of the problems of interpreting Raydist data are discussed on p. 13.

11. 6M-3207 (3 December 1954), p. 3; 6M-3457 (11 March 1955), p. 3; 6M-3482 (25 March 1955), pp. 4–5, 13.

12. 6M-3482, pp. 4–5.

13. 6M-3233, p. 3; 6M-3257, p. 3.

14. "1954 Cape Cod System Master Make-up and Display Program Specifications," 6M-2977, May 11, 1955, p. 1. Directed to C. R. Wieser (and Distribution List) and formally approved by Wieser (officially acting in this capacity for C. A. Zraket), this 169-page technical M-Note was the product of eight authors. These were, in the order listed, H. D. Benington, W. E. Ball, A. R. Chandler, L. B. Collins, O. T. Conant, F. F. Gucker, I. B. Hazel, A. R. Shoolman.

15. Ibid., p. 10.

16. Ibid.

17. Ibid. A disclaimer at the end of the introduction to this treatise, perhaps written by Benington to give him the intellectual "elbow room" he usually demanded, suggested that "the preparation of this memorandum was dictated more by time

schedule than by accuracy requirements. The reader is asked to overlook the many statements which are either not felicitous or are actually inaccurate."

18. Hqs, ADC, Operational Plan, p. 77; Jacobs 1986, pp. 82–85.

Chapter 28

1. For Wieser's statement see Division 6 QPR, March 15, 1955, pp. vii and 3; for Forrester and Everett's view, see p. iii. For Israel's February estimate see "Biweekly Report for 25 February 1955," 6M-3405, p. 11. Compare R. J. Horn's internal biweekly report of December 3, 1954 ("live testing of the 1954 Cape Cod System has now begun") with Israel's internal report of March 25, 1955 (6M-3482, p. 10)

2. Division 6 QPR, March 15, 1955, pp. iii and vii.

3. E. D. Lundberg, "SAGE System Meeting, 2 May 1955," Division 6 Memorandum 6M-3571, item 1; "Group Leaders' Meeting–August 8, 1955," Division 6 Memorandum 6L-215, item 7; Division 2 QPR, May 15, 1955, pp. iv, 8, and 9; Division 2 QPR, November 15, 1955, pp. 15 and 33; Division 6 QPR, December 15, 1954, pp. 3, 6; Division 6 QPR, March 15, 1955, pp. 3–12. The Air Force subsequently replaced the term "subsector" with "sector"; however, "subsector" will be used here in accordance with contemporary Lincoln Laboratory practice. The reader may think of the two terms as synonymous.

4. LSC minutes, April 11, 1955, p. 2; LSC minutes, May 2, 1955, p. 1; Division 2 QPR, May 15, 1955, p. iv. "With this realignment," wrote C. F. J. Overhage, the new Division Head, "the work of Division 2 is now focused more directly on problems relating to SAGE planning, testing, and equipment." Although Overhage's note is dated June 10, 1955, the date on the title page of this QPR is "15 May 1955," and the QPR itself covered "the period 1 February 1955 to 30 April 1955" (p. iii).

5. J. W. Forrester to Lincoln Steering Committee, "Organization and Tasks of Division 6," 16L-173, November 6, 1954, pp. 5–9, figs. A, B. "In the following sections and on the attached figures will be found the organization and duties of Division 6 staff members." (p. 1)

6. Division 6 QPR, June 15, 1955, p. ix. Compare the listings of subjects on p. 5 of this QPR with those on p. 5 of the QPR of September 15, 1955 and with the organizational breakdown described in 6L-173.

7. Division 2 QPR, May 15, 1955, p. 8.

8. See abstract of "Consolidated Weekly Operations Schedule," Lincoln Memorandum 2M-0073, issued by A. Wright of Bell Telephone Laboratories, March 12, 1956 (AC-139, MITRE archives).

9. Division 2 QPR, May 15, 1955, p. 8; "SAGE System meeting 2 May 1955," 6M-3571, item 1. See also "SAGE Test Committee Meeting No. 1," 6M-5002, issued

April 14, 1955 (AC-72, MITRE archives). For a wider perspective on these events, see the eyewitness account on pp. 86–89 of Jacobs 1986.

10. Division 2 QPR, May 15, 1955, p. 9.

11. Ibid.

12. D. R. Israel to R. R. Everett et al., "Results from 1954 Cape Cod System Tests," June 29, 1955.

13. H. W. Boehmer, C. R. Wieser, and J. F. Jacobs to G. M. Holloway and G. E. Valley, "SAGE Testing and Evaluation," July 14 1955. Copies of this memorandum went to Overhage, Forrester, Everett, Dodd, and B. J. Driscoll.

14. SAGE Test Committee Meetings 1, 2, 3, and 4, reported in, respectively, 6M-5002, 6M-5007, 6M-5023, and 6M-5031 (AC-72, MITRE archives).

15. Boehmer and Wieser to Forrester and Overhage, "Test Program for 1954 Cape Cod System," 6L-212 (Draft), July 28, 1955, pp. 1–5.

16. Ibid., p. 11.

17. Ibid., pp. 4 and 12.

18. LSC minutes, July 26, 1955, item 66, "Present Status of the SAGE System."

19. "SAGE Test Committee" and "The SAGE Experimental Test Policy Committee," Lincoln Laboratory Memoranda from M. G. Holloway to All Division Heads, Group Leaders; Resident Representatives: IBM, Burroughs, BTL, RAND; Lincoln Project Office Liaison Officers: USAF, Army, Navy, August 8, 1955 (ASC-23, MITRE archives). These two memoranda bore the same identifying code: "MGH:GEV:mb." See also LSC meeting, July 26, 1955, item 6; LSC meeting, August 8, 2955, item 12; "Group Leaders' Meeting–August 8, 1955" (6L-215), items 4 and 7. Holloway had replaced Albert G. Hill as Director of Lincoln Laboratory on May 5, as had been announced at the LSC meeting of May 2, 1955.

20. Division 2 QPR, November 15, 1955, pp. iii, 15, and 33; Division 6 QPR, December 15, 1955, pp. 3–14. See also "Group Leaders' Meeting–August 8, 1955," 6L-215, issued August 11, 1955, item 7; Jacobs 1986, pp. 111–113.

21. Division 6 QPR, December 15, 1955, p. 38.

22. Division 2 QPR, November 15, 1955, p. iii. The five groups were identified as follows: Group 21, Radar Data; Group 22, Analysis and Evaluation; Group 23, Operations; Group 24, Data Processing; and Group 25, Displays.

23. Division 6 QPR, June 15, 1956, p. 42. See also S. H. Dodd to Committee 236, May 2, 1957 (ASC-355, MITRE archives); J. W. Degan to C. F. J. Overhage, "Experimental SAGE Sector Manning," Lincoln Laboratory Memorandum 38-5805, June 10, 1958 (AC-139, MITRE archives), p. 3.

24. Division 2 QPR, November 15, 1955, pp. 14–15; Division 2 QPR, November 15, 1956, p. 12. See also Walter I. Wells, "Group 22 Responsibilities and Organization," Lincoln Laboratory memorandum 2A-0072-1, October 3, 1956.

25. 6L-212, p. 9.

26. Ibid., pp. 16–37.

27. W. Wells and V. A. Nedzel, "1954 Cape Cod Test Program Schedules," 6M-5049, August 8, 1955 (AC-16, MITRE archives).

28. Ibid.; 6L-212, pp. 16–25.

29. 6M-5049, p. 2; L-212, p. 16.

30. Division 2 QPR, August 15, 1956, p. 15.

31. Division 2 QPR, February 15, 1956, pp. 17 and 43; Division 2 QPR, May 15, 1956 (AC-21, MITRE archives), pp. iii, 3, and 69.

32. O. V. Fortier and C. W. Uskavitch to C. R. Wieser and Distribution List, "Test Specifications: Cape Cod System Live Exercises (SLEs)," 2M-0569, July 19, 1956 (AC-139, MITRE archives), p. 1. See also Division 2 QPR, November 15, 1956 (AC-21, MITRE archives), p. 13.

33. O. V. Fortier and C. W. Uskavitch to C. R. Wieser and Distribution List, Correction #1, "Cape Cod System Operation Exercises (SOEs)," 2M-0569, August 6, 1956 (AC-139, MITRE archives).

34. Robert N. Davis to All Laboratory Division Heads, Group Leaders of 21, 22, 23, 24, 61, 62, and 64, "Cape Cod System," 2M-0793, September 26, 1957 (ASC-23, MITRE archives); Group 22 to C. R. Wieser, "Mission Report: Experimental SAGE Sector, Evaluation Test No. 1, 2 October 1957," 2M-0800-1, issued November 15, 1957 (AC-41, MITRE archives). For dates of the SOTs, see the index to the "2M" series of memoranda (AC-139, MITRE archives).

35. G. Lewitzky to Distribution List, "Test Specifications: System Operation Tests, Series I," 6M-5073, October 10, 1955, pp. 1–3.

36. Division 2 QPR, May 15, 1956 (AC-21, MITRE archives), p. 19. See index to "2M" memoranda: 2M-0014-3 gives October 25, 1955 as the date of the first test, while 2M-0026-5 indicates that the eighth and last test in Series I was run on January 31, 1956.

37. Division 2 QPR, February 15, 1956, pp. 17–18; Division 2 QPR, May 15, 1956 (AC-21, MITRE archives), p. 19; G. Lewitzky, "Test Specifications: System Operation Tests, Series I," SAGE Operations Memorandum 6M-5073, October 10, 1955 (AC-16, MITRE archives), pp. 1–2; Lewitzky to V. A. Nedzel and Distribution List, "System Operation Test, Series 1, Test 5, 6 January 1956," 2M-0018, issued January 13, 1956 (AC-139, MITRE archives).

38. Division 2 QPR, May 15, 1956, table 22-111, p. 26; Division 2 QPR, August 15, 1956, table 22-1, p. 24 (AC-21, MITRE archives). The tables cited list 19 tests; however, the index to MITRE archives AC-139 lists reports for only 14, beginning with 2M-0069-1 (February 14) and ending with 2M-0069-14 (June 20).

39. Division 2 QPR, May 15, 1956, p. 24; R. C. Holland to V. A. Nedzel, "Preliminary Report: System Operation Test #1, Series 2," 2M-0069-1, February 16, 1956; Holland to Nedzel, "Preliminary Report, SOT #14, 20 June 1956," 2M-0069-14, July 6, 1956 (AC-139, MITRE archives).

40. Division 2 QPR, May 15, 1956, table 22-111, p. 26; Division 2 QPR, August 15, 1956, table 22-1 (AC-21, MITRE archives), p. 24.

41. Division 2 QPR, February 15, 1956, pp. 19–20.

42. Division 2 QPR, May 16, 1956, p. 19.

43. Division 2 QPR, May 15, 1956, pp. 3 and 21; Division 2 QPR, August 15, 1956, pp. iii, 5–6, and 25.

44. Division 2 QPR, August 2, 1956, p. 20.

45. C. W. Uskavitch to V. A. Nedzel, "Test Specifications: Cape Cod System Operation Tests, Series III," 2M-0562, June 29, 1956 (AC-139, MITRE archives), p. 3; Division 2 QPR, November 15, 1956, p. 13. For test dates, see index to 2M memoranda in AC-139, MITRE archives, beginning with 2M-0562-1 and ending with 2M-0562-25.

46. Division 2 QPR, August 15, 1956, p. 26.

47. F. W. Graham and Lt. C. E. Cannon to V. A. Nedzel, "Test Specifications: Cape Cod Operation Tests, Series IV," 2M-0659, February 12, 1957 (AC-139, MITRE archives), p. 3. See also"Test Specifications: Cape Cod System Operation Tests, Series III" (2M-0562), abstract and p. 3.

48. Division 2 QPR, November 15, 1956, pp. 13–14.

49. Division 2 QPR, February 14, 1956, p. 15. (Seenote at bottom of page.)

50. Division 2 QPR, May 15, 1956, p. 13, figure 22-1 and bottom note; Division 2 QPR, August 15, 1956, p. 21, figure 22-1.

51. W. I. Wells to C. F. J. Overhage and Distribution List, "Proposal for Future Cape Cod System Modifications," 2M-0525, June 18, 1956, p. 1; Division 2 QPR, August 15, 1956, p. iii; Division 2 QPR, November 15, 1956, p. 70.

52. Division 2 QPR, August 15, 1956, p. iii; Division 2 QPR, November 15, 1956, p. 70.

53. 2M-0525, p. 1. A "SAGE-type system" was defined on p. 17 of "Test Specifications: Cape Cod System Operation Tests, Series IV" (2M-0659, February

12, 1957) as "an area defense system, its object being to produce maximum attrition as fast as possible of any enemy air force entering the area of coverage."

54. 2M-0525, p. 1; D. R. Israel and W. I. Wells, eds., "Cape Cod and Sage Operational Problems," 6M-4532, August 9, 1956 (AC 44-10, MITRE archives), p. 1.

55. 6M-4532, p. 2.

56. Ibid., pp. 1–9.

57. Frank H. Ditto to V. A. Nedzel,"Cape Cod System Test Operating Procedures—Radar Mapping Supervisor," 2M-0538 Draft 1, September 4, 1956, p. 2.

58. Ibid., pp. 2–4.

59. Ibid., pp. 5–7. For a clear drawing of the Mapping Supervisor's Console, see p. 8.

60. "Proposal for Future Cape Cod System Modifications," 2M-0525, June 18, 1956, pp. 2–14; "Test Specifications: Cape Cod System Operation Tests, Series IV," 2M-0659, February 12, 1957, pp. 3–4.

61. Ibid. "IND" apparently refers to "Intercept Directors." It was proposed that a "Recovery Officer" would "relieve intercept directors of the necessity of returning all of the interceptors to base" (p. 11).

62. 2M-0659, pp. 4, 8, and 9; F. W. Graham to V. A. Nedzel, "The Test Design for the 1957 Cape Cod System Operation Tests, Series IV," 2M-0642, February 26, 1957, p. 2.

63. 2M-0642, p. 2.

64. Ibid., pp. 3, 12; 2M-0659, p. 9.

65. 2M-0659, p. 9.

66. 2M-0642, p. 3.

67. 2M-0659, p. 9.

68. See Index to 2M memoranda, beginning with 2M-0659-2 and ending with 2M-0659-13 (AC-139, MITRE archives).

69. See the "Preliminary Reports" for SOT 3 (2M-0659-3, March 20, 1957), SOT 5 (2M-0659-5, April 3, 1957), SOT 6 (2M-0659-6, April 10, 1957), SOT 7 (2M-0659-7, April 17, 1957), and SOT 8 (2M-0659-8, April 24, 1957). No report for the first test in this series is listed in the index.

70. Holland to Nedzel, "Preliminary Report: SOT 2, Series IV, 13 March 1957," 2M-0659-2, April 1, 1957, pp. 1 and 3. See also 2M-0659-4, March 27, 1957 (AC-139, MITRE archives), p. 3.

71. 2M-0659-9, May 20, 1957.

72. See "Preliminary Reports" 2M-0659-10 (May 8, 1957), 2M-0659-11 (May 15, 1957), 2M-0659-12 (May 22, 1957), and 2M-0659-13 (May 29, 1957).

73. W. I. Wells to C. R. Wieser, "Status of Cape Cod System Effective 1 August 1957," July 30, 1957 (ASC-23, MITRE archives).

74. Robert N. Davis to all Laboratory Division Heads, Group Leaders of 21, 22, 23, 24, 61, 62 and 64, "Cape Cod System," 2M-0793, September 26, 1957 (ASC-23, MITRE archives).

75. Group 22 to Wieser, "Mission Report: Experimental SAGE Sector, Evaluation Test No. 1, 2 October 1957," 2M-0800-1, November 15, 1957 (AC-41, MITRE archives).

Chapter 29

1. Jacobs 1986, p. 108.

2. Jay W. Forrester to Lincoln Steering Committee, "Organization and Tasks of Division 6," 6L-173, November 16, 1954. News of the rearrangement formally reached Division 6 engineers earlier, in "Biweekly Report for September 24, 1954 (6M-3075)." See also Division 6 QPR, December 15, p. 3.

3. 6L-173, pp. 5–9; 6M-3075, pp. 1–10; Division 6 QPR, December 15, 1954, pp. 3–9.

4. 6L-173, pp. 10–12.

5. Ibid., pp. 18–19.

6. Ibid., pp. 22–23 and Fig. H.

7. Ibid., pp. 1–4; see especially summary on p. 3.

8. 6M-3075, pp. 1–2; "Biweekly Report for 5 November 1954," 6M-3151, pp. 2, 8–11. The notices pointed out that of course Group 61 would continue to be preoccupied with making the 1954 Cape Cod System operational so that they could begin shakedown testing, and a footnote in the September Biweekly explained to Division 6 engineers that "Semi-Automatic Ground Environment System" was the new name for the Transition System.

9. "Biweekly Report for 31 December 1954," 6M-3257, p. 1.

10. For an eyewitness account in informed retrospect three decades later, see pp. 151–152 and chapters 20, 22, and 26 of Jacobs 1986. Overhage is first listed as Head of Division 2 in the Division 2 QPR of May 15, 1955.

11. Jacobs 1986, p. 112.

12. Division 6 QPR, September 15, 1955, pp. iii–iv; Division 6 QPR, December 15, 1955, pp. iii, 10–11, 18; Division 2 QPR, November 15, 1955, pp. iii, 15; Division 2 QPR, February 15, 1956, pp. iii, 15–16, 43.

13. See, e.g., the growing number of tasks described in the following: Division 6 QPR, March 15, 1956, pp. iii, 53–55; Division 6 QPR, June 15, 1956, pp. iii, ix, 3, 11–13, 41–44; Division 6 QPR, September 15, 1956, pp. iii, 13–15, 61–63.

14. Division 6 QPR, September 15, 1955, pp. iii, 82–85.

15. Division 2 QPR, February 15, 1956, pp. 15–16; W. I. Wells to C. R. Wieser, and H. W. Boehmer, "Introduction to the ESS Evaluation Program," 22M-0100, December 7, 1955. Members of the SAGE Experimental Test Policy Committee included Valley, Overhage, Forrester, Wieser, Everett, V. A. Nedzel, F. Frick, H. Sherman, Col. J. D. Lee (commander of the 4620th Air Defense Wing), Lt. Col. R. La Montagne (of the USAF Lincoln Project Office), M. J. Burger (Bell Labs, Lincoln), A. Herckmans (Bell Labs, Whippany branch), J. F. Mills (IBM, Lincoln), and F. Simon (Western Electric, Lincoln).

16. Division 2 QPR, February 15, 1956, p. iii.

17. Ibid. For further details, see two Inter-Office Memoranda: O. V. Fortier and K. E. McVicar, "Personnel of Teams Working on Test Plans for the ESS" as of October 5 and as of November 23, 1955. In October there were 36 members spread among five teams; by the end of November there were 57 spread among six teams: Gap Filler Inputs, Long Range Inputs, Height, Outputs (Ground-to-Air Data Links, TTY, Crosstell Outputs and Inputs), Ground-to-Air Voice Radio and Wire Communications, and an Equipment Programming Service Committee to provide computer programs required by the test teams. Committee personnel were drawn from the various divisions as follows: "October: 5 from Division 2, 2 from Division 3, 13 from Division 6, 11 from West. El., 5 from BTL. November: 6 from Division 2, 3 from Division 3, 16 from Division 6, 21 from West. El., 8 from BTL, 1 from IBM, 2 from RAND."

18. "X-Telling XD-1 to WW and Return; Future SAGE Tests with Whirlwind," 6M-3684, June 13, 1955, pp. 1–4.

19. Ibid., p. 3.

20. "Minutes of Meeting on XD-1—WWI Crosstelling Link, 5 Oct. 1955," 6M-3923, October 6, 1955, p. 1.

21. Division 6 QPR, December 15, 1955, issued May 18, 1956, p. iii.

22. J. F. Jacobs to Jay W. Forrester and Distribution List, "Rand-Lincoln Relationship," 6M-3713, July 7, 1955, p. 1; Division 6 QPR, September 15, 1955, p. 3. In addition to this detailed analysis of the tasks that lay ahead, see chapters 23 and 32 of Jacobs 1986. For a history of SDC, see Baum 1981)

23. J. J. Carson, "Draft 6M-3985, Experimental SAGE Subsector (ESS) Planning and Progress Report," November 14, 1955, p. 1 of covering memo (and first page of draft memo itself). In 45 pages the draft memo went into elaborate detail.

24. Division 6 QPR, March 15, 1956, p. iii; Division 6 QPR, December 15, 1955, p. iii.

25. H. D. Benington, H. K. Rising, and C. A. Zraket to J. F. Jacobs, "Operational and Mathematical Specifications for the Initial ESS Program," 6M-4061, December 16, 1955, pp. 1–2. The three software "specs" went into esoteric detail, relied heavily on technical M-Notes and related documents, and included "Master Specifications" (covering the "ultimate requirements of the production program"), "Initial-Program Specifications" (covering "initial ESS program preparation"), and "ESS Specifications" (to be used by "groups responsible for initial operation of the Experimental Subsector").

26. Division 6 QPR, 15 March 1956, p. 73. Compare Division 6 QPR, December 15, 1955, pp. 37–39; Division 6 QPR, March 15, 1956, pp. 53–55. For a description of the Duplex Central concept, see Division 6 QPR, September 15, 1955, pp. 15–17.

27. Division 6 QPR, June 15, 1956, pp. 70–71. In parameter testing, "each coded component subprogram is submitted to a parameter test which is performed on the computer. The parameter test uses an environment which simulates pertinent portions of the complete system of programs. Each parameter test is documented in a set of test specifications which detail the environment that was used and the outputs that were obtained."

28. Division 2 QPR, November 15, 1955, p. iii; Division 6 QPR, September 15, 1955, pp. iii–iv, xi.

29. Jacobs 1986, p. 108.

30. "Summary and Transmittal of SAGE Phasing Group Minutes, 13 June 1956," issued by ADES Project Office and Electronic Defense Systems Division, Air Materiel Command, USAF, p. 1.

31. Ibid., p. 2. See also "Transmittal of 18 July SAGE Phasing Group Minutes," issued 26 July 1956, pp. 10–14.

32. "Interim Report in Lieu of September 1956 SAGE Phasing Group Meeting," issued 12 September 1956 by ADES Project Office, pp. 1–2.

33. "Summary of SAGE Phasing Group Meeting 17 October 1956," issued by ADES Project Office, Electronics Defense Systems Division, Air Materiel Command, p. ii.

34. Ibid., p. iii.

35. Division 6 QPR, June 15, 1956, p. iii. This "Division 6 Summary" is dated "2 January 1957" and bears Forrester's and Everett's names at its end, as usual. Either it was their joint product or it was written, presumably by Everett, after Forrester's departure from Lincoln in June. For another view of Forrester's departure, see chapter 13 of Redmond and Smith 1980, especially p. 213.

36. Division 6 QPR, September 15, 1955, p. 4–5; Division 6 QPR, December 15, 1955, pp. iii, 3–5; Division 6 QPR, March 15, 1956, pp. iii, 3–4.

37. Division 6 QPR, June 15, 1956, pp. 41–44.

38. Ibid., pp. iii, 13–14.

39. R. E. Witherbee to S. H. Dodd, "Facilities of the Experimental SAGE Sector," 6M-4981, May 6, 1957 (AC-72, MITRE archives), pp. 1–3.

40. 6M-4981, pp. 3–5; Draft 6M-3985, pp. 23–24.

41. Draft 6M-3985, pp. 3–4, 27–28; 6M-4981, pp. 5–6.

42. Draft 6M-3985, pp. 6–11.

43. For more on The MITRE Corporation, see chapter 31.

44. E. S. Rich to C. F. J. Overhage, "ESS Shakedown Testing: Summary Report (3 December 1956–30 September 1957) (Material presented at ADES Status Meeting 8 October 1957)," 6M-5384, issued October 16, 1957 (AC 44-5, MITRE archives), p. 5. See also the following Division 6 QPRs: December 15, 1957, pp. 31–35; March 15, 1958, p. 27; June 15, 1958, p. 39; September 15, 1958, p. 23; December 15, 1958, p. 58.

45. ESS Shakedown Test Planning Section to E. S. Rich, "General Test Requirements for ESS System Shakedown," 6M-4281-1, April 11, 1957 (AC-72, MITRE archives).

46. 6M-5384, p. 1.

47. Ibid., pp. 1–5.

48. Division 6 QPR, 15 December 1957, p. 51; Baum 1981, pp. 26–28.

49. C. W. Uskavitch to V. A. Nedzel, "Evaluation Testing of the Experimental SAGE Sector," 2M-0280, September 30, 1958 (AC-139, MITRE archives), p. 5. This technical memorandum contains the text of a paper Uskavitch presented September 3, 1958 at a Symposium on the Prediction of Performance of Large-Scale Systems, sponsored by the Operational Research Department of Willow Run Laboratories (University of Michigan).

50. 2M-0280, pp. 6–8.

51. See ibid., pp. 5–6. See also technical memoranda of series 2M-0800, "Mission Report: Experimental SAGE Sector, Group 64 Shakedown and Evaluation Tests." The first of this series, reporting a shakedown test of August 21, 1957, was issued on October 4, 1957. See especially 2M-0800-18 (May 1, 1958)and 2M-0800-19 (June 11, 1958) (AC-41, MITRE archives).

52. 2M-0800-20, June 13, 1958, p. 2.

53. 2M-0800-23, June 30, 1958, p. 2.

54. Group 22 to C. R. Wieser, "Mission Report: Experimental SAGE Sector, Evaluation Test No. 1, 2 October 1957," 2M-0800-1, issued November 15, 1957, pp. iii, 3, 6–8.

55. 2M-0800-11, March 28, 1958, pp. 3, 7–8, 10–12.

56. 2M-0800-23, June 30, 1958, pp. 2–3, 10.

57. M. D. Rubin to C. R. Wieser, "Interim Report on ECM Missions: Experimental SAGE Sector Evaluation Tests No. 27 and 29, 16 April 1958 and 30 April 1958," 2M-0800-27, June 19, 1958 (AC-41, MITRE archives).

58. 2M-0280, p. 5.

59. 2M-0800, p. 5.

60. See 2M-0280, esp. pp. 9–16.

61. 2M-0280, pp. 11–12.

62. 2M-0280, pp. 12–13.

63. 2M-0280, pp. 14–16.

Chapter 30

1. Interview with R. R. Everett and J. W. Forrester by the authors, January 11, 1983.

2. See e.g. Jacobs 1986, pp. 111–113; Redmond and Smith 1980.

3. Division 6 QPR, 1 June 1952, pp. 6-28–6-69. (This was the first of the QPRs.) The quotation is taken from Linvill's discussion on p. 6-68.

4. Lt. Gen. E. E. Partridge to James R. Killian Jr., January 28, 1953 (Wood, Case History, document 25; Project Lincoln, volume 3, item 60). See also Partridge to Harlan Hatcher, January 28, 1953 (Hatcher Papers).

5. Maj. Gen. D. L. Putt to Cmdg. Gen., AFCRC, "Revision of Command Policy Pertaining to Lincoln Transition System," May 6, 1953 (Wood, Case History, document 29; Project Lincoln, volume 3, item 66); Jay W. Forrester, "Bomarc Meeting with AFCRC, WADC, and Boeing," L-102, June 23, 1953; Lincoln Steering Committee, Meetings of June 22, 1953, p. 2; July 6, 1953, p. 2; August 3, 1953, p. 2; Division 6 Memoranda L-147 (April 20, 1954), item 13; 6L-174 (November 29, 1954), item 8; 6L-175 (December 6, 1954), item 8.

6. Technical Memorandum No. 20, p. 58.

7. Ibid., p. 87.

8. J. Arnow, D. Israel, and R. Walquist to Distribution List, "Draft of appendix B of Revision of TM-20," M-2946, July 29, 1954 (AC-2, MITRE archives), p. 4. See also D. R. Israel, "Weapons Integration," draft memorandum, December 27, 1956.

9. See TM-20, pp. 57–58, 87–106.

10. In May 1953, ARDC had instructed its Cambridge Research Center "to promptly examine the relationship of the Bomarc ground control system to the Transition System, to see that the performance of the Bomarc system would be adequately provided for as economically as possible, and to recommend a plan to be followed by the Air force to provide proper Bomarc control equipment." See Maj. Gen. D. L. Putt to Cmdg Gen, AFCRC, "Revision of Command Policy Pertaining to Lincoln Transition System," May 6, 1953 (Wood, Case History, document 29); Project Lincoln, volume 3, item 66. See also Forrester, "Bomarc Meeting with AFCRC, WADC, and Boeing," L-102, June 23, 1953. Lincoln Steering Committee Meetings: June 22, 1953, p. 2; July 6, 1953, p. 2; August 3, 1953, p. 2. Minutes of Group Leaders Meetings: L-147, April 20, 1954, item 13; 6L-174, November 29, 1954, item 8; 6L-175, December , 1954, item 8.

11. L-102, p. 1. See also LSC minutes, June 22, 1953, p. 2.

12. L-102, p. 3 and second page of attached draft copy of "Recommendation Pertaining to Lincoln Transition System–Bomarc Weapon System Development." It appeared that Westinghouse had begun working on the problem of ground control of BOMARC in the summer of 1952, had been awarded a formal contract the following January, and had proposed to use a computer "specified to be along the lines of that in G-12 which is now being assembled by the University of Michigan. The G-12 model should be complete in three or four months and shipped to Florida in five or six months."

13. L-102, p. 4.

14. LSC minutes, June 22, 1953, item 5; Forrester, "Air Force Cancellation of Boeing Subcontract to Westinghouse," L-108, July 10, 1953, p. 1. On the day after the Steering Committee Meeting, Forrester drew up L-102, the record of the first BOMARC meeting (June 15–18, 1953), and on July 10 he issued report L-108. On the first page of L-108, Forrester noted that the action was "initiated by and decided by the Air Force as an economy measure and in order to obtain a more standardized and closely knit ground environment. Although the decision appears technically sound, it was reluctantly accepted by the Lincoln Laboratory because full information on the technical problems involved is not yet available and because of the additional load it will put on the Lincoln staff. Cancellation of the Westinghouse work may lead to some unhappiness toward MIT."

15. Lincoln Steering Committee meetings: June 22, 1953, p. 2; July 6, 1953, p. 2; October 26, 1953, p. 2. See also L-108.

16. George Valley, "Foreword," Division 2 QPR, October 15, 1953 (AC-21, MITRE archives), p. iii. See also LSC minutes, August 31, 1953, item 8; George Valley, "Lincoln-Boeing Meeting—24, 25 August 1953," August 31, 1953 (ASC 23-4, MITRE archives).

17. "Group Leaders' Meeting—October 24, 1955," 6L-225, item 7.

18. "Group Leaders' Meeting—October 31, 1955," 6L-226, item 11.

19. 6L-226, pp. 3–4. Those attending, in addition to Forrester and Everett, included J. A. Arnow, D. R. Brown, S. H. Dodd, C. W. Farr, D. R. Israel, J. F. Jacobs, B. E. Morriss Jr., J. A. O'Brien, W. N. Papian, J. C. Proctor, E. S. Rich, N. H. Taylor, and P. Youtz. "AICBM" refers to anti-intercontinental ballistic missile measures.

20. "Group Leaders' Meeting—November 14, 1955," 6L-227, item 8.

21. "Group Leaders' Meeting—November 21, 1955," 6L-228, items 8 and 11. "APG" refers to the Air Proving Ground at Eglin Air Force Base in Florida.

22. Lt. Gen. Donald Putt to Cmdr, ARDC, February 3, 1956.

23. Ibid.

24. Col. A. R. Shiely to M. G. Holloway, March 2, 1956. See Wood, Case History, document 68.

25. Forrester to Distribution List, "General Procedure for Integrating New Weapons with SAGE," Memorandum 6M-4048, January 30, 1956, pp. 1–3. Forrester's policy statement was forwarded as an enclosure, with a cover letter dated 2 February 1956, from M. G. Holloway (Director, Lincoln Laboratory) to Lt. Col. Ralph S. La Montagne (Deputy Director for Lincoln Operations, Electronics Research Directorate, Air Force Cambridge Research Center). Holloway's and Forrester's documents appear in Wood, Case History, as documents 72 and 73.

26. 6M-4048, pp. 2–4, 7.

27. Ibid., pp. 1–6

28. Ibid., pp. 1–4.

29. Ibid., pp. 3–4. See also David R. Israel, "Weapons Integration," draft memorandum, December 27, 1956 (ASC 23-4, MITRE archives).

30. 6M-4048, p. 4. The emphasis is Forrester's.

31. 6M-4048, pp. 5–8 ; Drawing No. B-75886 ("Proposed Integration Procedure').

32. Forrester to Distribution List, "Reorganization of Division 6" (Administrative Memorandum 6A-188), January 9, 1956. See also Division 6 QPR, March 15, 1956, pp. 3, 7.

33. These developments are reflected in the following: E. D. Lundberg to Distribution List, "SAGE System Meeting, 28 November 1955," 6M-4023, item 4; "SAGE System Meeting, 5 December 1955," M-4038, item 1; Division 6 QPR, March 15, 1956, pp. 7–8.

34. LSC minutes, October 17 1955, item 4; meeting of October 31, 1955, item 5; meeting of November 7, 1955, item 5.

35. Putt to Power, February 20, 1956.

36. Brig. Gen. Ivan L. Farman, "Integration of SAGE/Air Defense Weapon Systems," April 27, 1956, items 1–4. For Hq., ARDC views re integration, see interview with Brigadier General Ivan L. Farman by E. G. Schwiebert, J. T. Schwiebert, and T. M. Smith, February 28, 1957; interview with I. L. Farman by H. C. Jordan and W. B. Hughes, June 18, 1957 (ESDHO).

37. 1st Indorsement to "B/L-Hq-USAF ltr d/3 Feb 56, Subj: (Uncl) SAGE," Hq, AMC to DCS/D, HQ, USAF, March 27, 1956 (ESDHO).

38. G. E. Valley to Col. A. R. Shiely Jr., April 10, 1956 (ESDHO).

39. The AN/FST-2, developed at AFCRC, was manufactured by Burroughs.

40. Valley to Shiely, April 10, 1956.

41. Shiely to Valley, April 20, 1956.

42. Shiely to DCS/D, Hq, USAF, April 23, 1956 (ESDHO).

43. Shiely to Brig. Gen. H. M. Estes Jr., April 23, 1956 (ESDHO).

44. Shiely to Valley, April 20, 1956 (ESDHO).

45. "Proposed Work Statement (Submitted by ARDC)" appears as document 76 in Wood, Case History, and also among Everett papers in the MITRE archives. See also attachment to letter, dated June 29, 1956, from Valley to Col. Forrest G. Allen, Wright-Patterson AFB, Ohio.

46. "Memorandum re ARDC—Proposed Work Statement on SAGE, Proposed by Lincoln Laboratory July 5, 1956" (Wood, Case History, document 75O).

47. See Everett, "Group Leaders' Meeting—July 23, 1956," 6L-269. See also Wood, Case History, documents 83 and 86 (ESDHO).

48. "Group Leaders' Meeting—March 5, 1956," 6L-249, p. 2. Most of the attendees were Division 6 veterans: Forrester, Everett, J. A. Arnow, H. D. Benington, D. R. Brown, W. J. Canty, C. L. Corderman, S. H. Dodd, R. S. Fallows, C. W. Farr, D. R. Israel, J. F. Jacobs, K. E. McVicar, B. E. Morriss Jr., J. A. O'Brien, W. N. Papian, J. C. Proctor, E. S. Rich, N. H. Taylor, P. Youtz, C. A. Zraket.

49. "Group Leaders' Meeting—March 12, 1956," 6L-255, p. 2. (The minutes of this meeting were not issued until April 25, 1956.)

50. Group Leaders' Meeting—April 23, 1956," 6L-256, "p. 2.

51. LSC minutes, April 9, 1956, pp. 2–3.

52. Ibid, pp. 3–4.

Chapter 31

1. Jacobs 1986, pp. 39, 86, 89. See also Redmond and Smith 1980, p. 213.

2. See Pugh 1984, p. 127.

3. R. R. Everett to G. E. Valley, "Lincoln in SAGE," June 22, 1956 (ASC 24, MITRE archives), p. 1.

4. Ibid., p. 2. "Perhaps," Everett added, "we should not at this time make specific recommendations about who such a systems contractor should be since both RAND and BTL are nibbling and should not be frightened away."

5. Ibid., pp. 2–3.

6. Ibid., pp. 3–4.

7. "Group Leaders' Meeting—July 23, 1956," 6L-269, item 5.

8. Col. F. G. Allen to M. G. Holloway, August 24, 1956 (Wood, Case History, document 85). See also document 86, "Introduction," identified by Wood as "Memo on Weapon systems/SAGE integration," n.d. (at bottom of p. 1 in pencil: "August 20, 1956"). Both documents bear the identification number "56RDZ-26192."

9. LSC minutes, July 2, 1956, p. 2.

10. LSC minutes, August 6, 1956, p. 3; September 4, 1956, p. 2; September 10, 1956, p. 1.

11. "Memorandum re ARDC—Proposed Work Statement on SAGE, Proposed by Lincoln Laboratory July 5, 1956" (enclosure to a letter from G. E. Valley to Col. Forrest G. Allen, July 5, 1956), p. 6; Wood, Case History documents 74 and 75, respectively. For the details of a similar, earlier expression of these views, see Everett to Valley, "Lincoln in SAGE," June 22, 1956, pp. 2–3.

12. "Memorandum re ARDC," p. 4. Compare Everett's memorandum to Valley, pp. 2–3.

13. C. W. Farr to R. S. Fallows and H. E. Anderson, "Weapons Integration," September 20, 1956 (AC-59, MITRE archives), p. 1.

14. Ibid., pp. 2–3.

15. In November of 1956, Secretary of Defense Charles E. Wilson would place TALOS under the Army's control, on the ground that it was better suited to "local rather than area defense."

16. Ibid.; Goldberg 1957, p. 209; David Israel, "Weapons Integration," December 27, 1956, pp. 10–14.

17. M. G. Holloway to Col. F. G. Allen, September 24, 1956, in Wood, Case History document 87 (at head of page: "Text of letter taken from TWX, AFCRC {CRL-9-Z-E} to ARDC {Brig. Gen. I. L. Farman, RDZE}, September 26, 1956").

18. D. R. Israel, "Weapons Integration" (draft memorandum), December 27, 1956, pp. 3–5, 10–14. For a Division 6 perspective on technical details regarding use of the Nike missiles at the McGuire SAGE site, as well as of the Army's

AN/FSG-1 Missile Master fire direction center for antiaircraft weapons, see R. S. Fallows to Distribution List, "Study of Initial SAGE Weapon Control Capability," 6M-4885 (draft version dated February 11, 1957, final version dated May 23, 1957). For more on the manned interceptors discussed in 6M-4885, see F. Garth and P. Sinesi to Distribution List, "Detailed Specification for Manned Interceptor Section of 6M-4885," 6M-4470, July 12, 1957 (AC 40-2, MITRE archives).

19. Farr, p. 2; Israel, pp. 4–6; Division 6 QPR, September 15, 1956, pp. 7, 9–10; Division 6 QPR, December 15, 1956, p. 6.

20. "Group 61 Bi-Weekly Report (30 November–14 December 1956)," 6M-4836, issued December 28, 1956, pp. 2–3; Division 6 QPR, December 15, 1956, p. 5; Israel, "Weapons Integration," pp. 8–9.

21. R. R. Everett to C. F. J. Overhage and W. H. Radford, "Air Defense Systems Engineering," June 3, 1957 (AC-59, MITRE archives).

22. C. W. Farr to R. S. Fallows and H. E. Anderson, "Weapons Integration," September 20, 1956, p. 4.

23. H. E. Anderson et al. to G. E. Valley Jr., R. R. Everett, N. H. Taylor, "A Six-Point Proposal for System Revision and Weapons Integration," September 28, 1956.

24. R. R. Everett to G. E. Valley, "Weapons Integration in SAGE," October 5, 1956, item 2. See also Everett, "Proposal for Weapons Integration," September 28, 1956 (ASC-23, MITRE archives).

25. Everett to Valley, October 5, 1956, item 1.

26. Murphy, Early History of MITRE, volume 1, p. 57. See also 6L-280, "Group Leaders' Meeting 8 October 1956, item 10-8-B; Lincoln Steering Committee, minutes, October 15, 1956, item 4, and meeting of December 28, 1956, item 3.

27. For details, see Redmond and Jordan, Air Defense Management (ESDHO archives).

28. Ibid, p. 32.

29. For a view of these developments as seen by a key Lincoln engineer and administrator, see chapters 31 and 32 of Jacobs 1986.

30. Robert L. Perry to Cmdr. ARDC, Att'n: Kent C. Redmond, Deputy Command Historian, Hq. ARDC, March 13, 1959, with attached "Memo for the Record"; interview with Col. R. S. Carter by Robert L. Perry, March 13, 1959; interview with Maj. H. E. Walmer by Redmond, August 19, 1957; interviews with Brig. Gen. I. L. Farman by E. G. Schwiebert, J. T. Schwiebert, and T. M. Smith, February 28, 1957, and by Harry C. Jordan and William B. Hughes, June 18, 1957; Lincoln Steering Committee, Meetings of April 1, 1957, item 6, and April 8 1957, item 4.

31. Jacobs 1986, pp. 135–136.

32. Interview with R. R. Everett and J. W. Forrester by the authors, January 11, 1983.

33. R. R. Everett to C. F. J. Overhage and W. H. Radford, "Air Defense Systems Engineering," June 3, 1957.

34. Ibid., pp. 2–3. (Everett explicitly cited Forrester's memorandum 6M-4048.)

35. Ibid., p. 3.

36. Ibid., p. 4.

37. Ibid., pp. 4–5.

38. Ibid., p. 5.

39. Ibid., p. 6.

40. Redmond and Jordan, Air Defense Management, pp. 35–39.

41. Ibid, pp. 40–41, 47–51.

42. See J. A. Stratton to James H. Douglas, June 3, 1958, p. 1. (This letter opens: "This is in reply to your letter of March 3.")

43. Redmond and Jordan, Air Defense Management, pp. 54–55; Murphy 1972, pp. 58–61. In November 1957, SDD became its own master as the System Development Corporation.

44. C. F. J. Overhage and M. O. Kappler to J. McCormack Jr., April 16, 1958, pp. 1–5. For an insider's view of these events, see Jacobs 1986, pp. 137–141.

45. Murphy 1972, pp. 63–66; Redmond and Jordan, Air Defense Management, pp. 54–55.

46. Jacobs 1986, pp. 137–138.

47. Ibid., p. 130.

48. General Plan—DRAFT (C-61, MIT archives). The final entry reads "RRE:jk 5/26/58." In Everett's handwriting at the top of the first page is "R. R. Everett, 26 May 1958." Following Everett's opening words, the document continued as follows: "An approximation to its administrative workings should be established at once. . . . As soon as the above are reasonably well settled, the transfer of Lincoln Laboratory personnel and facilities to the new corporation should be effected. . . ." Then came a list of nine "steps to be taken at once." The remaining 6½ pages were devoted to correspondingly numbered comments on the steps, e.g., "1. Establish corporation and appoint officers. This matter is in MIT's hands and is going forward. In particular, an immediate, clear understanding should be reached with Halligan." This document made no mention of a name for "the new corporation."

49. Stratton to Douglas, June 3, 1958, p. 1.

50. Ibid., p. 2.

51. Ibid.

52. Ibid., p. 4.

Chapter 32

1. See, e.g., William H. Radford, "M. I. T. Lincoln Laboratory: Its Origin and First Decade," *Technology Review*, January 1962, p. 15; Augarten 1984; Baum 1981.

2. Interview with Lieutenant General Arthur C. Agan, Commander, Aerospace Defense Command, Colorado Springs, Colorado, conducted by Denys Volan, ADC Command Historian, February 1970 (U.S. Air Force Oral History Interview [K239.0512-899], Albert F. Simpson Historical Research Center, Air University), p. 69.

3. News release 609-58, Office of Public Information, Department of Defense (ESDHO).

4. J. C. Forbes to Distribution List, "SAGE Press Kit," January 10, 1962 (enclosed with the press kit that the Air Force distributed when the Sioux City site was turned over to the Air Defense Command on December 15, 1961). See news release for that occasion, Air Force Systems Command, Electronic Systems Division. See also *MITRE, The First Twenty Years* (MITRE Corp., 1979), pp. 25, 49.

5. Gordon Bell, "Field Trip to North Bay," *Computer Museum Report*, spring 1983, p. 13. See also U.S. Treaties and Other International Agreements, Canada—Defense—Sept. 27, 1961 (IAS 4859), pp. 1376–1380; *Ontario* (Department of National Defense, Ottawa, Ontario, Royal Canadian Air Force Headquarters, SAGE Installation, North Bay, , n.d. [October 1960] (ESDHO).

6. Agan-Volan interview, pp. 68–69.

7. Interview with Lieutenant General Arthur C. Agan (Rtd.) by Lieutenant Colonel Vaughn H. Gallacher, April 19–22, 1976 (U.S. Air Force Oral History Interview K239-0512-900, Albert F. Simpson Historical Research Center, Air University), pp. 307–308.. General Agan had overseen the incorporation of the SAGE site at McGuire AFB (the New York Air Defense Sector) into the continental air defense system between September 1958 and September 1959. In August of 1967, after several other assignments, he assumed command of the Aerospace Defense Command at Ent Air Force Base in Colorado.

8. A Report to the National Security Council by the Defense Committee on Continental Defense (NSC 159), July 22, 1953, p. 53.

9. Murphy 1972, volume I, pp. 43–44. The late Professor Murphy's unpublished history offers an informed, carefully researched, and cogent description and analysis of the emergence of The MITRE Corporation, its creation, and its early activities.

10. Thomas K. Glennan Jr., "Issues in the Choice of Development Policies," in Marschak et al. 1967, pp. 40–41.

11. *MITRE, The First Twenty Years,* pp. 52–55.

12. Goldstine 1972, p. 213.

13. Ibid., p. 310.

14. Watson and Petre 1990, p. 233. See also Ferguson and Morris 1993, pp. 6, 234–235.

15. Ibid., p. 243; Pugh 1984, pp. 123–124, 150, 307; Augarten 1984, p. 205.

16. Augarten 1984, pp. 208–209.

17. Pugh 1984, p. 262.

18. See Baum 1981, p. 25.

19. Ibid., pp. 26–29, 31–39.

20. Ibid., p. 36.

21. Ibid., pp. 12–13.

22. David Warsh, "Reaping the Whirlwind," *Boston Globe,* August 10, 1986.

23. Pugh 1984, pp. 127–128.

24. Goldstine 1972, pp. 213, 84–99.

25. See Redmond and Smith 1980, especially chapter 2.

26. MITRE Corporation, "Annual Report 1989," p. 3.

27. Ibid., p. 20; Warsh, "Reaping the Whirlwind."

28. Robert R. Everett, "Epilogue," *Annals of the History of Computing,* 5, no. 4 (1983): 403. For a list of SAGE "firsts," see Baum 1981, pp. 24–25.

29. Augarten 1984, p. 208.

Bibliography

Angoff, Charles, ed. 1968. *The Humanities in the Age of Science.* Fairleigh Dickinson University Press.

Augarten, Stan. 1984. *Bit by Bit: An Illustrated History of Computers.* Ticknor & Fields.

Baum, Claude. 1981. *The System Builders: The Story of SDC.* System Development Corporation.

Baxter, J. P. 1946. *Scientists against Time.* Little, Brown.

Bush, Vannevar. 1949. *Modern Arms and Free Men: A Discussion of the Role of Science in Preserving Democracy.* Simon and Schuster.

Bush, Vannevar. 1970. *Pieces of the Action.* Morrow.

Craven, Wesley Frank, and James Lea Cate, eds. 1948. *The Army Air Forces in World War II,* volume I, *Plans and Early Operations, January 1939 to August 1942.* University of Chicago Press.

Dupree, A. Hunter. 1957. *Science in the Federal Government.* Harvard University Press.

Fagen, M. D. ed. 1978. A History of Engineering and Science in the Bell System, National Service in War and Peace (1925–1975). Bell Telephone Laboratories,

Ferguson, Charles H., and Charles R. Morris. 1993. *Computer Wars: How the West Can Win in a Post-IBM World.* Times Books.

Futrell, Robert Frank. 1971. Ideas, Concepts, Doctrine: A History of Basic Thinking in the United States Air Force, 1907–1964. Air University Study AU-19, Maxwell Air Force Base, Alabama.

Getting, Ivan A. 1989. *All in a Lifetime: Science in the Defense of Democracy.* Vantage.

Gilpin, Robert. 1962. *American Scientists and Nuclear Weapons Policy.* Princeton University Press.

Gimbel, John. 1968. *The American Occupation of Germany; Politics and the Military, 1945–1949.* Stanford University Press.

Goldberg, Alfred, ed. 1957. *A History of the United States Air Force, 1907–1957.* Van Nostrand.

Goldstine, Herman H. 1972. *The Computer from Pascal to von Neumann.* Princeton University Press.

Gray, George W. 1943. *Science at War.* Harper.

Hartcup, Guy. 1970. *The Challenge of War: Britain's Scientific and Engineering Contributions to World War Two.* Taplinger.

Hartmann, Frederick H. 1965. *Germany between East and West: The Reunification Problem.* Prentice-Hall.

Holly, I. B., Jr. 1953. *Ideas and Weapons: Exploitation of Aerial Weapons by the United States during World War I.* Yale University Press.

Huntington, Samuel P. 1961. *The Common Defense: Strategic Programs in National Politics.* Columbia University Press.

Jacobs, John F. 1986. The SAGE Air Defense System, A Personal History. MITRE Corp.

Kevles, Daniel J. 1977. *The Physicists.* Knopf.

Koppes, Clayton R. 1982. *JPL and the American Space Program: A History of the Jet Propulsion Lab.* Yale University Press.

Lavington, Simon. 1980. *Early British Computers.* Digital Press.

Leslie, Stuart W. 1993. *The Cold War and American Science: The Military-Industrial-Academic Complex at MIT and Stanford.* Columbia University Press.

Liddell Hart, B. H. 1930. *The Real War, 1914–1918* (Little, Brown, 1964).

Major, John. 1971. *The Oppenheimer Hearing.* Batsford.

Marschak, Thomas, Thomas K. Glennan, Jr., and Robert Summers. 1967. *Strategy for R&D: Studies in the Microeconomics of Development.* Springer-Verlag.

Metropolis, N. J. Howlett, and G.-C. Rota, eds. 1980. *A History of Computing in the Twentieth Century.* Academic Press.

Millikan, Robert A. 1950. *Autobiography.* Prentice-Hall.

Morison, Samuel Eliot. 1961. *History of United States Naval Operations in World War II,* volume III: *Rising Sun in the Pacific.* Little, Brown.

Morrow, John H., Jr. 1993. *The Great War in the Air, Military Aviation from 1909 to 1921.* Smithsonian Institution Press.

Murphy, Howard R. 1972. The Early History of the MITRE Corporation, Its Background, Inception, and First Five Years. Unpublished draft.

Noyes, W. A., Jr., ed. 1948. *Chemistry: A History of the Chemistry Components of the National Defense Research Committee, 1940–1946.* Little, Brown.

Price, Don K. 1954. *Government and Science: Their Dynamic Relation in American Democracy.* New York University Press.

Pugh, Emerson W. 1984. *Memories That Shaped an Industry.* MIT Press.

Redmond, Kent C., and Thomas M. Smith. 1980. *Project Whirlwind: The History of a Pioneer Computer.* Digital Press.

Ropp, Theodore. 1959. *War in the Modern World.* Duke University Press.

Sapolsky, Harvey M. 1990. *Science and the Navy: The History of the Office of Naval Research.* Princeton University Press.

Schilling, Warner R., Paul Y. Hammond, and Glenn H. Snyder. 1962. *Strategy, Politics, and Defense Budgets.* Columbia University Press.

Smith, Charles J. 1964. History of the Electronic Systems Division, January–June 1964, volume I, SAGE: Background and Origins. Historical Division, Air Force Systems Command, Hanscom Field, Bedford, Massachusetts.

Stern, Nancy. 1981. *From ENIAC to UNIVAC, An Appraisal of the Eckert-Mauchly Computers.* Digital Press.

Stokesbury, James L. 1980. *A Short History of World War II.* Morrow.

Sturm, Thomas A. 1967. The USAF Scientific Advisory Board, Its First Twenty Years, 1944–1964. USAF Historical Division Liaison Office, Washington, D.C.

Turnbull, Archibald D., and Clifford L. Lord. 1949. *History of United States Naval Aviation.* Yale University Press.

Volan, Denys. Air Defense in Theory and Practice. ADC Historical Study No. 16. Historical Division, Office of Information Services, HQ, Air Defense Command (copy filed in Air University Archives, Maxwell Air Force Base, Alabama).

von Karman, Theodore, with Lee Edson. 1967. *The Wind and Beyond.* Little, Brown.

Wang, An, with Eugene Linden. 1986. *Lessons: An Autobiography.* Addison-Wesley.

Watson, Thomas J., Jr., and Peter Petre. 1990. *Father, Son & Co.: My Life at IBM and Beyond.* Bantam.

Watson-Watt, Robert. 1959. *The Pulse of Radar: The Autobiography of Sir Robert Watson-Watt.* Dial.

Welchman, Gordon. 1982. *Hut Six Story: Breaking the Enigma Codes.* McGraw-Hill.

Index